Richard H. Groshong, Jr.

3-D Structural Geology

Springer

Berlin
Heidelberg
New York
Barcelona
Hong Kong
London
Milan
Paris
Singapore
Tokyo

Richard H. Groshong, Jr.

3-D Structural Geology

A Practical Guide to Surface and Subsurface Map Interpretation

With 366 figures and 10 tables

Springer

Professor Dr. Richard H. Groshong, Jr.
Department of Geology
The University of Alabama
Tuscaloosa, Alabama 35487-0338
U.S.A.

rgroshon@wgs.geo.ua.edu

The cover illustrates the top of the Cretaceous Selma Chalk in the Gilbertown graben, which is located in the subsurface of southern Alabama, 3x vertical exaggeration. The surface is a triangulated irregular network, developed in GeoSec3D software and colored according to elevation; cool colors are lower elevations. The fault surfaces have been removed for clarity. Structure by G. Jin, J.C. Pashin, and RHG as part of research on the Gilbertown oil field, sponsored by the US Department of Energy Grants DE-AC22-94PC91008 and DE-FC02-91ER75678.

ISBN 3-540-65422-4 Springer-Verlag Berlin Heidelberg New York

Library of Congress Cataloging-in-Publication Data
Groshong, Richard H., Jr 1943–
3-D structural geology : a practical guide to surface and subsurface map interpretation / Richard H. Groshong, Jr.
p. cm.
Includes bibliographical references and index.
ISBN 3-540-65422-4 (hardcover : acid-free paper)
1. Geology, Structural – Maps – Data processing. 2. Geological mapping. I. Title.
QE601.3.D37676 1999 99-20758
551.8'022'3–dc21

Cover: Erich Kirchner
Typesetting: Data conversion by MEDIO, Berlin

SPIN: 10573136 32/3020 - 5 4 3 2 1 0 – Printed on acid-free paper.

Contents

Introduction

Many important decisions, ranging from locating an oil prospect or a land-fill site to determining the location and size of an earthquake-producing fault, are based on geological maps. Because a map-scale structure is never completely sampled in three dimensions, geological maps and the cross sections derived from maps are always interpretations. The interpretation may be complicated by direct structural observations, like bedding attitudes, that are misleading because they represent a local structure, not the map-scale structure. Some data may simply be wrong. The interpretation of the geometry of even a single horizon, therefore, always involves inferences about the validity and meaning of the observations themselves as well as the nature of the geometry between the observation points. How is an accurate interpretation to be constructed and how is it to be validated once complete?

The objective of this book is to demonstrate the concepts and techniques required to obtain the most complete and accurate interpretation of the geometry of structures at the map scale. The methods are designed primarily for interpretations based on outcrop measurements and subsurface information of the type derived from well logs and two-dimensional seismic reflection profiles. These forms of information all present a similar interpretive problem, which is to define the geometry from isolated and discontinuous observations. The underlying philosophy of interpretation is that structures are three-dimensional solid bodies and that data from throughout the body should be integrated into an internally consistent interpretation. The techniques presented here are drawn from the best and most practical methods of surface and subsurface geology and provide a single methodology intended to be appropriate for both. Because computers are widely used for interpretation and drafting, quantitative techniques and computer-oriented methods are emphasized. The techniques are developed using approaches that can be done with a pencil, a piece of paper, and a pocket calculator, because an understanding of the methods is best achieved by working through at least a few examples by hand. Most of the methods are also presented in analytical forms suitable for implementation on a computer. The final equations are given in the body of the text with the derivations placed at the end of the chapters so that the text can concentrate on the applications.

The presentation is directed at geoscience professionals who require practical and efficient techniques for interpreting real-world structural geometries at the map scale. Numerous methods are provided for the interpretation and validation of geologic maps, as well as for the creation of original maps. The techniques are designed to help identify and develop the best interpretation from incomplete data and to provide unbiased methods by which erroneous data and erroneous interpretations can be recognized and rejected. Attention is given to the effects of measurement errors and to the effects of different interpretive techniques applied to the same data. The goal is to squeeze the most information out of the data and to obtain the best possible interpretation of the geometry.

Recognizing that not all users of this book will have had a recent course in structural geology, the book begins with basic structural concepts and progresses systematically from data collection and presentation through increasingly advanced levels of data interpretation. Chapter 1 contains a short review of the elements of structural geology and the concepts useful in map interpretation. The mechanical interpretation of folds and faults and the relationships between the geometry and mechanics are emphasized. Even with abundant data, geologic map interpretation requires inferences, and the best inferences are based on both the hard data and on mechanical principles. The fundamental building blocks of structural interpretations are the attitudes of surfaces and the thicknesses of units, covered in Chapter 2, along with the necessary considerations for accurately locating observation points in xyz space. Structure contours provide the primary means for representing the geometry of structures in three dimensions. In Chapter 3, the contouring techniques that will be used to construct and display the geometry of both folds and faults are described and discussed. Chapter 4 is a discussion of folds, including the recognition of cylindrical and conical domains, methods for projecting folds along trend, dip-sequence analysis, the recognition and use of minor folds, and growth folding. Chapters 5 and 6 cover faults and unconformities. Chapter 5 discusses the recognition of faults, calculating heave and throw from stratigraphic separation, the geometric properties of faults, and growth faults. Chapter 6 is concerned with mapping fault surfaces and their intersections with each other, and with mapping faulted horizons. Cross sections are used to understand, illustrate, and validate map interpretations. The construction of cross sections, including the projection of data onto the line of section is given in Chapter 7. To be a valid interpretation, a structure must be restorable to the geometry it had before deformation and so the book concludes in Chapter 8 with an introduction to the most widely useful techniques for the restoration and validation of cross sections.

An advanced user might wish to begin with Chapter 3 on structure contouring, followed by Chapter 7 on cross section construction, and then cover the material in Chapters 4 to 6 on the geometry of folded and faulted surfaces. The necessary topics in Chapters 1 and 2 can be reviewed as they are cited in the later chapters. Chapter 8 can be covered any time after Chapter 7 has been completed.

The text contains numerous worked examples designed to explain and illustrate the techniques. Problems are also provided at the ends of Chapters 2 through 8. Many of these are based on the examples in the text, either exactly or approximately. If worked independently from the answers in the text, the problems provide the opportunity to review the methods and to understand the interpretation process. Many of the map interpretation problems are designed to provide just enough information to allow a unique solution. It is instructive to see what answers may be obtained by deleting a small amount of the information from the well or the map or by deliberately introducing erroneous data of a type commonly encountered, for example by transposing numbers in a measurement, reversing a dip direction, or by mislocating a contact. For additional problems, use the questions provided at the ends of the chapters to interpret other geologic maps.

Many thanks to my graduate students for numerous helpful suggestions, and especially to Bryan Cherry, Diahn Johnson, and Saiwei Wang, whose work has been utilized in some of the examples. I am extremely grateful to Denny Bearce, Lucian Platt, John Spang and Hongwei Yin for their reviews and for their suggestions which have led to significant improvements in the presentation. Additional helpful suggestions have been made by Jean-Luc Epard, Gary Hooks, Jack Pashin, George Davis, Jiafu Qi, Jorge Urdaneta, and the University of Alabama Advanced Map Interpretation class of 1997.

Tuscaloosa, Alabama
Spring 1999

Richard H. Groshong, Jr.

Chapter 1
Elements of Map-Scale Structure

1.1
Introduction

The primary objective of structural map making and map interpretation is to develop an internally consistent three-dimensional picture of the structure that agrees with all the data. This can be difficult or ambiguous because the complete structure is usually undersampled. Thus an interpretation of the complete geometry will probably require a significant number of inferences, as, for example, in the interpolation of a folded surface between the observation points. Constraints on the interpretation are both topological and mechanical. The basic elements of map-scale structure are the geometries of folds and faults, the shapes and thicknesses of units, and the contact types. This chapter provides a short review of the basic elements of the structural and stratigraphic geometries that will be interpreted in later chapters, reviews some of the primary mechanical factors that control the geometry of map-scale folds and faults, and examines the typical sources of data for structural interpretation and their inherent uncertainties.

1.2
Representation of a Structure in Three Dimensions

A structure is part of a three-dimensional solid volume that probably contains numerous beds and perhaps faults and intrusions (Fig. 1.1). An interpreter strives to develop a mental and physical picture of the structure in three dimensions. The best interpretations utilize the constraints provided by all the data in three dimensions. The most complete interpretation would be as a three-dimensional solid, an approach now possible with 3-D computer graphics programs. Two-dimensional representations of structures by means of maps and cross sections remain major interpretation and presentation tools. When the geometry of the structure is represented in two dimensions on a map or cross section, it must be remembered that the structure of an individual horizon or a single cross section must be compatible with those around it. This book presents methods for extracting the most three-dimensional interpretive information out of local observations, for example at wells (Fig. 1.1), and for using this information to build a three-dimensional interpretation of the whole structure.

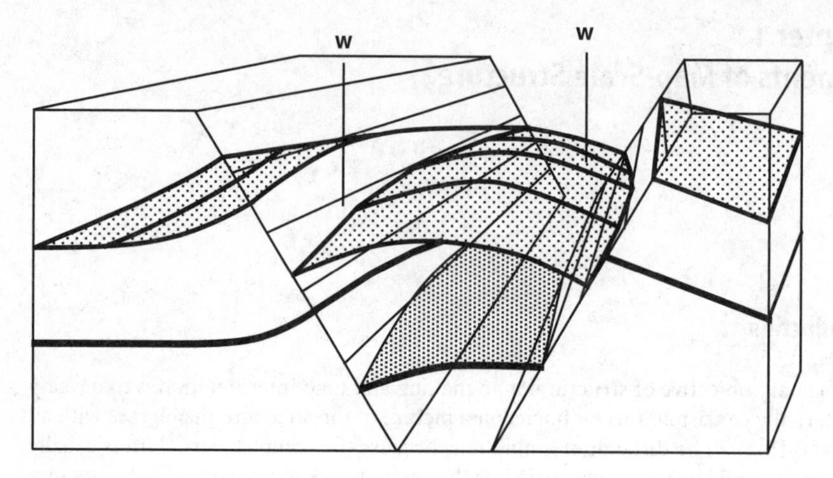

Fig. 1.1. Geometry of a structure in three dimensions. *w* Well location where the structure is sampled

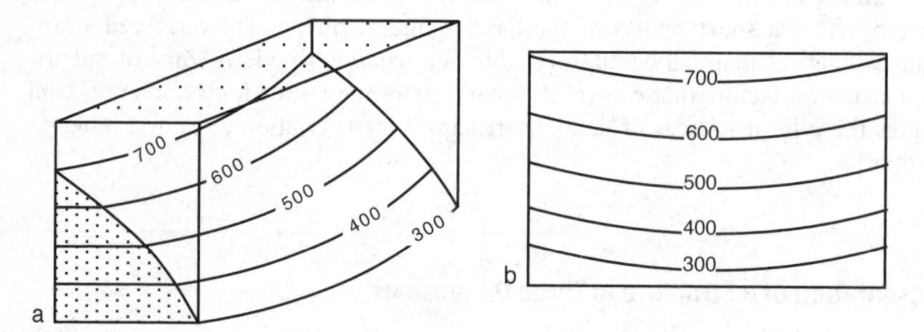

Fig. 1.2. Structure contours. **a** Lines of equal elevation on the surface of a map unit. **b** Lines of equal elevation projected onto a horizontal surface to make a structure contour map

1.2.1
Structure Contour Map

A structure contour is the trace of a horizontal line on a surface (e.g., on a formation top or a fault). A structure contour map represents a topographic map of the surface of a geological horizon (Fig. 1.2). The dip direction of the surface is perpendicular to the contour lines and the dip amount is proportional to the spacing between the contours. Structure contours provide an effective method for representing the three-dimensional form of a surface in two dimensions. Structure contours on a faulted horizon (Fig. 1.3) are truncated at the fault.

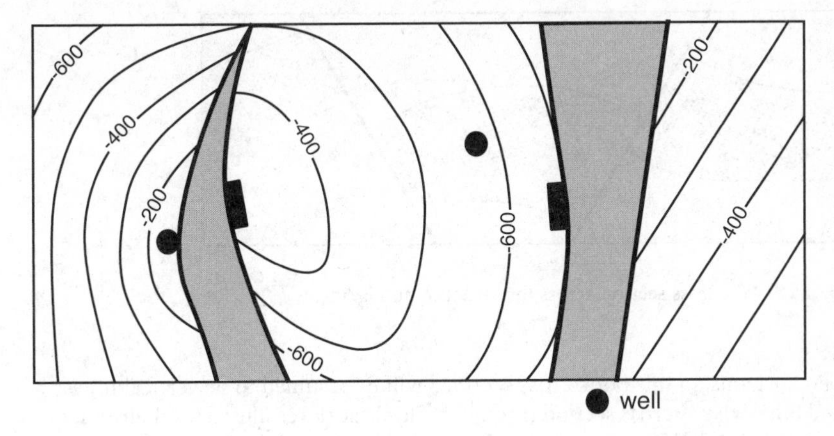

Fig. 1.3. Structure contour map of the faulted upper horizon in Fig. 1.1. Contours are at 100-unit intervals, with negative elevations being below sea level. Fault gaps, where the horizon is missing, are *shaded*

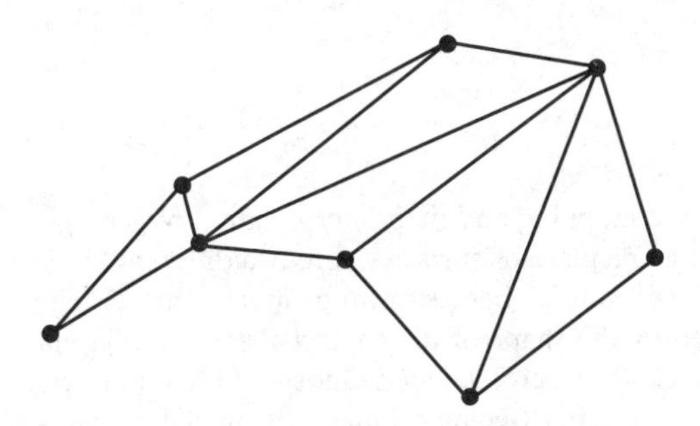

Fig. 1.4. Triangulated irregular network (TIN) of points. This could represent the map view of the points projected onto a horizontal plane or could be a perspective view of points plotted in three dimensions

1.2.2
Triangulated Irregular Network

A triangulated irregular network (TIN; Fig. 1.4) is an array of points that define a surface. In a TIN network, the nearest-neighbor points are connected to form triangles that form the surface (Banks 1991; Jones and Nelson 1992). If the triangles in the network are shaded, the three-dimensional character of the surface can be illustrated. This is an effective method for the rendering of surfaces by computer. The TIN can be contoured to make a structure contour map.

1.2.3
Cross Section

Even though a structure contour map or TIN represents the geometry of a surface in three dimensions, it is only two-dimensional because it has no thickness. To completely represent a structure in three dimensions, the relationship between different horizons must be illustrated. A cross section of the geometry that would be seen on the face of a slice through the volume is the simplest representation of the relationship

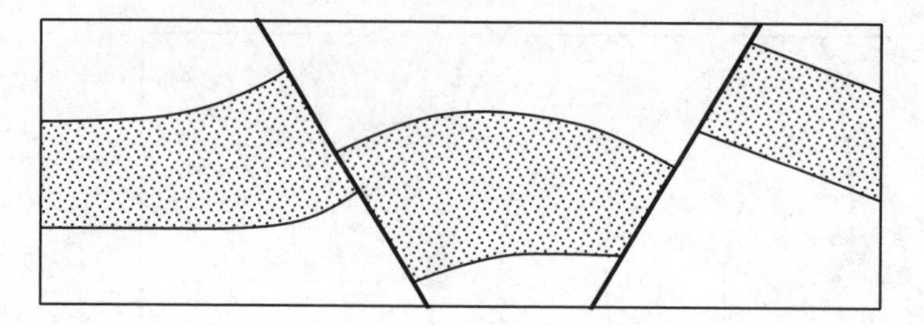

Fig. 1.5. East–west cross section across the structure in Fig. 1.1

between horizons. In this book cross sections will be assumed to be vertical unless it is stated otherwise. A cross section through multiple surfaces illustrates their individual geometries and defines the relationships between the surfaces (Fig. 1.5). The geometry of each surface provides constraints on the geometry of the adjacent surfaces. The relationships between surfaces forms the foundation for many of the techniques of structural interpretation that will be discussed.

1.3 Map Units and Contact Types

The primary concern of this book is the mapping and map interpretation of geologic contacts and geologic units. A contact is the place or surface where two different kinds of rocks come together. A unit is a closed volume between two or more contacts. The geometry of a structure is represented by the shape of the contacts between adjacent units. Dips or layering within a unit, such as in a crossbedded sandstone, are not necessarily parallel to the contacts between map units. Geological maps are made for a variety of purposes and the purpose typically dictates the nature of the map units. It is important to consider the nature of the units and the contact types in order to distinguish between geometries produced by deposition and those produced by deformation. Units may be either right side up or overturned. A stratigraphic horizon is said to face in the direction toward which the beds get younger. If possible, the contacts to be used for structural interpretation should be parallel and have a known paleogeographic shape. This will allow the use of a number of powerful rules in the construction and validation of map surfaces (Chaps. 3 and 4), in the construction of cross sections (Chap. 7), and should result in geometries that can be restored to their original shapes as part of the structural validation process (Chap. 8). Contacts that were originally horizontal are preferred. Even if a restoration is not actually done, the concept that the map units were originally horizontal is implicit in many structural interpretations.

1.3.1 Depositional Contacts

A depositional contact is produced by the accumulation of material adjacent to the contact (after Bates and Jackson 1987). Sediments, igneous or sedimentary extrusions, and air-fall igneous rocks have a depositional lower contact which is parallel to the pre-existing surface. The upper surface of such units is usually, but not always, close

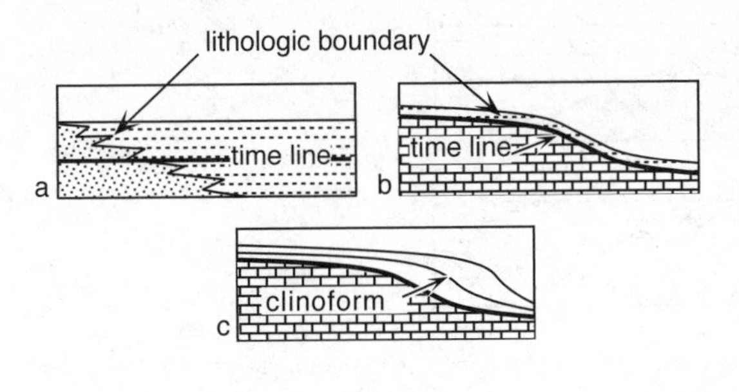

Fig. 1.6. Cross sections showing primary depositional lithologic contacts that are not horizontal. **a** Laterally equivalent deposits of sandstone and shale. The depositional surface is a time line, not the lithologic boundary. **b** Draped deposition parallel to a topographic slope. c Primary topography associated with clinoform deposition

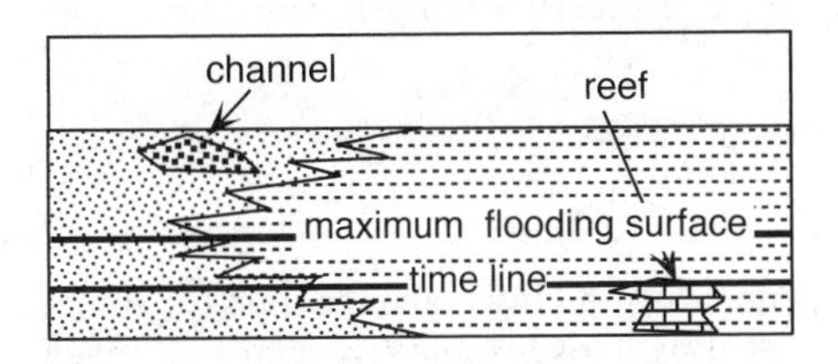

Fig. 1.7. Cross sections showing primary sedimentary facies relationships and maximum flooding surface. All time lines are horizontal in this example

to horizontal. A conformable contact is one in which the strata are in unbroken sequence and in which the layers are formed one above the other in parallel order, representing the uninterrupted deposition of the same general type of material, e.g., sedimentary or volcanic (after Bates and Jackson 1987).

Lithologic boundaries that represent lateral facies transitions (fig 1.6a), were probably not horizontal to begin with. Certain sedimentary deposits drape over pre-existing topography (Fig. 1.6b) while others are deposited with primary depositional slopes (Fig. 1.6c). The importance of the lack of original horizontality depends on the scale of the map relative to the magnitude of the primary dip of the contact. Contacts that dip only a few degrees might be treated as originally horizontal in the interpretation of a local map area, but the depositional contact between a reef and the adjacent basin sediments may be close to vertical (Fig. 1.7), for example, and could not be considered as originally horizontal at any scale. Depositional contacts that had significant original topographic relief (Fig. 1.7) should be restored to their original depositional geometry, not to the horizontal.

1.3.2 Unconformities

An unconformity is a surface of erosion or nondeposition that separates younger strata from older strata. An angular unconformity (Fig. 1.8a) is an unconformity between two groups of rocks whose bedding planes are not parallel. An angular unconformity with a low angle of discordance is likely to appear conformable at a local scale. Distinguishing between conformable contacts and low-angle unconformities is difficult but can be extremely important to the correct interpretation of a map. A buttress unconformity (Fig. 1.8b; Bates and Jackson 1987) is a surface on which onlapping strata abut against a steep topographic scarp of regional extent. A disconformity (Fig. 1.8c) is an unconformity in which the bedding planes above and below the break are essentially parallel, indicating a significant interruption in the

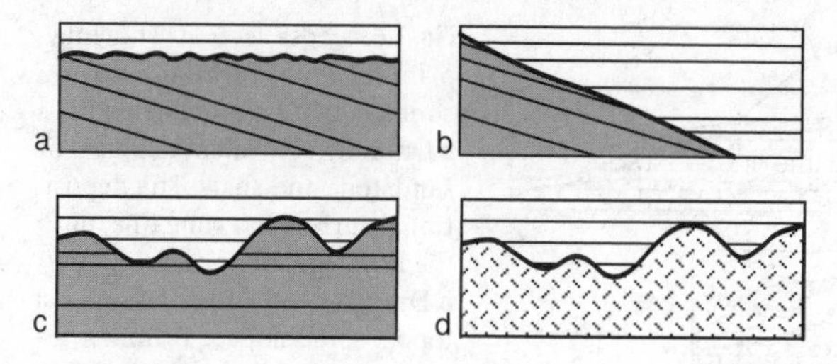

Fig. 1.8. Unconformity types. The unconformity (*heavy line*) is the contact between the older, underlying *shaded* units and the younger, overlying *unshaded* units. **a** Angular unconformity. **b** Buttress or onlap unconformity. **c** Disconformity. **d.** Nonconformity. The *patterned* unit is plutonic or metamorphic rock

orderly sequence of sedimentary rocks, generally by an interval of erosion (or sometimes of nondeposition), and usually marked by a visible and irregular or uneven erosion surface of appreciable relief. A nonconformity (Fig. 1.8d) is an unconformity developed between sedimentary rocks and older plutonic or massive metamorphic rocks that had been exposed to erosion before being covered by the overlying sediment.

1.3.3 Time-Equivalent Boundaries

The best map-unit boundaries for regional structural and stratigraphic interpretation are time-equivalent across the map area. Time-equivalent boundaries are normally established using fossils or radiometric age dates and may cross lithologic boundaries. It can be difficult to establish time-equivalent map horizons because of the absence or inadequate resolution of the paleontologic or radiometric data, lithologic and paleontologic heterogeneity in the depositional environment, and because of the occurrence of time-equivalent nondeposition or erosion in adjacent areas. Time-equivalent map-unit boundaries may be based on certain aspects of the physical stratigraphy. A *sequence* is a conformable succession of genetically related strata bounded by unconformities and their correlative conformities (Mitchum 1977; Van Wagoner et al. 1988). A parasequence is a subunit within a sequence that is bounded by marine flooding surfaces (Van Wagoner et al. 1988) and the approximate time equivalence along flooding surfaces makes them suitable for structural mapping. A maximum flooding surface (Fig. 1.7; Galloway 1989) can be the best for regional correlation because the deepest-water deposits can be correlated across lithologic boundaries. At the time of maximum flooding, the sediment input is at a minimum and the associated sedimentary deposits are typically condensed sections, seen as radioactive shales or thin, very fossiliferous carbonates. Volcanic ash fall deposits, which become bentonites after diagenesis, are excellent time markers. Because an ash fall drapes the topography and is relatively independent of the depositional environment, it can be used for regional correlation and to determine the depositional topography (Asquith 1970).

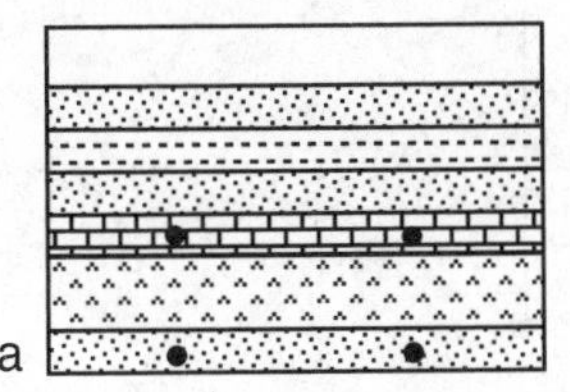

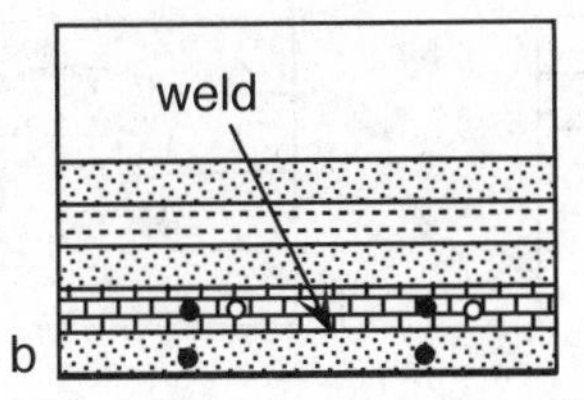

Fig. 1.9. Cross sections illustrating the formation of a welded contact. *Solid dots* are fixed material points above and below the unit which will be depleted. **a** Sequence prior to depletion. **b** Sequence after depletion: *solid dots* represent the final positions of original points in a hangingwall without lateral displacement; *open circles* represent final positions of original points in a hangingwall with lateral displacement

1.3.4 Welds

A weld joins strata originally separated by a depleted or withdrawn unit (after Jackson 1995). Welds are best known where a salt bed has been depleted by substratal dissolution or by flow (Fig. 1.9). If the depleted unit was deposited as part of a stratigraphically conformable sequence, the welded contact will resemble a disconformity. If the depleted unit was originally an intrusion, like a salt sill, the welded contact will return to its original stratigraphic configuration. A welded contact may be recognized from remnants of the missing unit along the contact. Lateral displacement may occur across the weld before or during the depletion of the missing unit (Fig. 1.9b). Welded contacts may crosscut bedding in the country rock if the depleted unit was originally crosscutting, as, for example, salt diapir that is later depleted.

1.3.5 Intrusive Contacts and Veins

An intrusion is a rock, magma, or sediment mass that has been emplaced into another distinct unit. Intrusions (Fig. 1.10) may form concordant contacts that are parallel to the layering in the country rock, or discordant contacts that crosscut the layering in the country rock. A single intrusion may have contacts that are locally concordant and discordant.

A vein is a relatively thin, normally tabular, rock mass of distinctive lithologic character, usually crosscutting the structure of the host rock. Many veins are depositional and represent the filling of a fracture, whereas others are the result of replacement of the country rock. Veins are mentioned here with intrusions because the contact relationships and unit geometries may be similar to those of some intrusions.

1.3.6 Other Boundaries

Many other attributes of the rock units and their contained fluids can be mapped, for example, the porosity, the oil-water contact, or the grades of mineral deposits. Most of the mapping techniques to be discussed will apply to any type of unit or contact. Some interpretation techniques, particularly those for fold interpretation, depend on the

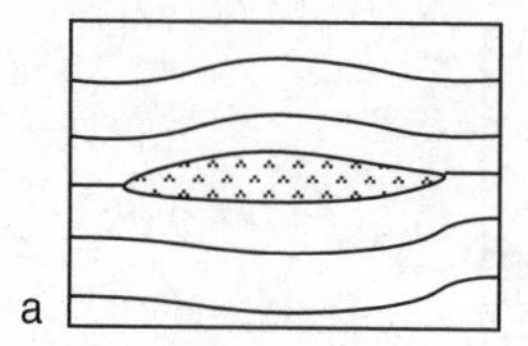

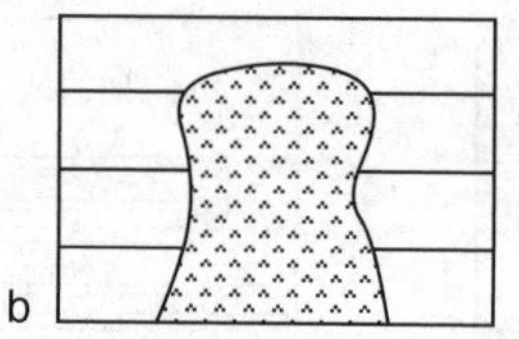

Fig. 1.10. Cross sections of intrusions. Intrusive material is *patterned* and *lines* represent layering in the country rock. **a** Concordant. **b** Discordant

contacts being originally planar boundaries and so those methods may not apply to nonstratigraphic boundaries.

1.4 Thickness

The thickness of a unit is the perpendicular distance between its bounding surfaces (Fig. 1.11a). The true thickness does not depend on the orientation of the bounding surfaces. If a unit has variable thickness, various alternatives might be used, such as the shortest distance between upper and lower surfaces or the distance measured perpendicular to either the upper or lower surface. The definition used here is based on the premise that if the unit was deposited with a horizontal surface but a variable thickness, then the logical measurement direction would be the thickness measured perpendicular to the upper surface, regardless of the structural dip of the surface (Fig. 1.11b).

Thickness variations can be due to a variety of stratigraphic and structural causes. Growth of a structure during the deposition of sediment typically results in thinner stratigraphy on the structural highs and thicker stratigraphy in the lows. Both growth folds and growth faults occur. A sedimentary package with its thickness influenced by an active structure is known as a growth unit or growth sequence. The upper part of a growth structure may be erosional at the same time that the lower parts are depositional. Thickness variations may be the result of differential compaction during and after deposition. If the composition of a unit undergoes a facies change from relatively uncompactable (i.e. sand) to relatively compactable (i.e. shale) then after burial and compaction the unit thickness will vary as a function of lithology. Deformation-related thickness changes are usually accompanied by folding, faulting, or both within the unit being mapped. Deformation-related thickness changes are likely to correlate to position within a structure or to structural dip.

1.5 Folds

A fold is a bend due to deformation of the original shape of a surface. An antiform is convex upward; an anticline is convex upward with older beds in the center. A synform is concave upward; a syncline is concave upward with younger beds in the center. Original curves in a surface, for example grooves or primary thickness changes, are not considered here to be anticlines or synclines.

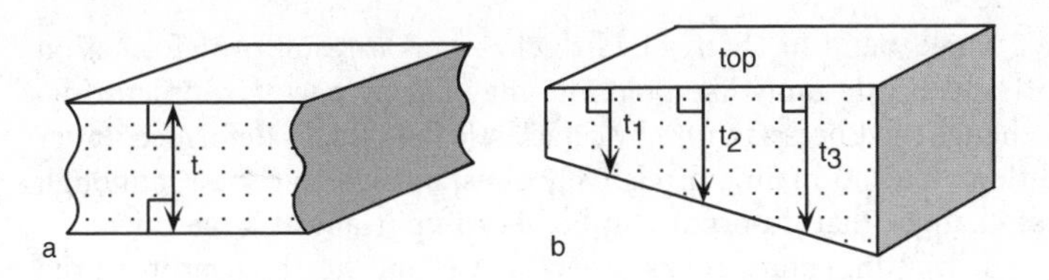

Fig. 1.11. Thickness (*t*). **a** Unit of constant thickness. **b** Unit of variable thickness

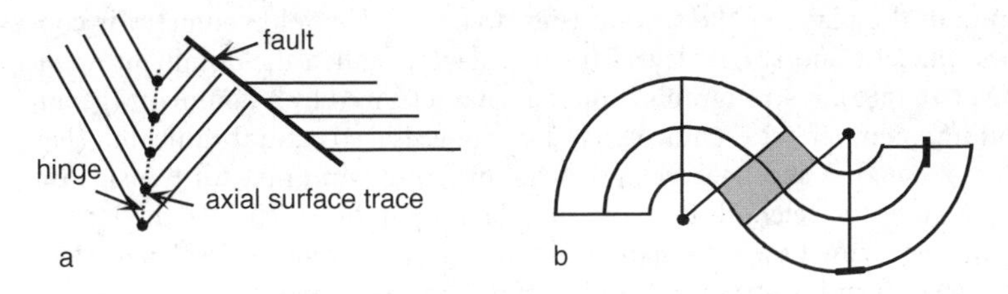

Fig. 1.12. Regions of uniform dip properties. **a** Dip domains. **b** Concentric domains separated by a planar dip domain (*shaded*)

1.5.1 Styles

Folds may be characterized by domains of uniform dip, by a uniform variation of the dip around a single center of curvature, or may combine regions of both styles. Regions of uniform dip (Fig. 1.12a) are called dip domains (Groshong and Usdansky 1988). Dip domains are separated from one another by axial surfaces or faults. Dip domains have also been referred to as kink bands, but the term kink band has mechanical implications that are not necessarily appropriate for every structure. An axial surface (fig 1.12a) is a surface that connects fold hinge lines, where a hinge line is a line of maximum curvature on the surface of a bed (Dennis 1967). A hinge (Fig. 1.12a) is the intersection of a hinge line with the cross section. A concentric domain (Fig. 1.12b) is defined here as a region in which beds have the same center of curvature. Dips in a concentric domain are everywhere perpendicular to a radius through the center of curvature. The center of curvature is determined as the intersection point of lines drawn perpendicular to the dips (Busk 1929; Reches et al. 1981). A circular curvature domain does not possess a line of maximum curvature and thus does not strictly have an axial surface. Dips within a domain may vary from the average values. If the dips are measured by a hand-held clinometer, variations of a few degrees are to be expected due to the natural variation of bedding surfaces and the imprecision of the measurements.

As a generality, the structural style is controlled by the mechanical stratigraphy and the directions of the applied forces. Mechanical stratigraphy is the stratigraphy described in terms of its physical properties. The mechanical properties that control the fold geometry are the stiffness contrasts between layers, the presence or absence of layer-parallel slip, and the relative layer thicknesses. Stiff (also known as competent) lithologies (for example, limestone, dolomite, cemented sandstone) tend to maintain constant bed thickness, and soft (incompetent) lithologies (for example, evaporites, overpressured shale, shale) tend to change bed thickness as a result of deformation (Fig. 1.13). Very stiff and brittle units like dolomite may fail by pervasive fracturing, however, and then change thickness as a unit by cataclastic flow. Rocks deformed under sedimentary conditions tend to maintain relatively constant bed thickness, although the small thickness changes that do occur can be very important. If large thickness changes are observed in sedimentary rocks other than evaporites or overpressured shale, primary stratigraphic variations should be considered as a strong possibility.

In cross section, folds may be harmonic, with all the layers nearly parallel to one another (Fig. 1.14), or disharmonic, with significant changes in the geometry between different units in the plane of the section (figs. 1.13, 1.15). The fold geometry is controlled by the thickest and stiffest layers (or multilayers) called the dominant members (Currie et al. 1962). A stratigraphic interval characterized by a dominant (geometry-controlling) member between two boundary zones is a structural-lithic unit (Fig. 1.15; Currie et al. 1962). A short-wavelength structural lithic unit may form inside the boundary of a larger-wavelength unit and be folded by it, in which case the longer-wavelength unit is termed the dominant structural-lithic unit and the shorter-wavelength unit is a conforming structural-lithic unit (Currie et al. 1962).

The dip changes within structural-lithic units may obscure the map-scale geometry. For example, the regional or map-scale dip in figure 1.15 is horizontal, although few dips of this attitude could be measured. Where small-scale folds exist, the map-scale geometry may be better described by the orientation of the median surface or the enveloping surface (Fig. 1.16; Turner and Weiss 1963, Ramsay 1967). The median surface (median line in two dimensions) is the surface connecting the inflection points of a folded layer. The inflection points are located in the central region of the fold limbs where the fold curvature changes from anticlinal to synclinal. An enveloping surface is the surface that bounds the crests (high points) of the upper surface or

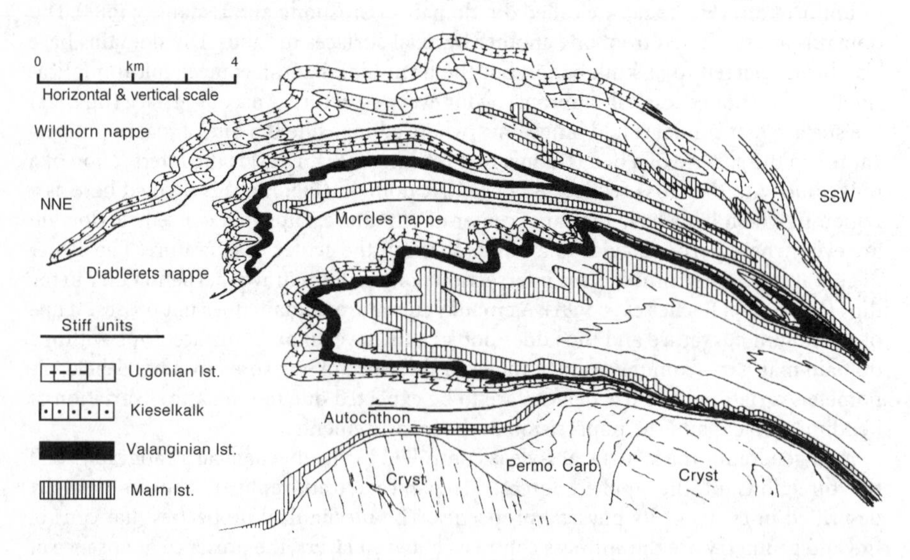

Fig. 1.13. Cross section of the Helvetic Alps, central Switzerland. Mechanical stratigraphy consists of thick carbonate units (stiff) separated by very thick shale units (soft). The folds, especially at the hinges, are circular in style. (After Ramsay 1981)

Fig. 1.14. Dip-domain style folds in an experimental model having a closely spaced multilayer stratigraphy. The *black* and *white layers* are plasticene and are separated by grease to facilitate layer-parallel slip. The *black layers* are slightly stiffer than the *white layers*. (After Ghosh 1968).

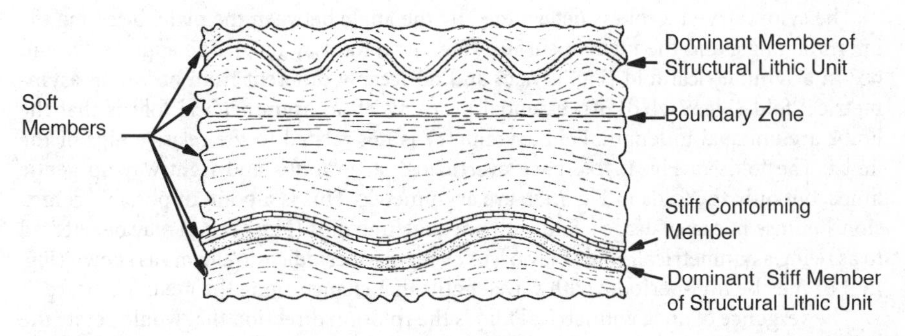

Fig. 1.15. Structural-lithic units. (After Currie et al. 1962)

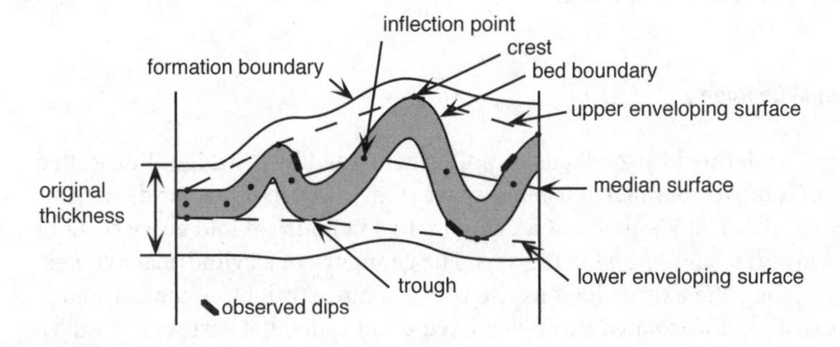

Fig. 1.16. Terminology for folded surfaces

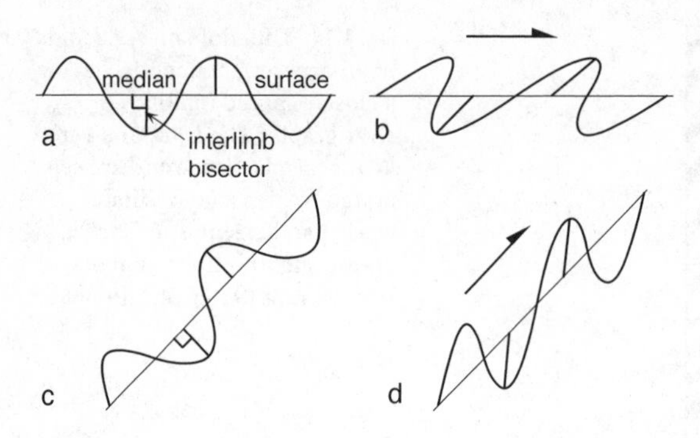

Fig. 1.17. Fold symmetry. **a** Symmetrical, upright. **b** Asymmetrical, overturned. **c** Symmetrical, overturned. **d** Asymmetrical, upright. *Arrow* gives direction of vergence

the troughs (low points) of the lower surface on a single unit. Figure 1.16 shows that the dip observed at a single location need not correspond to the dip of either the median surface, the enveloping surface, or to the trace of a formation boundary. The dip of the median surface may be more representative of the dip required for map-scale interpretation than the locally observed bedding dips.

The symmetry of a fold is determined by the angle between the plane bisecting the interlimb angle and the median surface (Fig. 1.17a; Ramsay 1967). The angle is close to 90° in a symmetrical fold (Fig. 1.17a,c) and noticeably different from 90° in an asymmetrical fold (Fig. 1.17b,d). An essential property of an asymmetrical fold is that the limbs are unequal in length. Fold asymmetry is not related to the relative dips of the limbs. The folds in Fig. 1.17b,c have overturned steep limbs and right-way-up gentle limbs, but only the folds in Fig. 1.17b are asymmetric. This is a point of possible confusion, because in casual usage a fold with unequal limb dips (Fig. 1.17 b,c) may be referred to as being asymmetrical. Folds may occur as regular periodic waveforms as shown (Fig. 1.17) or may be non-periodic with wavelengths that change along the median surface.

The vergence of an asymmetrical fold is the rotation direction that would rotate the axial surface of an antiform from an original position perpendicular to the median surface to its observed position at a lower angle to the median surface. The vergence of the folds in Fig. 1.17b,d is to the right.

1.5.2 Three-Dimensional Geometry

A cylindrical fold is defined by the locus of points generated by a straight line, called the fold axis, that is moved parallel to itself in space (Fig. 1.18a). In other words, a cylindrical fold has the shape of a portion of a cylinder. In a cylindrical fold every straight line on the folded surface is parallel to the axis. The geometry of a cylindrical fold persists unchanged along the axis as long as the axis remains straight. A conical fold is generated by a straight line rotated through a fixed point called the vertex (Fig. 1.18b). The cone axis is not parallel to any line on the cone itself. A conical fold changes geometry and terminates along the trend of the cone axis.

The crest line is the trace of the line which joins the highest points on successive cross sections through a folded surface (Figs. 1.18, 1.19a; Dennis 1967). A trough line is the trace of the lowest elevation on cross sections through a horizon. The plunge of a cylindrical fold is parallel to the orientation of its axis or a hinge line (Fig. 1.19b). The

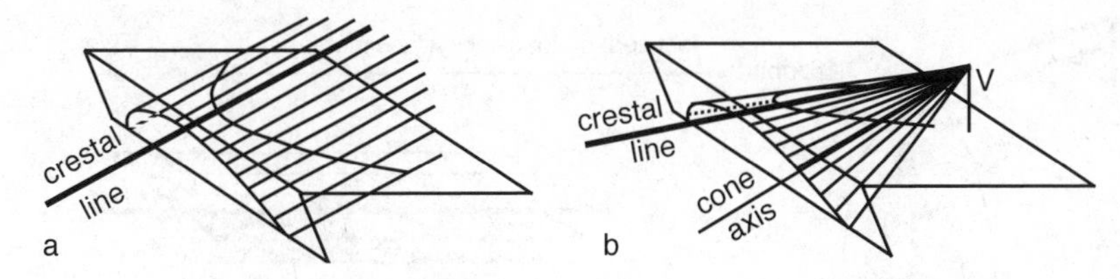

Fig. 1.18. Three-dimensional fold types. **a** Cylindrical. All straight lines on the cylinder surface are parallel to the fold axis and to the crestal line. **b** Conical. *V* vertex of the cone. Straight lines on the cone surface are not parallel to the cone axis

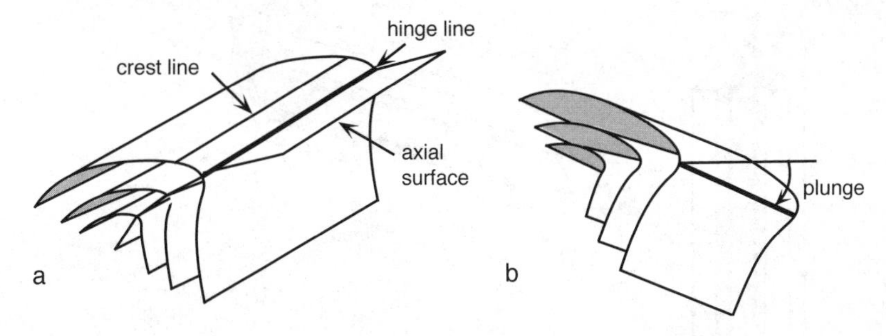

Fig. 1.19. Cylindrical folds. **a** Non-plunging. **b** Plunging

most useful measure of the plunge of a conical fold is the orientation of its crest line or trough line (Bengtson 1980).

The complete orientation of a fold requires the specification of the orientation of both the fold axis and the axial surface (Fig. 1.20). In the case of a conical fold, the orientation can be specified by the orientation of the axial surface and the orientation of the crestal line on a particular horizon. Common map symbols for folds are given in Fig. 1.21.

1.5.3 Mechanical Origins

The fundamental mechanical types of folding are based on the direction of the causative forces relative to layering (Ramberg 1963; Gzovsky et al. 1973; Groshong 1975), namely longitudinal contraction, transverse contraction, and longitudinal extension (Fig. 1.22). If the stratigraphy is without mechanical contrasts, forces parallel to layering produce either uniform shortening and thickening or uniform extension and thinning. If some shape irregularity is pre-existing, then it is amplified by layer-parallel shortening to give a passive fold. If the stratigraphy has significant mechanical contrasts, then a mechanical instability can occur that leads to buckle folding in contraction and pinch-and-swell structure (boudinage) in extension. If the forces are not equal vertically, then a forced fold is produced, regardless of the mechanical stratigraphy. Longitudinal contraction, transverse contraction, and longitudinal extension are end-member boundary conditions; they may be combined to produce folds with combined properties.

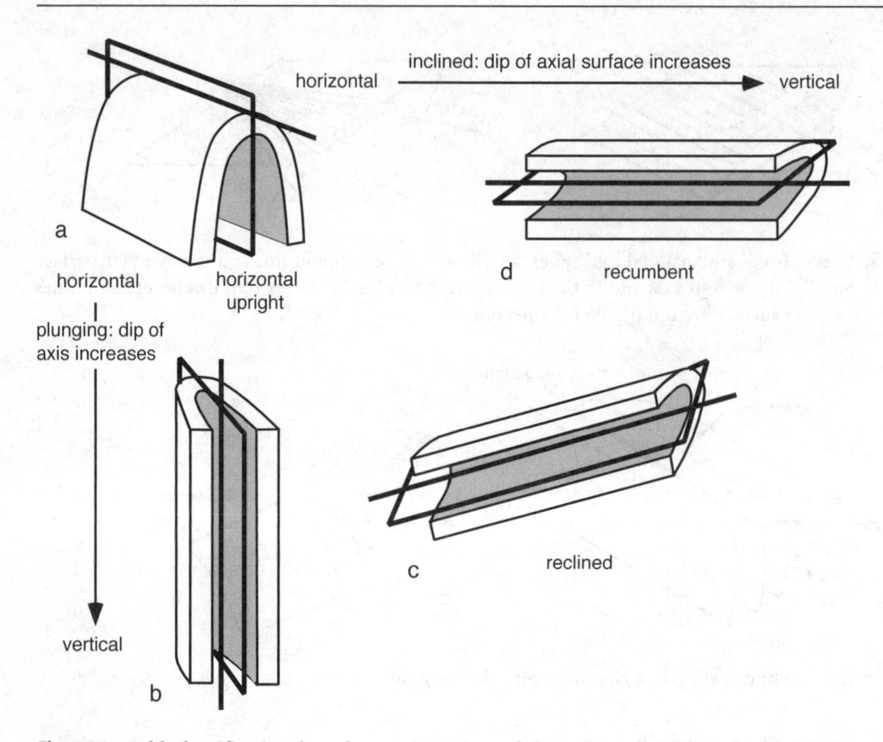

Fig. 1.20. Fold classification based on orientation of the axis and axial surface. **a** Horizontal upright: horizontal axis and vertical axial surface. **b** Vertical: vertical axis and vertical axial surface. **c** Reclined: inclined axis and axial surface. **d** Recumbent: horizontal axis and axial surface. (After Fleuty, 1964)

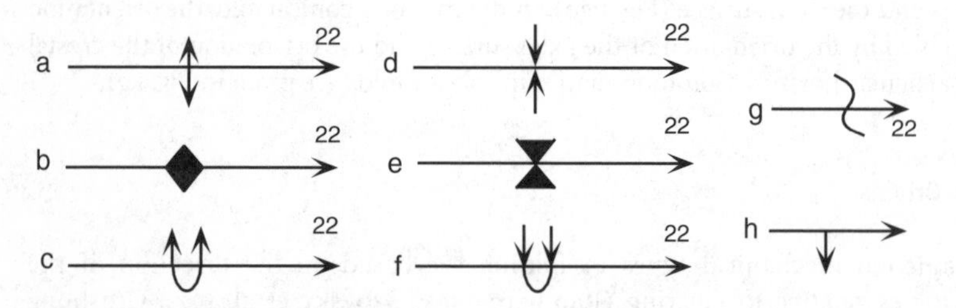

Fig. 1.21. Common map symbols for folds. Fold trend is indicated by the *long line*, plunge by the *arrowhead*, with amount of plunge given if known. The fold trend may be the axial trace, crest or trough line or a hinge line. **a, b** Anticline. **c** Overturned anticline; both limbs dip away from the core. **d, e** Syncline. **f** Overturned syncline; both limbs dip toward the core. **g** Minor folds showing trend and plunge of axis. **h** Plunging monocline with only one dipping limb

Buckle folds normally form with the fold axes perpendicular to the maximum principal compressive stress, σ_1. The folds are long and relatively unchanging in geometry parallel to the fold axis but highly variable in cross section. Buckle folds are characterized by the presence of a regular wavelength that is proportional to the thickness of the stiff

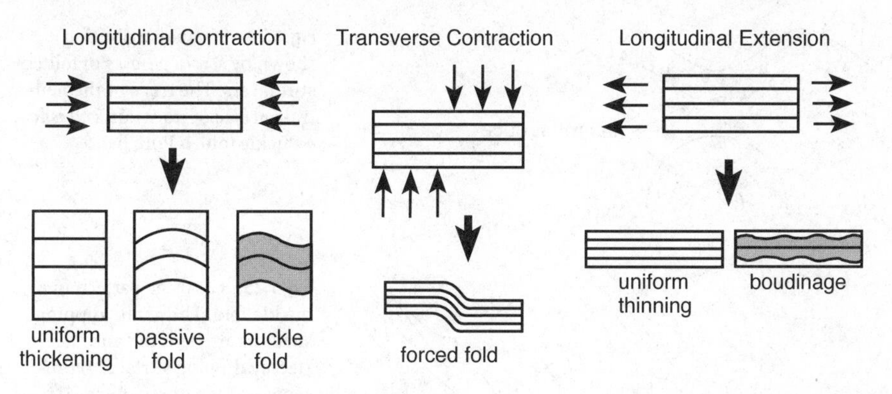

Fig. 1.22. End–member displacement boundary conditions showing responses related to the mechanical stratigraphy. *Shaded* beds are stiff lithologies, *unshaded* beds are soft lithologies

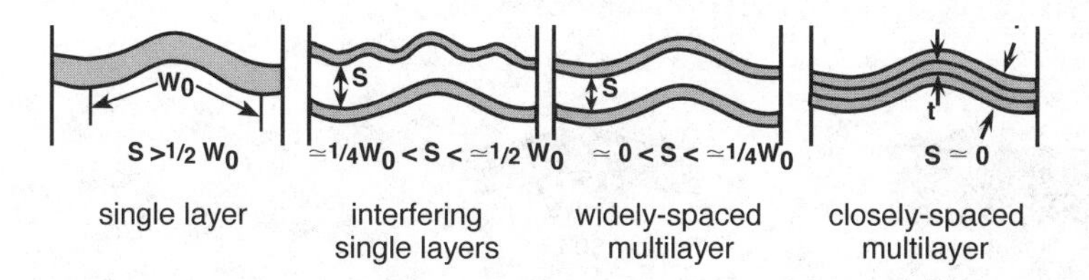

Fig. 1.23. Transition from single layer folding to multilayer folding as space between stiff layers decreases. The stiff layers are *shaded*

unit(s). A single-layer buckle fold consists of a stiff layer in a surrounding confining medium. The dip variations associated with a given stiff layer die out into the regional dip within the softer units at a distance of about one-half arc length away from the layer (Fig. 1.15). In the author's experience, buckle folds in sedimentary rocks typically have arc-length to thickness ratios of 5 to 30, with common values in the range of 6 to 10.

As buckled stiff layers become spaced closer together, the wavelengths begin to interfere (Fig. 1.23) resulting in disharmonic folds. Once the layers are sufficiently closely spaced, they fold together as a multilayer unit. A multilayer unit has a much lower buckling stress and an appreciably shorter wavelength than a single layer of same thickness (Currie et al. 1962). Stiff units, either single layers or multilayers, tend to produce circular to sinusoidal folds if enclosed in a sufficient thickness of softer material (Fig. 1.13). Stratigraphic sections made up of relatively thin-bedded multilayer units result in folds that tend to have planar dip-domain styles (Fig. 1.14).

The strain distribution in the stiff layer of a buckle fold (Fig. 1.24a) is approximately layer-parallel shortening throughout most of the fold. The neutral surface separates regions of layer-parallel extension from regions of layer-parallel contraction. Only in a pure bend (Fig. 1.24b) are the areas of extension and contraction about equal and the neutral surface in the middle of the layer. The strain in thick soft layers between stiff layers is shortening approximately perpendicular to the axial surface. The strain in thin soft layers between stiff layers may be close to layer-parallel shear strain. The pure bending strain distribution is more closely approached in transverse contraction folds, for example above a salt dome, than in buckle folds.

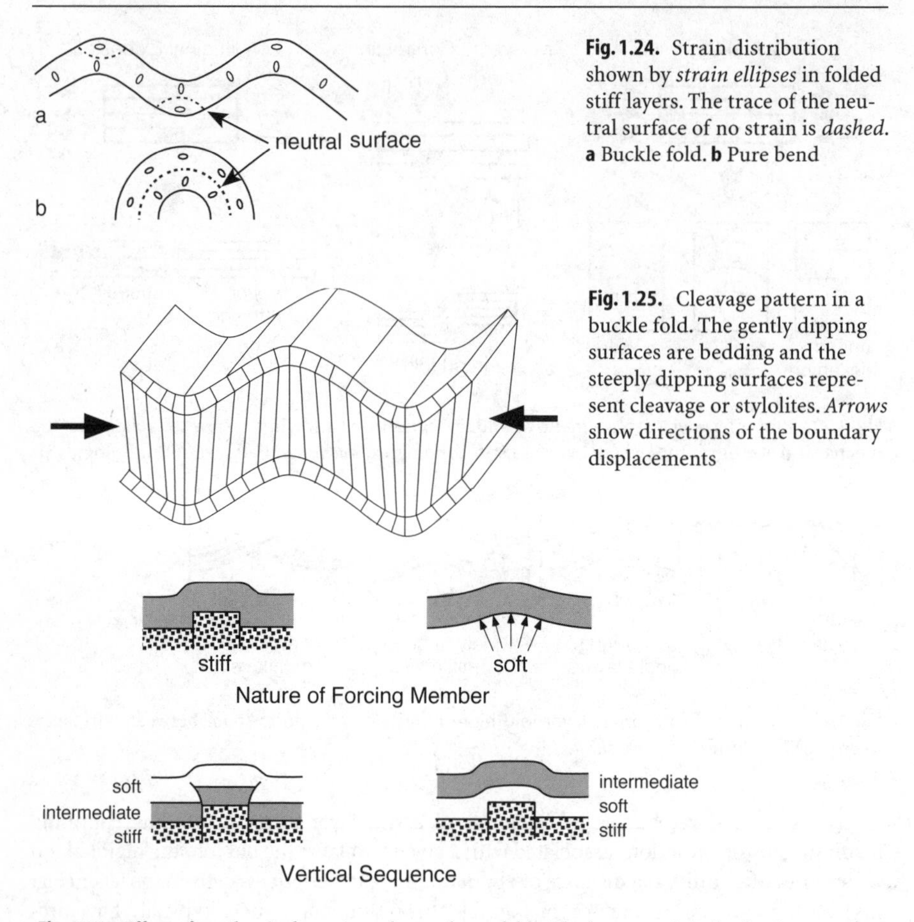

Fig. 1.24. Strain distribution shown by *strain ellipses* in folded stiff layers. The trace of the neutral surface of no strain is *dashed*. **a** Buckle fold. **b** Pure bend

Fig. 1.25. Cleavage pattern in a buckle fold. The gently dipping surfaces are bedding and the steeply dipping surfaces represent cleavage or stylolites. *Arrows* show directions of the boundary displacements

Fig. 1.26. Effect of mechanical stratigraphy on drape folds. The lowest unit is the forcing member

Cleavage planes and tectonic stylolites in a fold can indicate the mechanical origin of the fold because they form approximately perpendicular to the maximum shortening direction by processes that range from grain rotation to pressure solution (Groshong 1988). Cleavage in a buckle fold is typically at a high angle to bedding (Fig. 1.25), being more nearly perpendicular to bedding in stiff units and more nearly parallel to the axial surface in soft units. Cleavage that is approximately perpendicular to bedding produces a cleavage fan across the fold. The line of the cleavage-bedding intersection is approximately or exactly parallel to the fold axis and can be used to help determine the axis.

Folds produced by an unequal distribution of forces in transverse contraction (Fig. 1.22) are termed drape folds or forced folds (Stearns 1978). Drape folds tend to be round to blocky or irregular in map view. The major control on the form of the fold is the rheology of the forcing member (Fig. 1.26). A stiff and brittle forcing member (i.e., crystalline basement) leads to narrow fault boundaries at the base of the structure and strain that is highly localized in the zone above the basement fault. A soft unit between

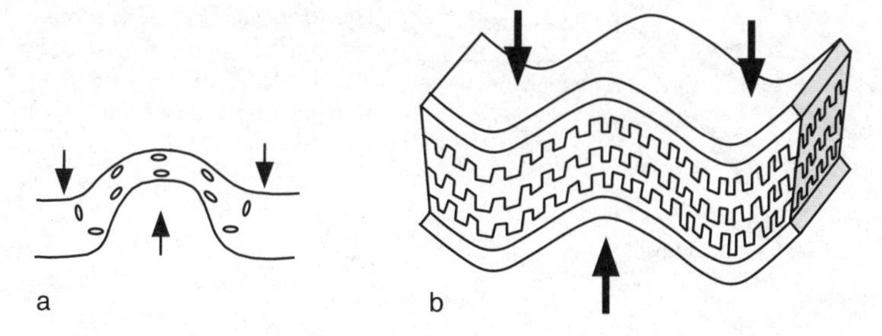

Fig. 1.27. Strain and cleavage patterns in transverse contraction folds produced by differential vertical displacement. **a** Strain distribution above a model salt dome (after Dixon 1975). **b** Cleavage or stylolites parallel to bedding. *Arrows* show directions of the boundary displacements

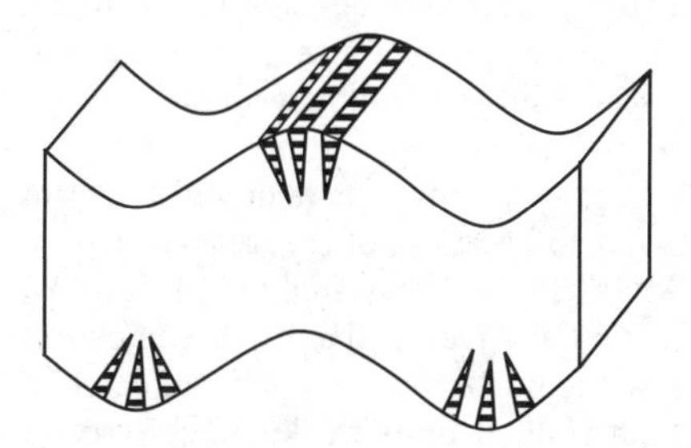

Fig. 1.28. Veins due to outer–arc bending stresses

a stiff forcing member and the cover sequence will cause the deformation to be disharmonic. A soft forcing member (like salt) typically produces round to elliptical structures with deformation widely distributed across the uplift.

Little strain need occur in the uplifted or downdropped blocks associated with a stiff forcing member. Nearly all the strain is localized in the fault zone between the blocks or in the steep limb of the drape fold over the fault zone. A soft forcing member (i.e., salt) will distribute the curvature and strain widely over the uplifted region (Fig. 1.27a). Because cleavage and stylolites form perpendicular to the shortening direction, in folds produced by displacements at a high angle to bedding, the expected cleavage and stylolite direction is parallel to bedding (Fig. 1.27b). In highly deformed rocks, cleavage parallel to bedding might be the result of deformation caused by a large amount of layer-parallel slip or by isoclinal refolding of an earlier axial-plane cleavage.

Extension fractures and veins may form due to the bending stresses in the outer arc of a fold (Fig. 1.28). Such features should become narrower and die out toward the neutral surface. The fracture plane is expected to be approximately parallel to the axis of the fold and the fracture-bedding line of intersection should be parallel to the fold axis. Bending fractures might occur in any type of fold.

1.6 Faults

A fault (Fig. 1.29) is a surface or narrow zone across which there has been relative displacement of the two sides parallel to the zone (after Bates and Jackson 1987). The term displacement is the general term for the relative movement of the two sides of

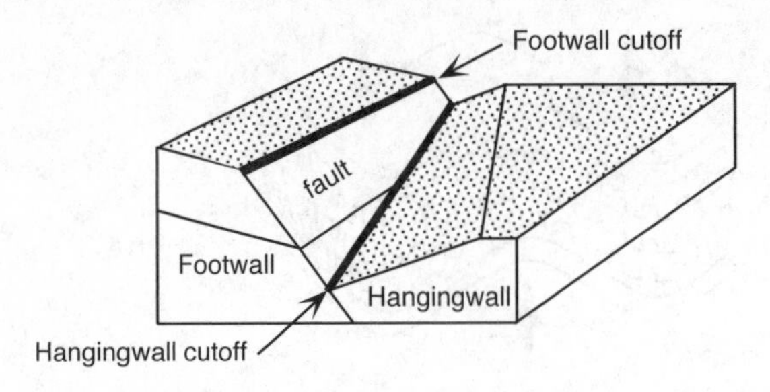

Fig. 1.29. General terminology for a surface (*patterned*) offset by a fault. *Heavy lines* are hangingwall and footwall cutoff lines

the fault, measured in any chosen direction. A shear zone is a general term for a relatively narrow zone with subparallel boundaries in which shear strain is concentrated (Mitra and Marshak 1988). As the terms are usually applied, a bed, foliation trend, or other marker horizon maintains continuity across a shear zone but is broken and displaced across a fault. It may be difficult to distinguish between a shear zone and a fault zone on the basis of observations at the map scale, and so here the term fault will be understood to include both faults and shear zones.

The term hangingwall refers to the strata originally above the fault and the term footwall to strata originally below the fault (Fig. 1.29). Because of the frequent repetition of the terms, hangingwall and footwall will often be abbreviated to HW and FW, respectively. A cutoff line is the line of intersection of a fault and a displaced horizon (Fig. 1.29). The HW and FW cutoff lines of a single horizon were in contact across a fault plane prior to displacement. Across a fault zone of finite thickness or across a shear zone, the HW and FW cutoffs were originally separated by some width of the offset horizon that is now in the zone.

1.6.1 Slip

Fault slip is the relative displacement of formerly adjacent points on opposite sides of the fault, measured along the fault surface (Fig. 1.30; Dennis 1967). Slip can be subdivided into horizontal and vertical components, the strike slip and dip slip components, respectively. A fault in which the slip direction is parallel to the trace of the cutoff line of bedding can be called a trace-slip fault. In horizontal beds a trace-slip fault is a strike-slip fault. Measurement of the slip requires the identification of the piercing points of displaced linear features on opposite sides of the fault. Suitable linear features at the map scale might be a channel sand, a facies boundary line, a fold hinge line, or the intersection line between bedding and a dike (Fig. 1.30). The displacement of dipping beds on faults oblique to the strike of bedding leads to complex relationships between the displacement and the slip in a specific cross-section direction, such as parallel to the dip of bedding. A strike-slip component of displacement is never visible on a vertical cross section.

1.6.2 Separation

Fault separation is the distance between any two parts of an index plane (e.g. bed or vein) disrupted by a fault, measured in any specified direction (Dennis 1967). The sep-

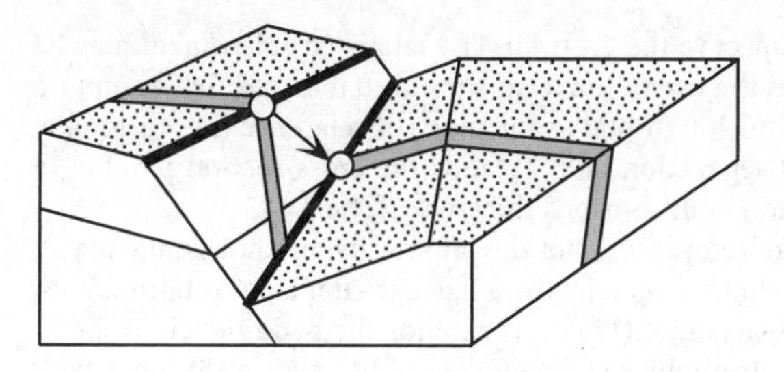

Fig. 1.30. Fault slip is the displacement of points (*open circles*) that were originally in contact across the fault. Here the correlated points represent the intersection line of a dike and a bed surface at the fault plane

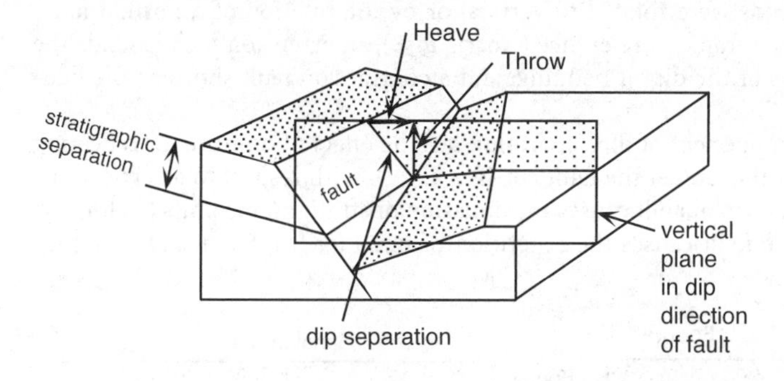

Fig. 1.31. Fault separation terminology

aration directions commonly important in mapping are parallel to fault strike, parallel to fault dip, horizontal, vertical and perpendicular to bedding. It should be noted that the definitions of the terms for fault separation and the components of separation are not always used consistently in the literature. Stratigraphic separation (Fig. 1.31) is the thickness of strata that originally separated two beds brought into contact at a fault (Bates and Jackson 1987) and is the stratigraphic thickness missing or repeated at the point, called the fault cut (Tearpock and Bischke 1991), where the fault is intersected. The amount of the fault cut is always a stratigraphic thickness.

Throw and heave (Fig. 1.31) are the components of fault separation most obvious on a structure contour map. Both are measured in a vertical plane in the dip direction of the fault. Throw is the vertical component of the dip separation measured in a vertical plane normal to the strike of the fault (Dennis 1967). Stratigraphic separation is not equal to the fault throw unless the marker horizons are horizontal (see Sect. 5.5.3). Heave is the horizontal component of the dip separation measured in a vertical plane normal to the strike of the fault (Dennis 1967).

1.6.3 Geometrical Classifications

A fault is termed normal or reverse on the basis of the relative displacement of the hangingwall with respect to the footwall (Fig. 1.32). For a normal fault, the hangingwall is displaced down with respect to the footwall, and for a reverse fault the hangingwall is displaced up with respect to the footwall. The relative displacement may be either a slip or a separation and the use of the term should so indicate, for example, a *normal-separation* fault. Using the horizontal as the plane of reference (i.e., originally horizontal bedding), a normal-separation fault extends the cross section parallel to bedding and a reverse-separation fault shortens the cross section.

Using bedding as the frame of reference is not the same as using a horizontal plane, as illustrated by Fig. 1.33 which shows the faults from Fig. 1.32 after a 90° rotation. With bedding vertical, a reverse displacement (Fig. 1.33) extends the bedding while shortening a horizontal line. The fault might have been caused by reverse slip on a fault formed after the beds were rotated to vertical or by the rotation of a normal fault. Using bedding as the frame of reference (Norris 1958), an extension fault extends the bedding, regardless of the dip of bedding, and a contraction fault shortens the bedding.

A fault with a component of dip separation has the effect of omitting or repeating stratigraphy across the fault at the fault cut (Fig. 1.34a). With respect to a vertical line or a vertical well, a normal fault causes the omission of stratigraphic units (well 1, Fig. 1.34b) and a reverse fault causes the repetition of units (well 1, Fig. 1.34c). Complex

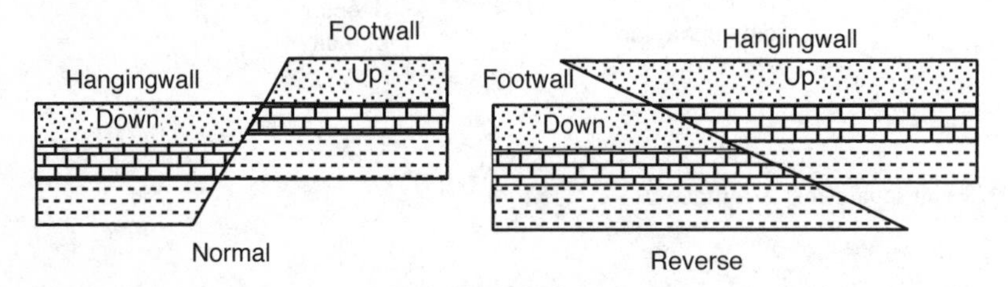

Fig. 1.32. Vertical cross section showing the relative fault displacement terminology with horizontal as the reference plane

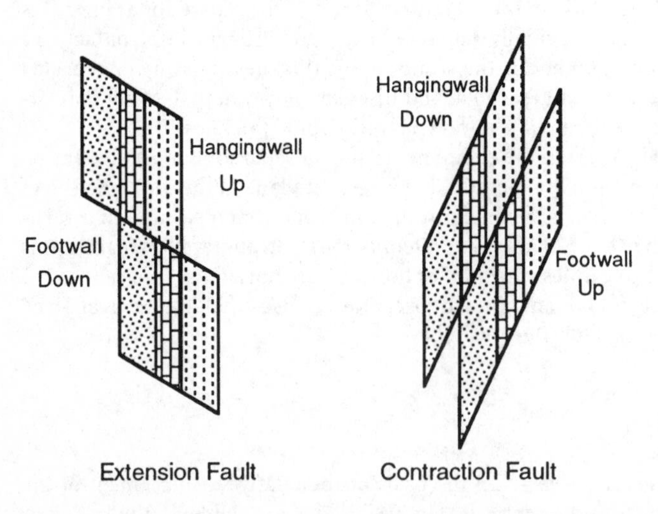

Fig. 1.33. Vertical cross section showing the relative fault displacement terminology with bedding as the reference plane

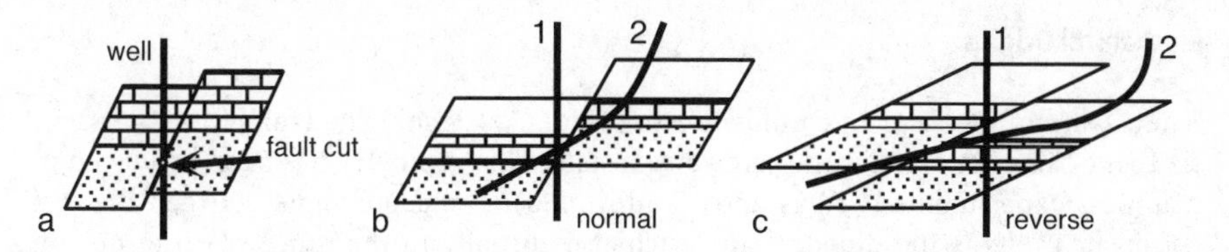

Fig. 1.34. Effect of well orientation on occurrence of missing or repeated section. All units are right side up and cross sections are vertical. **a** Point of observation is the position of the fault cut. **b** Wells penetrating a normal fault: *1* has missing section, *2* has repeated section. **c** Wells penetrating a reverse fault: *1* has repeated section, *2* has missing section

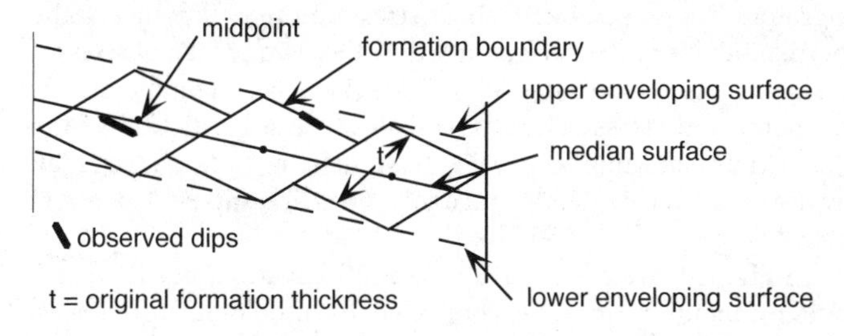

Fig. 1.35. Terminology for faulted surfaces

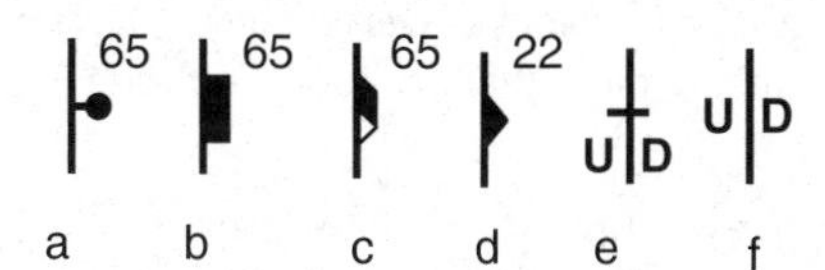

Fig. 1.36. Map symbols for faults indicating separation. **a–c** Normal separation, symbol on hangingwall. **d** Reverse separation, triangle on hangingwall. **e** Vertical fault, vertical separation indicated (*U* up; *D* down). **f** Fault of unknown dip, vertical separation indicated

stratigraphic omissions and repetitions may occur in a well that is not vertical (Mulvany 1992). For example, a well drilled from the footwall to the hangingwall of a normal fault will show repeated section down the well (well 2, Fig. 1.34b) and will show missing section down the well if the fault is reverse (well 2, Fig. 1.34c).

The median surface of a faulted unit (median line in two dimensions) is the surface connecting the midpoints of the blocks in the middle of the reference unit (Fig. 1.35). An enveloping surface is the surface that bounds the high corners or the low corners on a single unit. Dips within the fault blocks may all be different from the dip of either the median surface or the enveloping surface. Within a fault block the original thickness may remain unaltered by the deformation (Fig. 1.35), although the entire unit has been thickened or thinned as indicated by the changed thickness between the enveloping surfaces. Common map symbols for faults are given in Fig. 1.36.

1.6.4 Mechanical Origins

Faults commonly occur in conjugate pairs (Fig. 1.37). Conjugate faults form at essentially the same time under the same stress state. This geometry has been produced in countless experiments (Griggs and Handin 1960). The acute angle between the two conjugate faults is the dihedral angle which is usually in the range of 30 to 60°. In experiments the maximum principal compressive stress, σ_1, bisects the dihedral angle. The least principal compressive stress, σ_3, bisects the obtuse angle, and the intermediate principal stress, σ_2, is parallel to the line of intersection of the two faults. The slip directions are directly related to the orientation of the principal stresses (Fig. 1.37), with one set being right lateral (dextral) and the other set left lateral (sinestral).

The surface of the ground is a plane of zero shear stress and therefore one of the principal stresses is perpendicular to the surface and the other two principal stresses lie in the plane of the surface (Anderson 1905, 1951). From the experimental relationship between fault geometry and stress (Fig. 1.37), this leads to a prediction of the three most common fault types and their dips (Fig. 1.38). Relative to the horizontal, normal faults typically dip 60°, reverse faults average 30°, and strike-slip faults are vertical.

The predicted dips in Fig. 1.38 are good for a first approximation, but there are many exceptions. Fault orientations may be controlled by lithologic differences, changes in the orientations of the stress field below the surface of the ground, and by the presence of pre-existing zones of weakness. True triaxial stress states can result in the formation of two pairs of conjugate faults having the same dips but slightly different strikes, forming a rhombohedral pattern of fault blocks (Oertel 1965). Oertel faults are likely to be arranged in low-angle conjugate pairs that are 10-30° oblique to

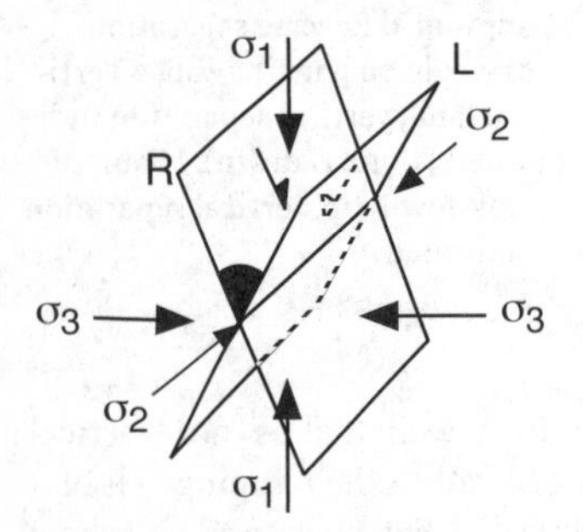

Fig. 1.37. Conjugate pair of faults related to stress orientation. The dihedral angle is *shaded*. The principal stresses are s_1, s_2, and s_3, in order of greatest to least compressive stress. *R* right-lateral fault plane; *L* left-lateral fault plane; *half-arrows* show the sense of shear on each plane

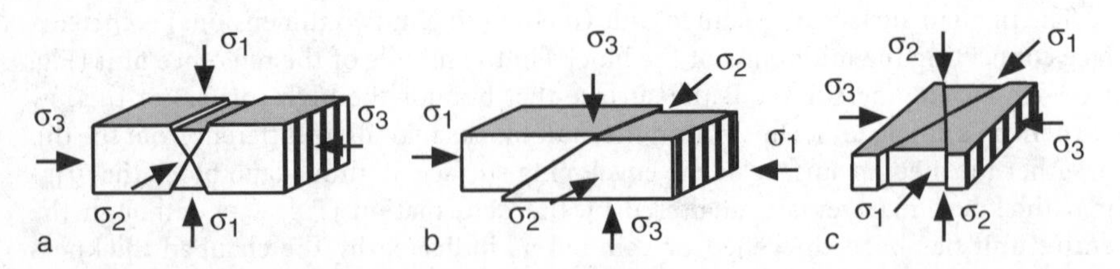

Fig. 1.38. Fault orientations at the surface of the earth predicted from Andersonian stress theory. **a** Normal. **b** Reverse. **c** Strike slip

each other. Faults will rotate to different dips as the enclosing beds rotate. Even with all the exceptions, it is still common for faults to have the approximate orientations given in Fig. 1.38.

1.6.5 Fault–Fold Relationships

A planar fault with constant displacement (Fig. 1.39a) is the only fault geometry that does not require an associated fold as a result of its displacement. Of course, all faults eventually lose displacement and end. A fault that dies out without reaching the surface of the earth is called blind, and a fault that reaches the present erosion surface is emergent, although whether it was emergent at the time it moved may not be known. Where the displacement ends at the tip of a blind fault, a fold must develop. Displacement on a curved fault will cause the rotation of beds in the hangingwall and perhaps in the footwall and will produce a fold (Fig. 1.39b,c). A generic term for the fold is a ramp anticline. The fold above a normal fault is commonly called a rollover anticline if the hangingwall beds near the fault dip toward the fault.

Fault dips may be controlled by the mechanical stratigraphy to form ramps and flats, although at the scale of the entire fault, the average dip may be maintained (i.e., 30° for a reverse fault). A flat is approximately parallel to bedding, at an angle of say, 10° or less. A ramp crosses bedding at an angle great enough to cause missing or repeated section at the scale of observation, say 10° or more. Characteristically, but not exclusively, ramps occur in stiff units such as limestone, dolomite, or cemented sandstone, whereas flats occur in soft units, such as shale or salt. Both normal and reverse faults may have segments parallel to bedding and segments oblique to bedding (Fig. 1.40). After displacement, hangingwall ramps and flats no longer necessarily match across the fault (Fig. 1.40b,c).

Another common fault geometry is listric (Fig. 1.41a) for which the dip of the fault changes continuously from steep near the surface of the earth to shallow or horizontal at depth. Both normal and reverse faults may be listric. The lower detachment of a listric fault is typically in a weak unit such as shale, overpressured shale or salt. Faults that flatten upward are comparatively rare, except in strike-slip regimes, and may be

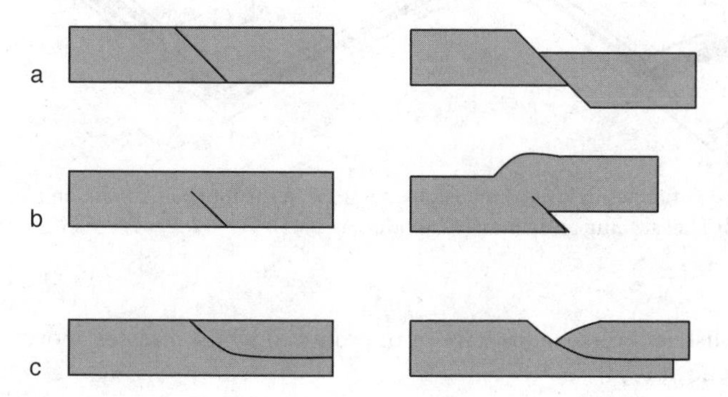

Fig. 1.39. Relationships between folds and faults. **a** Constant slip on a planar fault does not cause folding. **b** Slip on either a plane or curved fault that dies out produces a fold in the region of the fault tip. **c** Slip on a curved fault causes folding in the hangingwall

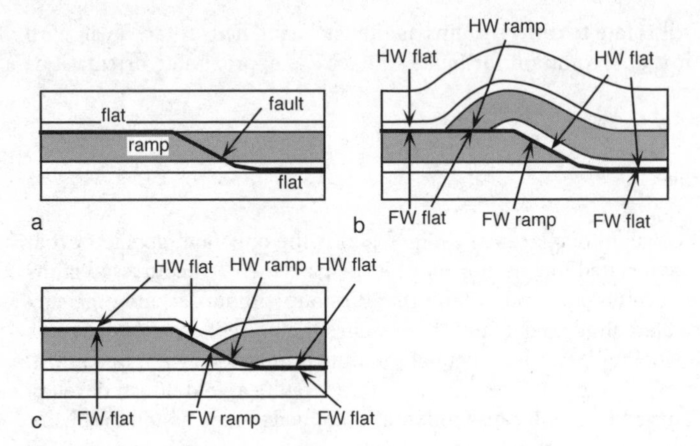

Fig. 1.40. Ramp–flat fault terminology. *HW* is hangingwall; *FW* is footwall. **a** Before displacement. **b** After reverse displacement. **c** After normal displacement. (After Woodward 1987)

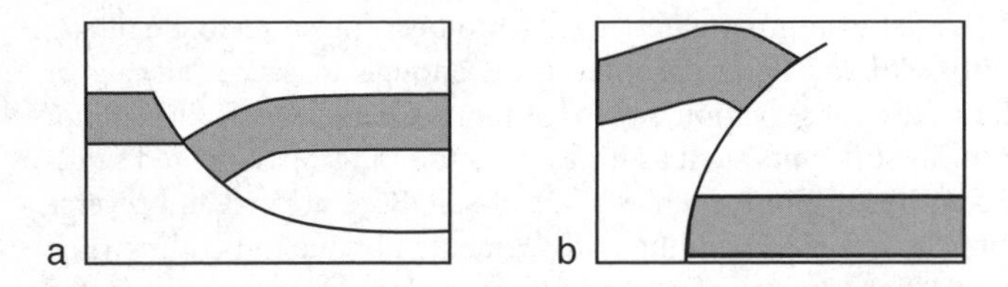

Fig. 1.41. Typical curved fault shapes in cross section. **a** Listric. **b** Antilistric

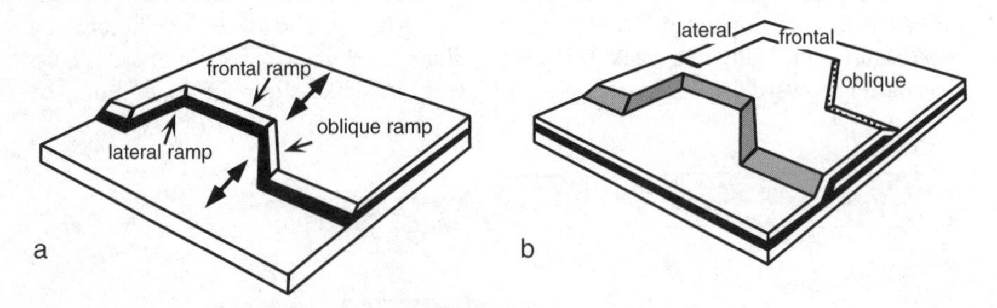

Fig. 1.42. Ramps and ramp anticlines in three dimensions. **a** Ramps in the footwall. *Arrows* indicate transport direction. **b** Thrust ramp anticlines. (After McClay 1992)

termed antilistric. Antilistric reverse faults have been produced by the stresses above a rigid block uplift (Sanford 1959).

In three dimensions, ramps are named according to their orientation with respect to the transport direction (Dahlstrom 1970; McClay 1992). Frontal ramps are approximately perpendicular to the transport direction, lateral ramps are approximately parallel to the transport direction and oblique ramps are at an intermediate angle (Fig. 1.42a). Displacement of the hangingwall produces ramp anticlines having orientations that correspond to those of the ramps (Fig. 1.42b). The ramp terminology was developed for thrust-related structures but is equally applicable to normal-fault structures. Lateral and oblique ramps necessarily have a component of strike slip and lateral ramps may be pure strike-slip faults. A lateral or oblique ramp separates the faulted interval into compartments in which the structures may be different and may evolve independently of one another.

1.7 Sources of Structural Data and Related Uncertainties

The fundamental information generally available for the interpretation of the structure in an area is the attitude of planes and locations of the contacts between units. The primary sources of this information are direct observations of exposures, well logs, and seismic reflection profiles. These data are never complete and may not be correct in terms of the exact locations or attitudes. Before constructing or interpreting a map, it is worth considering the uncertainties inherent in the original data.

1.7.1 Direct Observations

Outcrop- and mine-based maps are constructed from observations of the locations of contacts and the attitudes of planes and lines. Good practice is to show on the working map the areas of exposure at which the observations have been made (Fig. 1.43a). Exposure is rarely complete and so uncertainties typically exist as to the exact locations of contacts. Surface topography is usually directly related to the underlying geology and should be used as a guide to contact locations. Contacts that control topography may be traced with a reasonable degree of confidence, even in the absence of exposure,

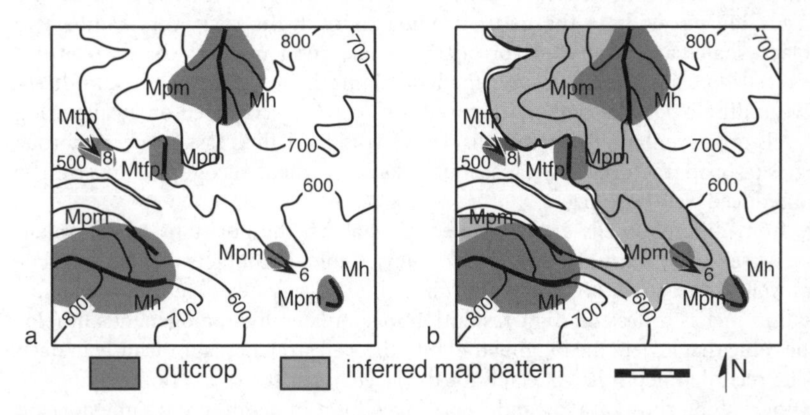

Fig. 1.43. Geologic map on a topographic base. Contours are in feet and the scale bar is 1000 ft. Three formations are present, from oldest to youngest: *Mtfp*, *Mpm*, *Mh*. Attitude of bedding is shown by an *arrow* pointing in the dip direction, with the dip amount indicated. **a** Outcrop map showing locations of direct observations (*shaded*). **b** Completed geologic map, contacts *wide* where observed, *thin* where inferred. *Lighter shading* is the interpreted outcrop area of the Mpm

at least where the structure is simple. The assignment of a particular exposure to a specific stratigraphic unit may be in doubt if diagnostic features are absent. Bedding attitudes measured in small outcrops might come from minor folds, minor fault blocks, or cross beds and not represent the attitude of the formation boundaries. The connectivity of the contacts in the final map is usually an interpretation, not an observation (Fig. 1.43b).

1.7.2 Wells

Wells provide subsurface information on the location of formation boundaries and the attitude of planes. Measurements of this information are made by a variety of techniques and recorded on well logs. Sample logs are made from cores or cuttings taken from the well as it is drilled. In wells drilled with a cable tool, cuttings are collected from the bottom of the hole every 5 or 10 feet and provide a sample of the rock penetrated in that depth interval. In wells that are rotary drilled, drilling fluid is circulated down the well and back to remove the cuttings from the bottom of the hole. The drilling fluid is sampled at intervals as it reaches the surface to determine the rock type and fossil content of the cuttings. Depths are calculated from the time required for the fluid to traverse the length of the hole and are not necessarily precise.

A wire-line log is a continuous record of the geophysical properties of the rock and its contained fluids that is generated by instruments lowered down a well. Lithologic units and their contained fluids are defined by their log responses (Asquith 1983; Jorden and Campbell 1986). Two logs widely used to identify different units are the spontaneous potential (SP) and resistivity logs (Fig. 1.44). In general, more permeable units show a larger negative SP value. The resistivity value depends on presence of a pore fluid and its salinity. Rocks with no porosity or porous rocks filled with oil generally have high resistivities and porous rocks containing saltwater have low resistivities. A variety of other log types is also valuable for lithologic interpretation, including gamma-ray, neutron density, sonic and nuclear magnetic resonance logs. The gamma-ray log responds to the natural radioactivity in the rock. Very radioactive (hot) black shales are often widespread and make good markers for correlations between wells. A caliper log measures the hole diameter in two perpendicular directions. Weak lithologies like coal or fractured rock can be recognized on a caliper log by intervals of hole enlargement. In wells drilled with mud, fluid loss into very porous lithologies or open fractures may cause mud cakes that will be recognized on a caliper log by a reduced well-bore size.

Logs from different wells are correlated to establish the positions of equivalent units (Levorsen 1967; Tearpock and Bischke 1991). Geologic contacts may be correlated from well to well to within about 30 ft in a lithologically heterogeneous sequence or to within inches or less on high-resolution logs in laterally homogeneous lithologies. The cable that lowers the logging tool into the well stretches significantly in deep wells. The recorded depth is corrected for the stretch, but the correction may not be exact. Different log runs, or a log and a core, may differ in depth to the same horizon by 20 ft at 10, 000 ft. Normally, different log runs will duplicate one of the logs, for example the SP, so that the runs can be accurately correlated with each other.

The orientation of the well bore is measured by a directional survey. Some wells, especially older ones, may be unintentionally deviated from the vertical and lack a

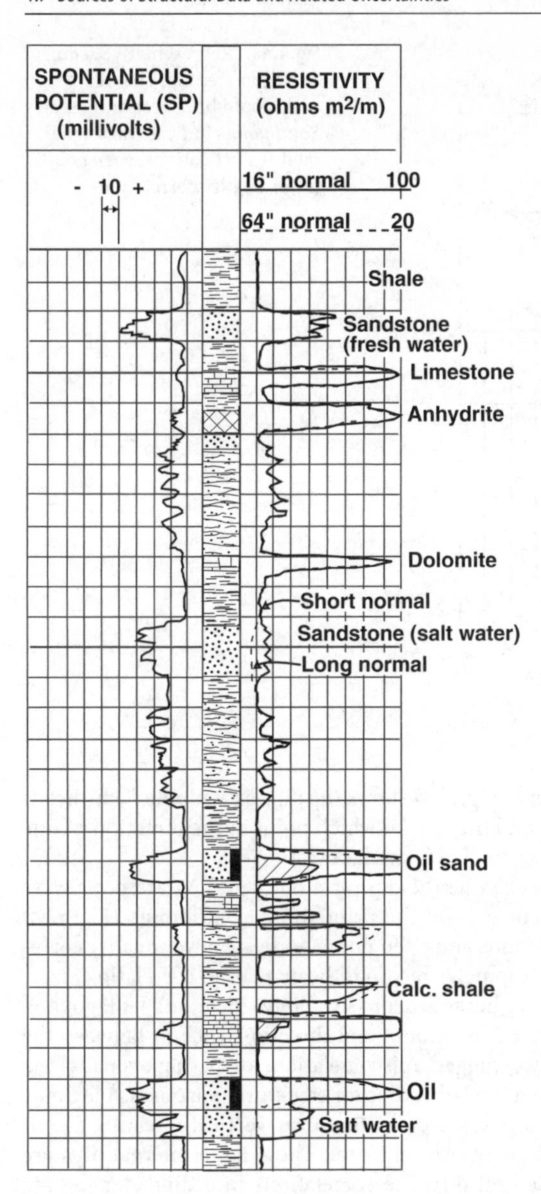

Fig. 1.44. Typical example of electric logs used for lithologic and fluid identification. The interpreted lithologic column is in the *center*. Short normal (16 in.) and long normal (64 in.) refer to spacing between electrodes on the resistivity tool. (After Levorsen 1967)

directional survey, resulting in spatial mislocation of the boundaries recorded by the well logs (if interpreted as being from a vertical well) which will lead to errors in dip and thickness determinations. The most common effect is for a well to wander down dip with increasing depth.

A dipmeter log is a microresistivity log that simultaneously measures the electrical responses of units along three or more tracks down a well (Schlumberger 1986). The responses are correlated around the borehole, and the dip of the unit is determined by a version of the 3-point method (Sect. 2.4). Dips may be calculated for depth intervals

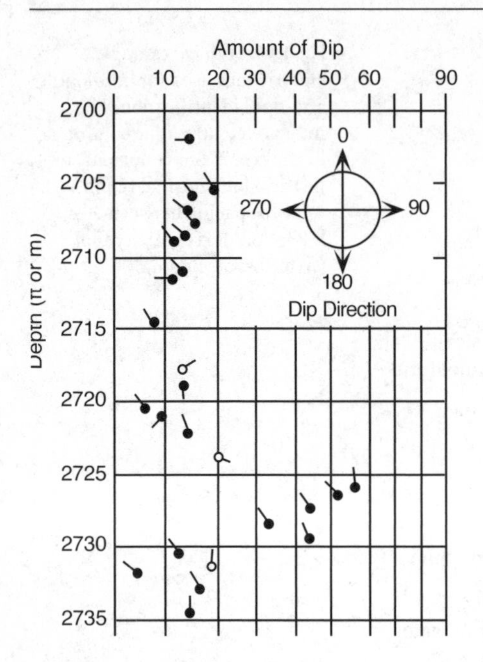

Fig. 1.45. Representative segment of a dipmeter log. The depth scale could be in feet or meters. *Solid points* indicate the higher quality correlations, *open points* lower quality correlations

as small as 8–16 cm. A typical record (Fig. 1.45) shows the dips as "tadpoles", the heads of which mark the amount of dip and the tails of which point in the direction of dip. The dips presented on the log are corrected for well deviation.

The correlations required to determine the dip for a dipmeter log are not always possible and may not always be correct. On the printed log, solid points (Fig. 1.45) indicate the highest quality correlations and open points indicate lower quality correlations. Sparse data or gaps on the dipmeter record indicate that no correlations were possible, a likely occurrence in a very homogeneous lithology (including fault gouge). Closely spaced dips that are scattered in amount and direction, such as between the depths of 2715 and 2725 in Fig. 1.45, suggest miscorrelations or perhaps small-scale bedding features, and are probably not reliable dips for structural purposes. A log may use a special symbol to show dips that are consistent over vertical intervals five or more times that of the minimum correlation interval. These large-interval dips are more likely to represent the structural dip. The correlations in a dipmeter log are made by scanning some distance (the scan angle) up and down the individual tracks to look for correlations. If the angle between the well bore and bedding is small (equivalent to a steep dip in a vertical well), the correlative units may lie outside the search interval. Thus dipmeter logs rarely show dips that are at angles of less than 30–40° from the well bore (50–60° dip in a vertical well) unless they were specifically programmed to look for them. If the dipmeter interpretation program is unable to make good correlations across the well it will probably show either no dips in the interval or may have made false correlations and so show low-quality dips at a high angle to the well bore.

1.7.3 Seismic Reflection Profiles

Many interpretations of subsurface structure are based on seismic reflection profiles. Sound energy generated at or near the earth's surface is reflected by various layer boundaries in the subsurface. The time at which the reflection returns to a recorder at the surface is directly related to the depth of the reflecting horizon and the velocity of sound between the surface and the reflector. Seismic data are commonly displayed as maps or cross sections in which the vertical axis is the two-way travel time (Fig. 1.46).

The geometry of a structure that is even moderately complex displayed in travel time is likely to be significantly different from the true geometry of the reflecting boundaries because of the distortions introduced by steep dips and laterally and vertically varying velocities (Fig. 1.47). Reflections from steeply dipping units may return to the surface beyond the outer limit of the recording array and so are not represented on the seismic profile. The structural interpretation of seismic reflection data requires the conversion of the travel times to depth. This requires an accurate model for the velocity distribution, something not necessarily well known for a complex structure. The most accurate depth conversion is controlled by velocities measured in nearby wells (Harmon 1991).

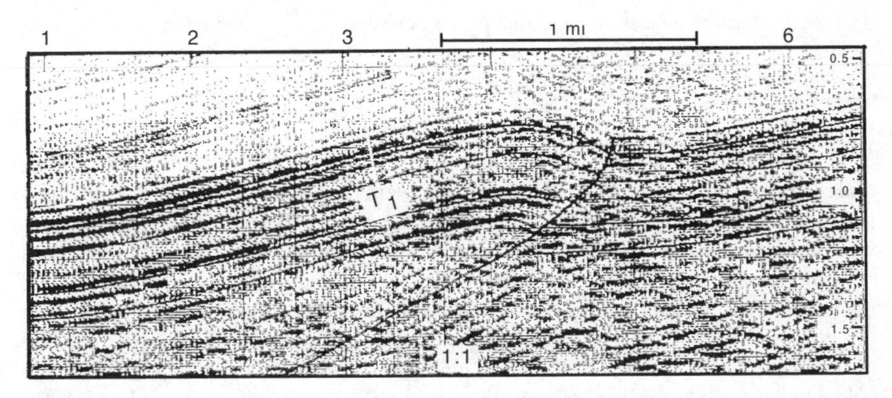

Fig. 1.46. Time-migrated seismic profile from central Wyoming (Stone 1991), displayed with approximately no vertical exaggeration. The vertical scale is two-way travel time in seconds. T_1 is a unit that can be correlated across the profile

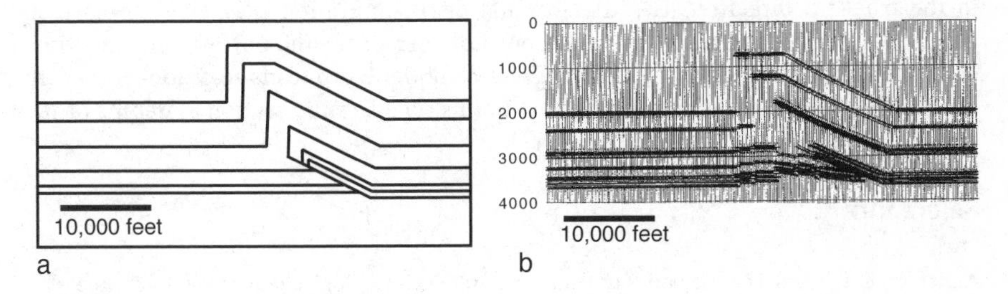

Fig. 1.47. Seismic model of a faulted fold. **a** Geometry of the model, no vertical exaggeration. **b** Model time section based on normal velocity variations with lithology and depth. Vertical scale is two-way travel time in milliseconds. (After Morse et al. 1991)

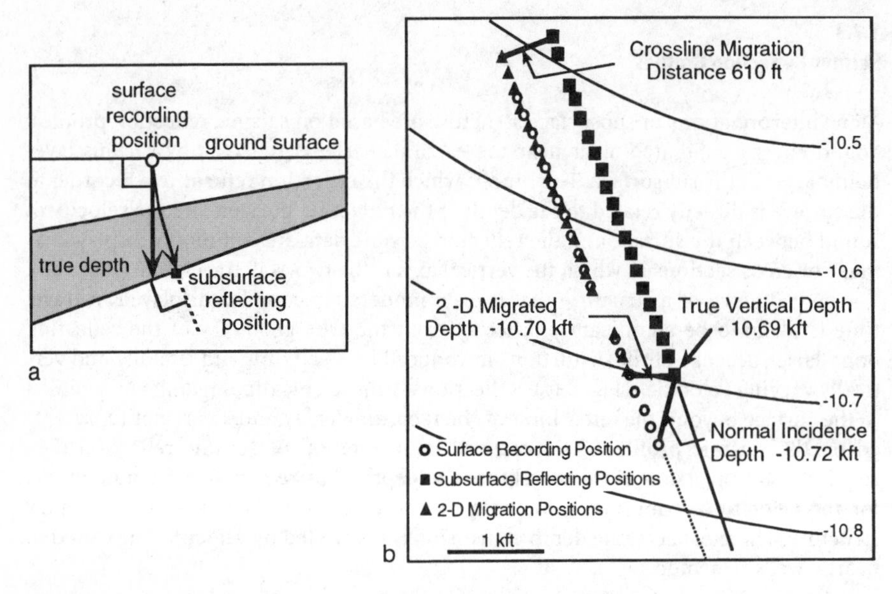

Fig. 1.48. Mislocation of seismic reflection points caused by dip of the reflector. **a** Ray path end points in vertical cross section. **b** Structure contours on a seismic reflector, depths subsea in kilo-feet, showing actual and interpreted locations of the reflecting points on a seismic line. (After Oliveros 1989)

If the trend of a seismic line is oblique to the dip of the reflector surface, two-dimensional reflection data have location problems similar to those of unknowingly deviated wells. This is in addition to the location problems associated with the conversion of travel time to depth. Reflections are interpreted to originate along ray paths that are normal to the reflector boundaries (Fig. 1.48a). A normal-incidence seismic ray is deflected up the dip (Fig. 1.48a). The true location of the reflecting point is up the dip and at a shallower depth than a point directly below the surface recording position. When the locations of reflecting points are plotted as if they were vertically below the surface recording stations, there will be a decrease in the calculated depth (or two-way travel time) relative to the true depth. Two-dimensional time migration is a standard processing procedure that corrects for the apparent dip of reflectors in the plane of the seismic line, but does not correct for the shift of the reflector positions in the true dip direction. Two-dimensional depth migration may give the correct depth to the reflecting point, but still does not correct for the out-of-plane position shift. For example (Fig. 1.48b), four degrees of oblique dip leads to a 400- to 600-ft shift in the true position of the reflection points on a seismic section at depths of 10, 500-10, 800 ft.

References

Anderson EM (1905) The dynamics of faulting: Edinburgh Geological Society, v. 8, 387–402
Anderson EM (1951) The dynamics of faulting, 2nd edn. Oliver and Boyd, London, 206 pp
Asquith DO (1970) Depositional topography and major marine environments, Late Cretaceous, Wyoming. Am. Assoc. Pet. Geol. Bull. 54: 1184–1224

Asquith G (1983) Basic well log analysis for geologists. American Association of Petroleum Geologists, Methods in Exploration Series, no 3, 216 pp
Banks R (1991) Contouring algorithms. Geobyte 6: 15–23
Bates RL, Jackson JA (1987) Glossary of geology, 3 rd edn. American Geological Institute, Alexandria, Virginia, 788 pp
Bengtson CA (1980) Structural uses of tangent diagrams. Geology 8: 599–602
Busk HG (1929) Earth flexures. Cambridge University Press, London, 106 pp
Currie JB, Patnode HW, Trump RP (1962) Development of folds in sedimentary strata. Geol. Soc. Am. Bull. 73: 655–673
Dahlstrom CDA (1970) Structural geology in the eastern margin of the Canadian Rocky Mountains. Bull. Can. Pet. Geol. 18: 332–406
Dennis JG (1967) International tectonic dictionary. American Association of Petroleum Geologists, Mem. 7, 196 pp
Dixon JM (1975) Finite strain and progressive deformation in models of diapiric structures. Tectonophysics 28: 89–124
Fleuty MJ (1964) The description of folds: Proc. Geol. Assoc. 75: 61–492
Galloway WE (1989) Genetic stratigraphic surfaces in basin analysis I: architecture and genesis of flooding-surface bounded depositional units. Am. Assoc. Pet. Geol. Bull. 73: 125–142
Ghosh SK (1968) Experiments of buckling of multilayers which permit interlayer gliding. Tectonophysics 6: 207–249
Griggs D, Handin J (1960) Observations on fracture and a hypothesis of earthquakes. Geol. Soc. Am. Mem. 79: 347–364
Groshong RH Jr. (1975) Strain, fractures and pressure solution in natural single-layer folds. Geol. Soc. Am. Bull. 86: 1363–1376
Groshong RH Jr (1988) Low-temperature deformation mechanisms and their interpretation. Geol. Soc. Am. Bull. 100: 1329–1360
Groshong RH Jr, Usdansky SI (1988) Kinematic models of plane-roofed duplex styles. Geological Society of America Special Paper 222, 97–206
Gzovsky MV, Grigoryev AS, Gushehenko OI, Mikhailova AV, Nikonov AA., Osokina DN (1973) Problems of the tectonophysical characteristics of stresses, deformations, fractures, and deformation mechanisms of the earth's crust. Tectonophysics 18: 167–205
Harmon C (1991) Integration of geophysical data in subsurface mapping. In: Tearpock D J, Bischke RE (eds) Applied subsurface geological mapping. Prentice Hall, Englewood Cliffs, pp 94–133
Jackson MPA (1995) Retrospective salt tectonics. In: Jackson MPA, Roberts, DG, Snelson S (eds) Salt tectonics: a global perspective. Am. Assoc. Pet. Geol. Mem. 65: 1–28
Jones NL, Nelson J (1992) Geoscientific modeling with TINs. Geobyte 7: 44–49
Jorden JR, Campbell FL (1986) Well logging II – electric and acoustic logging. Society of Petroleum Engineers, New York, 182 pp
Levorsen AI (1967) Geology of petroleum. WH Freeman, San Francisco, 724 pp
McClay KR (1992) Glossary of thrust tectonics terms. In: McClay KR (ed) Thrust tectonics. Chapman and Hall, London, pp 419–433
Mitchum RM (1977) Seismic stratigraphy and global changes of sea level, Part I. Glossary of terms used in seismic stratigraphy. In Payton CE (ed) Seismic stratigraphy – applications to hydrocarbon exploration. Am. Assoc. Pet. Geol. Mem. 26: 205–212
Mitra G, Marshak S (1988) Description of mesoscopic structures. In: Marshak S, Mitra G (eds) Basic methods of structural geology. Prentice Hall, Englewood Cliffs, New Jersey, pp 213–247
Morse PF, Purnell GW, Medwedeff DA (1991) Seismic modeling of fault-related structures. In: Fagan SW (ed) Seismic modeling of geologic structures. Society of Exploration Geophysicists, Tulsa, Oklahoma, pp 127–152
Mulvany PS (1992) A model for classifying and interpreting logs of boreholes that intersect faults in stratified rocks. Am. Assoc. Pet. Geol. Bull. 76: 895–903
Norris DK (1958) Structural conditions in Canadian coal mines. Geol. Surv. Can., Bull. 44. 54 pp
Oertel G (1965) The mechanism of faulting in clay experiments. Tectonophysics 2: 343–393

Oliveros RB (1989) Correcting 2–D seismic mis–ties. Geobyte 4: 43–47
Ramberg H (1963) Strain distribution and geometry of folds. Upps. Univ. Geol. Inst. Bull. 42: 1–20
Ramsay JG (1967) Folding and fracturing of rocks. McGraw–Hill, New York, 568 pp
Ramsay JG (1981) Tectonics of the Helvetic nappes. In: McClay KR, Price J (eds) Thrust and nappe tectonics. Blackwell, Boston, pp 293–309
Reches A, Hoexter DF, Hirsch F (1981) The structure of a monocline in the Syrian arc system, Middle East– surface and subsurface analysis: J. Pet. Geol. 3: 413–425
Sanford AR (1959) Analytical and experimental study of simple geologic structures. Geol. Soc. Am. Bull. 70: 19–51
Schlumberger (1986) Dipmeter interpretation. Schlumberger Limited, New York, 76 pp
Stearns DW (1978) Faulting and forced folding in the Rocky Mountains foreland. Geol. Soc. Am. Memoir 151: 1–37
Stone DS (1991) Analysis of scale exaggeration on seismic profiles. Am. Assoc. Pet. Geol. Bull. 75: 1161–1177
Tearpock DJ, Bischke RE (1991) Applied subsurface geological mapping. Prentice Hall, Englewood Cliffs, 648 pp
Turner FJ, Weiss LE (1963) Structural analysis of metamorphic tectonites. McGraw Hill, New York, 545 pp
Van Wagoner JC, Posamentier HM, Mitchum RM, Vail PR, Sarg JF, Loutit TS, Hardenbol J (1988) An overview of the fundamentals of sequence stratigraphy and key definitions. In: Wilgus CK, Hastings BS, Posamentier H, Van Wagoner J, Ross CA, Kendall CG St C (eds) Sea–level changes: an integrated approach. Soc. Econ. Paleontol. Mineral. Spec. Publ. 42: 39–45
Woodward NB (1987) Stratigraphic separation diagrams and thrust belt structural analysis. 38 th Field Conf. 1987, Jackson Hole, Wyoming. Wyoming Geological Association Guidebook, pp 69–77

Chapter 2
Location, Attitude, and Thickness

2.1
Introduction

This chapter covers methods for locating points in three dimensions on a map or in a well, for determining the attitude of a plane, the orientation of a line, and for finding the thickness of a unit from a map or in a well. The effects of measurement and location errors on attitudes and thicknesses are considered.

2.2
Location

The locations of data points recorded on maps and well logs must be converted to a single, internally consistent coordinate system in order to be used in the interpretive calculations given in this book. The positions of points in three dimensions will be described in terms of a right-handed Cartesian coordinate system with $+x$ = east, $+y$ = north, and $+z$ = up. Dimensions will be given in feet and kilofeet or meters and kilometers, depending on the units of the original source of the data. Parts of a foot will be expressed as a decimal fraction. Unit conversions are a common source of error which are largely avoided by retaining the original units of the map or well log. The conversions between locations on a topographic map or well log and the xyz coordinate system are given next.

2.2.1
Map Coordinate Systems, Scale, and Accuracy

The true locations of points on or in the earth are given in the spherical coordinate system of latitude, longitude, and the position along an earth radius. Maps are converted from the spherical coordinate system to a plane Cartesian coordinate system by some form of projection. Some distortion of lengths or angles or both is inherent in every projection technique, the amount of which depends on the type of projection and the scale of the map (Greenhood 1964; Robinson and Sale 1969) but is not significant at the scale of normal field mapping.

In many regions, maps are based on the transverse Mercator or polar stereographic projections and contain a superimposed rectangular grid called the Universal Transverse Mercator (UTM) or the Universal Stereographic Projection (USP) grid (Robinson and Sale 1969; Snyder 1987). The UTM grid is used at lower latitudes and the USP grid is used in the polar regions. The UTM grid divides the earth into 6 x 8° quadrilaterals that are identified by reference numbers and letters. UTM coordinates along the margin of a United States Geological Survey (USGS) map are given at inter-

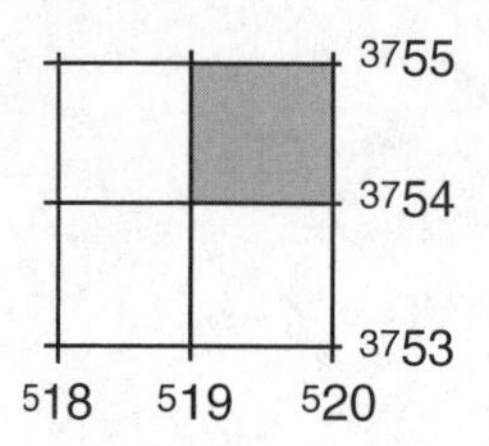

Fig. 2.1. UTM grid in north-central Alabama as given on a USGS topographic map. Each square block is 1000 m on a side. Block 1954 is *shaded*

vals of 1000 m (Fig. 2.1). On the map, the first digit or digits of the UTM coordinates are shown as a superscript and the last three zeros are usually omitted. Locations within the grid are given as the coordinates of the southwest corner of a block within the grid system. The x value is called the easting, and the y value is called the northing. For example, the lower left coordinate in Fig. 2.1, ^{5}18, is 518, 000 m east of the origin. Any block can be subdivided into tenths in both the x and y directions, adding one significant digit to the coordinates of the sub-block. Sub-blocks may be similarly subdivided. UTM coordinates are commonly written as a single number, easting first, then the northing, with the superscript and the trailing zeros omitted. For example, a grid reference of 196542 in the map of Fig. 2.1 represents a block 100 m on a side with its southwest corner at x = 196, y = 542. A grid reference of 19605420 represents a block 10 m on a side with its southwest corner at x = 1960, y = 5420. This coordinate system is internally consistent over large areas and is convenient for maintaining large databases. The United Kingdom uses a similar metric system called the National Grid (Maltman 1990).

The surface locations of wells are commonly recorded according to the coordinates given on a cadastral map, which is a map for officially recording property boundaries, land ownership, political subdivisions, etc. In much of the United States the cadastral system is based on the Land Office grid of Townships and Ranges (Fig. 2.2). The grid is aligned with latitude at base lines and with longitude at guide meridians. Every 24 miles the grid is readjusted to maintain the 6 mile dimensions of the blocks. An individual township is located according to its east-west coordinate (Township) and its north-south coordinate (Range). The township that has been subdivided is T.2 N., R.3 E. A township is subdivided into 36 sections, each 1 mile on a side and numbered from 1 to 36 as shown in Fig. 2.2. For a more precise location, each section is subdivided into quarter sections (may be called corners) as in the northeast quarter of section 7, abbreviated NE$^1/4$ sec. 7. Quarter sections may themselves be divided into quarters, as in the northwest quarter of the northeast quarter of section 7, abbreviated NW$^1/4$NE$^1/4$ sec. 7, T. 2 N., R. 3 E. Locations within a quarter are given in feet measured from a point that is specified. The surveys were not always done perfectly and were sometimes forced into irregular shapes by the topography. It is necessary to see the local survey map to be certain of the locations.

Map scales are expressed as a ratio in which the first number is the length of one unit on the map and the second number is the number of units of the same length on the ground. The larger the scale of the map, the smaller the second number in the ratio. Geological maps suitable for detailed interpretation are typically published at scales ranging from 1:63, 300 (1 in. to the mile: 1 in. = 63, 300 in. = 1 mile) to 1:24, 000 (1 in. = 2, 000 inches, the 7.5 minute quadrangles of the USGS). Larger scales are useful for making very detailed maps. Base maps should always contain bar scales which make it easy to enlarge or reduce the map while preserving the correct scale.

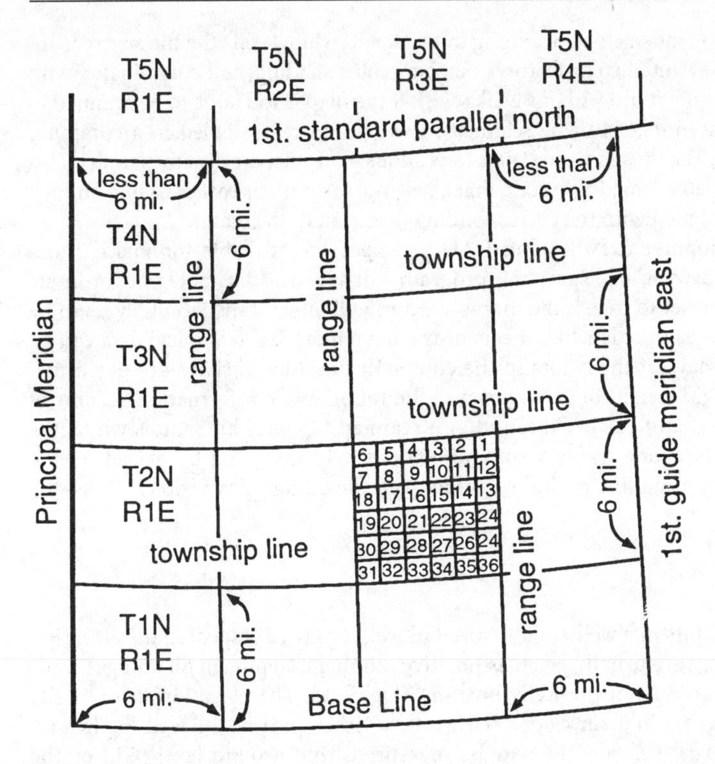

Fig. 2.2. The Township – Range grid system. The basic unit is a 24-mile block of 6-square-mile townships. Township T2N, R3E is divided into sections. (After Greenhood 1964)

Most governmentally produced base maps have an accuracy equivalent to the US Class 1 map standard (Fowler 1997). This standard states that the horizontal position of 90% of the points must be within 0.5 mm of their true location at the scale of the map. For example, at the 1:24, 000 scale, a point on the map must be within 0.5 mm x 24, 000 = 12 m (39.4 ft) of its true location. The vertical position of 90% of all contours must be within half the contour interval in open areas and spot elevations must be within one-quarter of a contour of their true elevation. A 20-ft contour (characteristic of 1:24, 000 USGS topographic maps) can be expected to be accurate to ± 10 ft (3 m). For comparison, the width of the thinnest contour line is about 0.01 in. on a 1:24, 000 USGS topographic map and represents 20 ft (6.01 m) at the scale of the map. The accuracy of an enlarged (or reduced) map is no better than it was at the original scale. Ground surveys of point locations such as wells can be expected to be accurate to within 1 m and satellite surveys can be accurate to 0.1 m within the survey area (Aitken 1994), significantly greater accuracies than that of many base maps.

Many of the techniques to be presented are designed for map interpretation by computer and are most convenient if the base map is stored in the computer. A simple technique for producing a computer base map is to scan the paper copy of a topographic map. Inasmuch as neither the topographic contours nor the cultural features are to be changed by the geological interpretation, the bitmapped version of the map produced by a scan should be satisfactory. Scanning in black and white keeps the file size reasonable and allows unwanted color features (such as color shading) to be eliminated with the scanner contrast control (Adams et al. 1996). If the map is too large to be scanned in one pass of the scanner, it can be scanned in strips and reassembled in a computer graphics package. The most critical step is to produce scans that are exactly parallel. This can be accomplished by making certain that a true north-south or east-west line on each map segment is aligned with the corresponding direction in the scanner.

Numerous computer drawing and CAD packages are available for making and working with maps. A suitable graphics program will have an adjustable xy coordinate system that can be set to match the map scale and will allow transparent layers to be placed over or under the map, like sheets of tracing paper. The geological data can be added to the overlays. If the colors in the computer graphics package are not transparent, the geologic formations will obscure the topographic information. Adams et al. (1996) suggest that if the background of a scanned topographic map is white, the background can be made transparent and colored map units can be placed on the layer below the map, making both the geology and the topography visible.

2.2.2 Wells

The location of points in a well are measured in well logs with respect to the elevation of the wellhead and are usually given as positive numbers. Depths in oil and gas wells are usually measured from the Kelly bushing (Fig. 2.3a). The elevation of the Kelly bushing (KB) is given in a surveyor's report included as part of the well-log header information. Alternatively, depths may be measured from ground level (GL) or the derrick floor (DF).

2.2.2.1 *Datum*

The coordinates of points in a well need to be corrected to a common datum elevation, normally sea level. The depths should be adjusted so that they are positive above sea level and negative below. The log depths are converted to a sea-level datum by subtracting the log depths from the elevation from which the depths were measured.

2.2.2.2 *Deviated Well*

Many wells are purposely drilled to deviate from the vertical. This means that points in the well are not directly below the surface location. The shape of the well is determined by a deviation survey, the terminology of which is given in Fig. 2.3a,b. The primary information from a deviation survey is the azimuth and inclination of the borehole and the downhole position of the measurement, for a number of points down the well. This information is converted by some form of smoothing calculation into the xyz coordinates of selected points in well-log coordinates, known as true vertical depth (TVD, Fig. 2.3a) and is given in the log of the survey. The TVD must be corrected for the elevation of the Kelly bushing to give the locations of points with respect to the datum. The calculated position of the points determined from a deviation survey

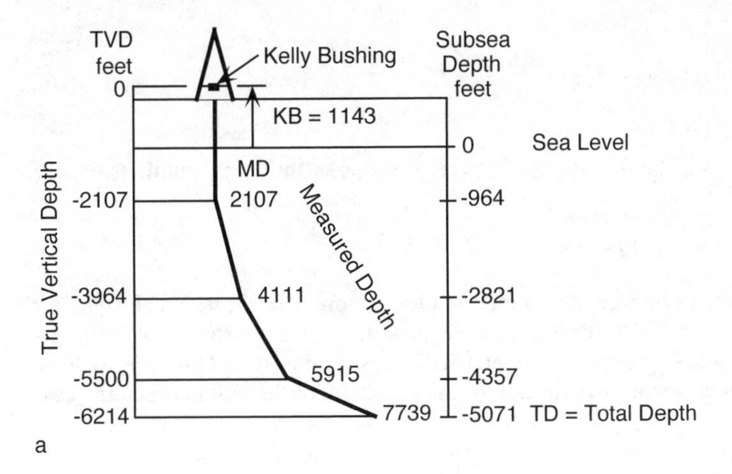

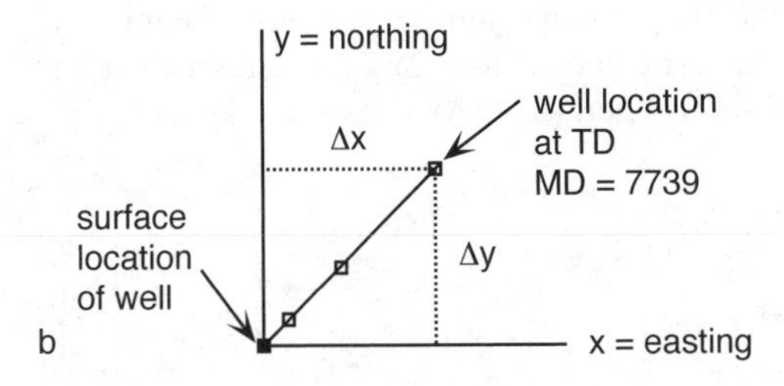

Fig. 2.3. Location in a deviated well. **a** Vertical section through a deviated well, in the northeast-southwest direction. Depths in the well are measured downward from the Kelly bushing *(KB)*. True vertical depths *(TVD)* are calculated from the borehole deviation survey. **b** Map view of the deviated well. Locations of points down hole are given by their distance from the surface location. *TD* total depth; *MD* measured depth

depends on the spacing between the measurement points and the particular smoothing calculation used to give the TVD locations. If the measurement points are spaced tens of meters or a hundred or more feet apart, the positions of points toward the bottom of a 3000-meter (10,000-ft) well might be uncertain by tens of meters or a hundred feet or so. This is because the absolute location of a point depends on the accuracy of the location of all the points above it in the well. Small errors accumulate. The relative positions of points spaced a small distance apart along the well should be fairly accurate. Points will be accurately located if the deviation survey is based on points spaced only a meter or less apart.

Locations in a deviated well are commonly given as the xyz coordinates of points in the well relative to the surface location, for example, P_1 and P_2 in Fig. 2.4a. Here z is the vertical axis, positive upward, the subscript 1 will always denote the upper point and the subscript 2 will denote the lower point. A unit boundary is likely to be located between control points, making it necessary to calculate its location.

To find the coordinates of an intermediate point, P, between P_1 and P_2, let r be the distance along the well from the upper control point to the point P. The coordinates of the boundary are (from Eq. D2.3; see Sect. 2.9 "Derivations"):

$$x = (rx_2 - rx_1 + Lx_1) / L; \tag{2.1a}$$

$$y = (ry_2 - ry_1 + Ly_1) / L; \tag{2.1b}$$

$$z = (rz_2 - rz_1 + Lz_1) / L, \tag{2.1c}$$

where L = the straight-line distance between the upper and lower points (from Eq. D2.2):

$$L = [(x_2 - x_1)^2 + (y_2 - y_1)^2 + (z_2 - z_1)^2]^{1/2}. \tag{2.2}$$

If the well is straight between the upper and lower points, L will be equal to the log distance. If L is not equal to the log distance, then the well is not straight and the calculated value of L will give the more internally consistent answer. More precise location of the point will require definition of the curvature of the well between the control points.

It may be required to locate a contact below the last control point given by the deviation survey (Fig. 2.4b). A simple linear extrapolation can be used, based on the assumption that the well continues in a straight line below the last control point with the same orientation it had between the last two control points. The subscripts 1 and 2 again represent the upper and lower control points. Let $\Delta x = (x_2 - x_1)$, $\Delta y = (y_2 - y_1)$, and $\Delta_z = (z_2 - z_1)$, then

$$x = x_2 + r\ \Delta x / L; \tag{2.3a}$$

$$y = y_2 + r\ \Delta y / L; \tag{2.3b}$$

$$z = z_2 + r\ \Delta z / L, \tag{2.3c}$$

where x, y, and z = the coordinates of a point at a well-log distance r below the last control point, P_2, having the coordinates (x_2, y_2, z_2), and L (Eq. 2.2) = the straight-line distance between the last two control points.

As an example of the location of points in a deviated well, consider the following information from a well (KB at 660 ft) on which the formation boundary of interest is at a log depth of 1225 ft. The deviation survey gives the coordinates of points above and below the boundary as 1200 ft TVD, 50 ft northing, -1050 ft easting and 1400 ft TVD, 150 ft northing, -1150 ft easting. What are the coordinates of the formation

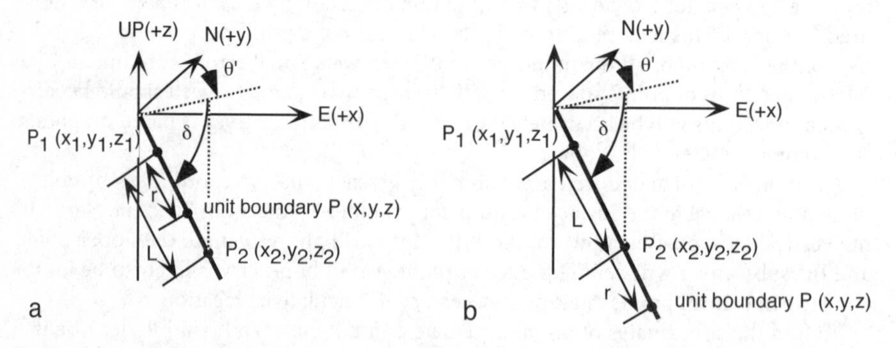

Fig. 2.4. Calculation of position of a unit boundary in a deviated well. **a** Boundary located between two control points. **b** Boundary located below the control points

boundary? The subsea depths of the upper and lower points, found by subtracting the log depths from the elevation of the Kelly bushing are $P_1 = -540$ ft and $P_2 = -740$ ft, respectively. From Eq. (2.2), the straight-line distance between the two points is $L = 225$ ft. From Eq. (2.1), the coordinates of a point $r = 25$ ft down the well from the upper point are −560 ft subsea, 60 ft northing, −1060 ft easting. Note that negative northing is to the south and negative easting is to the west.

2.3 Orientation of Lines and Planes

The basic structural measurements at a point are the orientations of lines and planes. The attitude of a plane is its orientation in three dimensions. The attitude may be given as the strike and dip (Fig. 2.5). Strike is the orientation of a horizontal line on the plane and the dip is the angle between the horizontal and the plane, measured perpendicular to the strike in the downward direction. Compass directions will be given here as the *trend*, which is the azimuth on a 360° compass (Ragan 1985), and will be indicated by numbers always containing three digits. Strike and dip are written in text form as strike, dip, dip direction; for example 340, 22NE and represented by the map symbols of Fig. 2.6a-d. The degree symbol may be written after each angle (i.e., 340°, 22°NE) or may be omitted for the sake of simplicity, as will be done here (Rowland and Duebendorfer 1994). The alternative form on a quadrant compass is the *bearing* and the dip (Ragan 1985), for which the same attitude would be written as N20W, 22NE. In subsurface geology, where the frame of reference may be a vertical shaft or well bore, the inclination of a plane may be given as the hade, which is the angle from a downward-pointing vertical line and the plane. The hade angle is the complement of the dip ($90° - \delta$). The attitude of a plane may also be given as the azimuth and plunge of the dip vector, written as dip amount, dip direction. By this convention, the previous attitude is given as 22, 070, indicated by the map symbol of Fig. 2.6e. This is the form that is usually used in this book because it is short and convenient for numerical calcula-

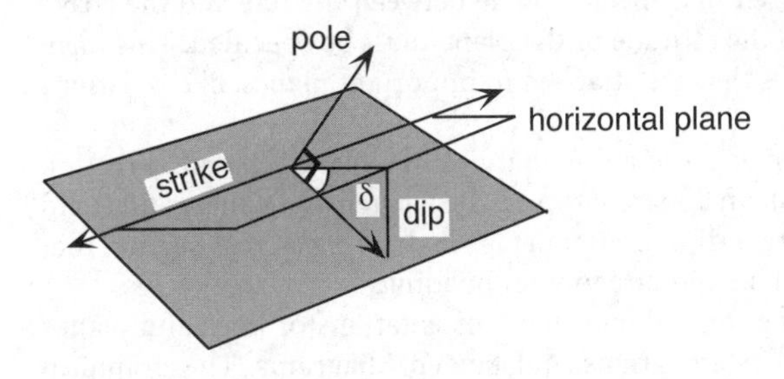

Fig. 2.5. Attitude of the *shaded* plane

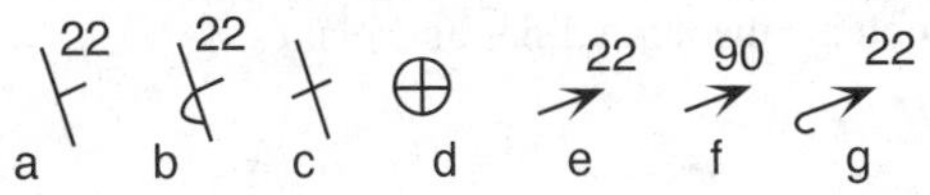

Fig. 2.6. Map symbols for the attitude of **a** plane. a strike and dip. **b** Strike and dip of overturned bed. **c** Strike of vertical bed. **d** Horizontal bed. **e** Azimuth and plunge of dip. **f** Facing (stratigraphic up) direction of vertical bed. **g** Azimuth and plunge of dip, overturned bed

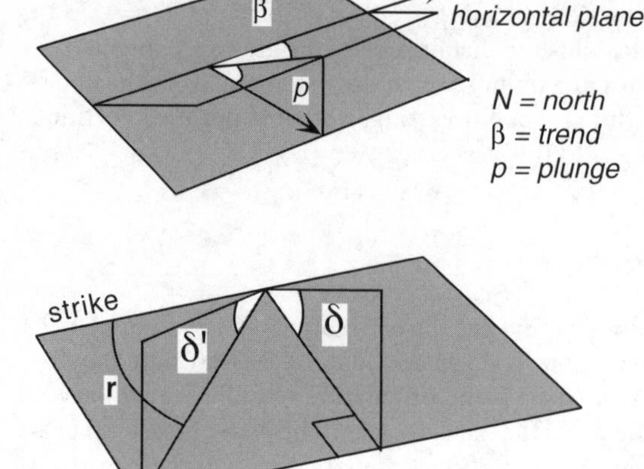

Fig. 2.7. Trend and plunge of a line in the *shaded* surface

Fig. 2.8. True dip, δ, apparent dip, δ′, and rake (or pitch), r

tions. The dip vector will occasionally be written in short form as **δ**, where the bold type indicates a vector. The attitude of a plane may also be represented by the orientation of its pole, a line perpendicular to the plane (Fig. 2.5).

The orientation of a line is given by its trend and plunge. The trend is the angle, β, between north and the vertical projection of the line onto a horizontal plane (Fig. 2.7). The plunge is the angle, p, in the vertical plane between the line and the horizontal. The orientation of a line is written as plunge amount, azimuth of trend, for example 30, 060 is a line plunging 30° toward the azimuth 060°. On a map the orientation of a line is represented by the same type of symbol as the azimuth and plunge of the dip (Fig. 2.6b) which is also a line. The two symbols could be differentiated on a map by means of line weight. The orientation of a line can also be represented by its pitch or rake, which is the angle measured in a specific plane between the line and the strike of the plane. Both the rake and the attitude of the plane must be specified. This form is convenient for recording lines that are attached to important planes, like striations on a fault plane.

Apparent dip, δ′, is the orientation of a line that lies in a plane in some direction other than the true dip (Fig. 2.8). An apparent dip can be written as a plunge and trend or as a rake in the plane. Apparent dips are important in drawing a cross section that is in a direction oblique to the true dip direction of bedding.

It is possible to represent the three-dimensional orientations of lines and planes quantitatively on graphs called stereograms and tangent diagrams. The graphical techniques aid in the visualization of geometric relationships and allow for the rapid solution of many three-dimensional problems involving lines and planes.

2.3.1 Stereogram Representation

A stereogram (Dennis 1967) is the projection of the latitude and longitude lines of a hemisphere onto a circular graph. Two different projections are commonly used in structural interpretation, the equal-area net (Schmidt or Lambert net) and the equal-angle stereographic net (Wulff net). The different projection techniques result in the preservation of different properties of the original sphere (Greenhood 1964). The equal-area net is required if points are to be contoured into spatially meaningful concentrations. A equal-angle net produces false concentrations from an evenly spaced distribution of points, although the angular relationships are correct. The lower-hemisphere equal-area stereograms (Fig. 2.9a) will be used throughout this book (an enlarged copy for use in working problems is given as Fig. 2.41 at the end of the chapter).

The outer or primitive circle of the equal-area stereogram is the projection of a horizontal plane (Fig. 2.9a). The compass directions are marked around the edge of the primitive circle. The projections of longitude lines form circular arcs that join the north and south poles of the graph and represent great circles. The trace of a plane is a great circle. A line plots as a point. The projections of latitude lines are elliptical curves, concentric about the north and south poles and represent small circles. The center of the graph is a vertical line. Planes and lines plotted on a stereogram can be visualized as if they were intersecting the surface of a hemispherical bowl (Fig. 2.9b).

A stereogram is used with a transparent overlay on which the data are plotted and which can be rotated about the center of the graph. To begin using a stereogram, on the overlay, mark the center and the north, east, south, and west directions, and the primitive circle (Fig. 2.10a). To plot the attitude of a plane from the strike and dip (for example, 60, 32SE), mark the strike on the primitive circle. Then rotate the overlay so that the strike direction lies on the N-S axis (Fig. 2.10b), find the great circle corresponding to the dip by counting down (inward) along the E-W axis from the primitive circle (zero dip) the dip amount in the dip direction. Draw a line along the great circle and then return the overlay to its original position to see the plane in its correct orientation (Fig. 2.10c). Find the pole to the plane by placing the overlay in the starting position for drawing bedding, that is, with the strike direction N-S. Along the E-W axis, count in from the primitive circle an amount equal to the dip, then count another 90° to find the pole (Fig. 2.10b) or, equivalently, count the dip amount up (outward) from the center point of the diagram. Mark the position of the E-W line on the primitive circle and return the overlay to its original position to find the trend of the pole

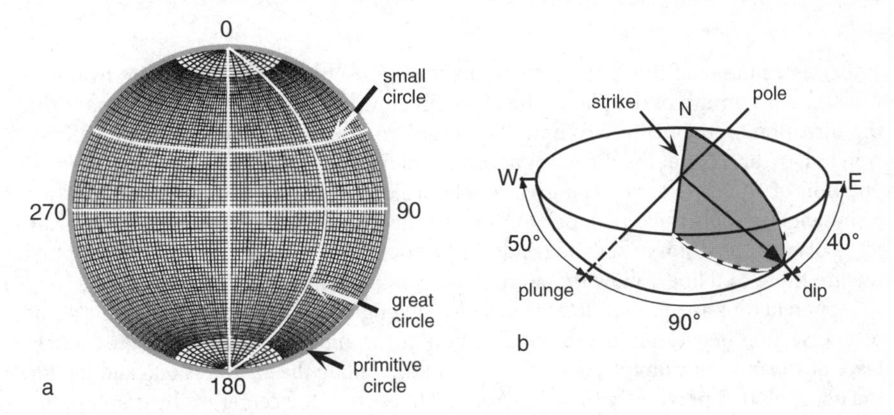

Fig. 2.9. Stereogram. **a** Equal-area stereogram (Schmidt or Lambert net), lower hemisphere projection. **b** Visualization of a plane and its pole in a lower hemisphere. (After Rowland and Duebendorfer 1994)

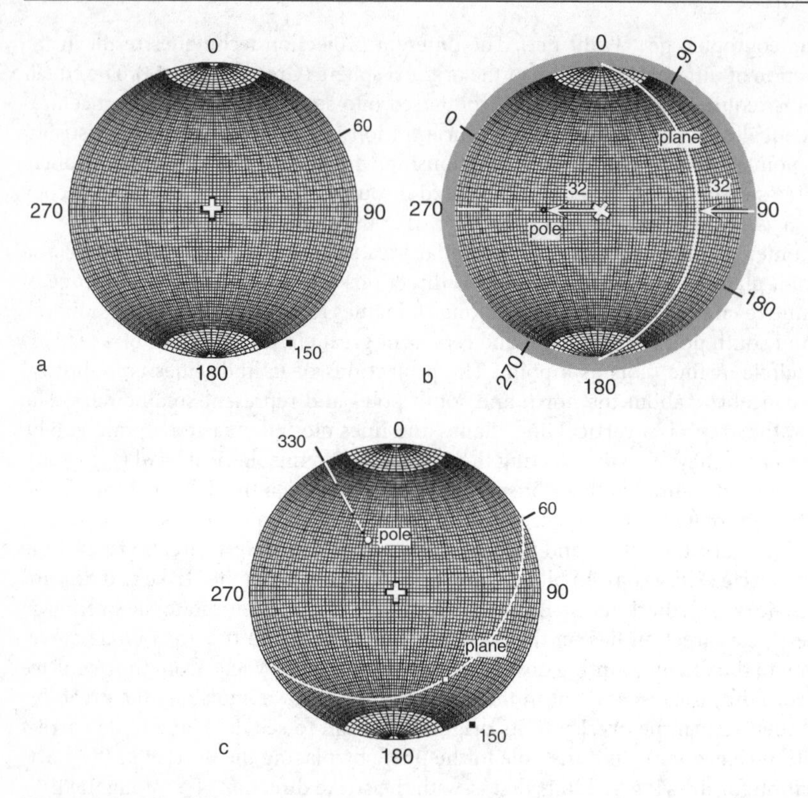

Fig. 2.10. Plotting the orientation of the plane and its pole on an equal-area lower-hemisphere stereogram. The plane has a dip of 32SE, and a strike of 060. **a** Overlay lined up with coordinates of graph. The *tic* marks the strike direction and the *black square* marks the dip direction. **b** Overlay (indicated by the *shaded* ring) rotated to bring the strike to north and the dip to east. **c** Overlay returned to its original position. The plunge and trend of the pole is 58, 330. The *black square* is the dip direction; the *white square* is the dip vector.

(330°). The plunge of the pole is 90° minus the dip. To plot planes and poles from the dip and dip azimuth of the plane (the plane previously plotted is $\boldsymbol{\delta} = 32, 150$) mark the dip direction on the primitive circle (Fig. 2.10a), rotate the overlay to bring this direction to E-W, and count the dip amount inward to find the point that represents the orientation of the dip vector. The great circle projection of the bed goes through this point. Plot 90° plus the dip along E-W to find the pole (Fig. 2.10b). Return the overlay to its original position to see the result (Fig. 2.10c). If the plunge and trend to be plotted are those of a line, follow the same steps as in plotting a dip vector.

Apparent dips are quickly determined on a stereogram as the orientation of the point of intersection between a line in the direction of the apparent dip and the great-circle trace of a plane. For example, find the apparent dips along the azimuths 080 and 180 for the plane plotted previously ($\boldsymbol{\delta} = 32, 150$). Plot lines from the center of the graph in the azimuth directions (Fig. 2.11a). The intersections of the azimuth lines with the great circle are the apparent dips. Dips can be read from the N-S axis as well as the E-W and so the 180° azimuth is in measurement direction. The angle measured inward from the

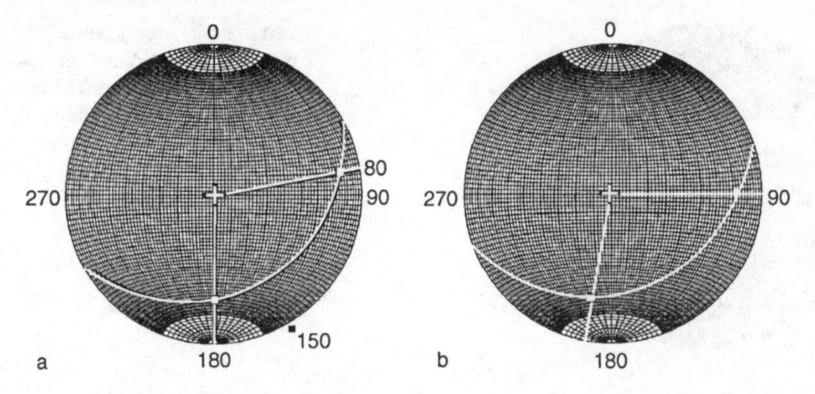

Fig. 2.11. Apparent dip in the plane given by the dip vector 32, 150 on an equal-area, lower-hemisphere stereogram. **a** Plane plotted with apparent dips shown as *white squares*. The south-trending apparent dip is 28, 180. **b** Overlay rotated 10° to bring the 080° trending apparent dip into measurement position; apparent dip is 15, 080

primitive circle to the intersection is the apparent dip, 28°. The overlay is rotated into measurement position for the 080° azimuth (Fig. 2.11b) to find the apparent dip of 15°.

Apparent dip problems can be worked backwards to find the true dip from two apparent dips. Plot the points representing the apparent dips, then rotate the overlay until both points fall on the same great circle. This great circle is the true dip plane.

2.3.2 Natural Variation of Dip and Measurement Error

The effect of measurement error or of the natural irregularity of the measured surface on the determination of the attitude of a plane is most readily visualized on a stereogram. The plane is represented by its pole (Fig. 2.12). Irregularities of the measured surface and measurement errors should produce a circular distribution of error around this pole (Cruden and Charlesworth 1976). An error of 4° around the true dip is probably the maximum expected for routine field measurements on a normally smooth bed surface. Measurements on a rough surface may show an even greater variability. A good average attitude from a rough surface can be obtained by making several measurements and then separately averaging the strikes and dips or the trends and plunges of the dips. A good field measurement procedure is to lie a flat field notebook or square of rigid plastic on a rough bed surface to average out the irregularities.

The effect of the error is related to the attitude of the plane. If the bed is horizontal, the pole is vertical (Fig. 2.12, point a) and the error means that the azimuth of the dip could be in any direction, even though the true three-dimensional orientation of the plane is rather well constrained. Small irregularities on a bed surface have the same effect (Woodcock 1976; Ragan 1985). On a steeply dipping plane (Fig. 2.12, point b), the same amount of error causes little variation in either the azimuth or the dip. Conversely, the measurement of the trend of a line on a gently dipping surface is accurate to within a few degrees, but the direction measured on a steeply dipping plane may show significantly greater error (Woodcock 1976). A precision of about 2° is about nor-

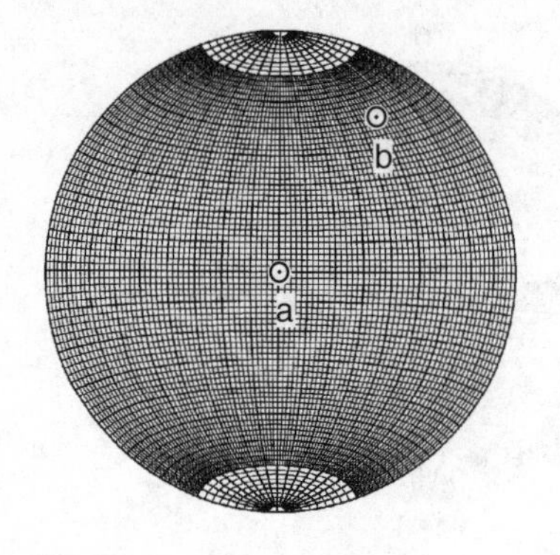

Fig. 2.12. Equal-area, lower-hemisphere stereogram. Measurement error of 4° (radius of circles) around true bedding poles (points *a* and *b*)

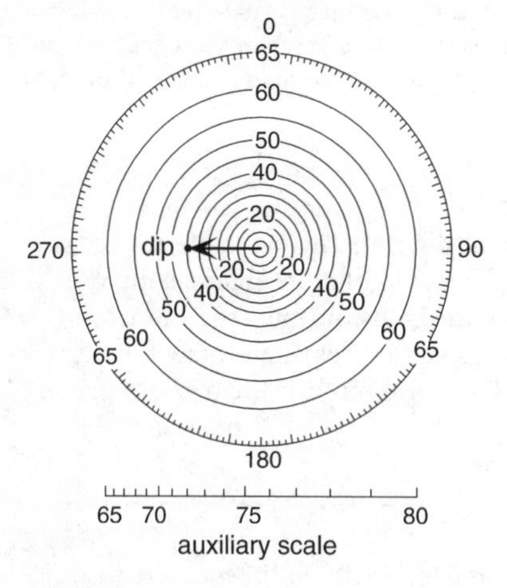

Fig. 2.13. Tangent diagram. Arrow represents a plane dipping 40° to the west (40, 270). (After Bengtson 1980)

mal for calculations done using a stereogram. For greater precision, the calculations should be done analytically by methods that will be presented later in the chapter.

2.3.3 Tangent Diagram Representation

The other useful diagram for representing the attitudes of planes is the tangent diagram (Fig. 2.13; an enlarged copy is given at the end of the chapter as Fig. 2.42). Developed by Hubbert (1931), it has been popularized by Bengtson (1980, 1981) who showed its value in the analysis of folds and faults. The concentric circles on the diagram represent the dip magnitude and their spacing is proportional to the tangent of the dip, hence the name tangent diagram. The center of the diagram is zero dip. The azimuth is marked around the margin of the outer circle. The attitude of a plane is represented by a vector from the origin in the direction of the azimuth and having a length equal to the dip amount (Fig. 2.13). The attitude can be shown with the complete vector or as a point plotted at the location of the tip of the vector. A major convenience of the tangent diagram is that no overlay is required and that certain problems are solved very quickly and without rotations (as are required with the stereogram). A major drawback of the tangent diagram is that very steep dips require a very large diagram. The dia-

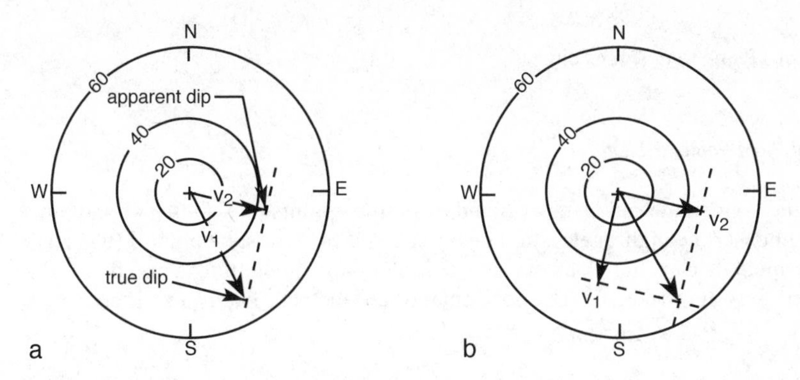

Fig. 2.14. Apparent dip on a tangent diagram. **a** Apparent dip from true dip. **b** True dip from two apparent dips. (after Bengtson, 1980)

gram in Fig. 2.13 extends to a dip of 65°. A calibrated scale that can be used to plot dips from 65° to 80° is given at the bottom of this figure. To use it, plot the vector along the appropriate radius and use the auxiliary scale to find the added length of the vector beyond the outer circle of the diagram. The diagram is not practical for manipulations where a large percentage of the dips are over 70°, for which a stereogram should be used. The tangent diagram can be used as a circular histogram, even for steep dips, by plotting the steep dips with their correct azimuths along the outer circle.

A tangent diagram is a convenient tool for finding the true or apparent dip. The apparent dip in a given direction (Fig. 2.14a) is the vector in the appropriate direction. For example, given a dip vector of a bed V_1, find the apparent dip in the direction of V_2. Project the tip of V_2 onto the direction of V_1. The projection is along a line perpendicular to the apparent dip. The length of the projected bedding vector in the direction of the apparent dip is the amount of the apparent dip. The true dip can be found from two apparent dips. The perpendiculars from the apparent dips (v_1 and v_2) intersect at the tip of the true dip vector (Fig. 2.14b).

2.4 Finding the Orientations of Planes

The attitude of a plane measured by hand with a compass or given by a dipmeter log is effectively the value at a single point. Measured over such a small area, the attitude is very sensitive to small measurement errors, surface irregularities, and the presence of small-scale structures. The following sections describe how to find the attitude from three points that can be widely separated and from structure contours. Both methods provide an average attitude representative of the map-scale structure.

2.4.1 Attitude of a Plane from Three Points

2.4.1.1 *Graphical Three-Point Problem*

The attitude of a plane can be determined from three points (a, d, f) that are not on a straight line. Let the highest elevation be point a and the lowest be point d (Fig. 2.15). The intermediate elevation, f, must also occur along the line joining a and d as point e. The line fe is the strike line. The horizontal (map) distance from a to e is ab:

$$ab = (ac \times be) / cd. \tag{2.4}$$

Plot the length ab on the map (Fig. 2.15b) and join point f and b to obtain the strike line. The dip vector lies along the perpendicular to the strike, directed from the high point to the intermediate elevation along the strike line. The dip amount is:

$$\delta = \arctan (v / h), \tag{2.5}$$

where v = the elevation difference between the highest and the lowest points and h = the horizontal (map) distance between the highest point and a strike-parallel line through d. The azimuth of the dip is measured directly from the map direction of the dip.

A typical example of a 3-point problem is seen on the map of Fig. 2.16a. The map shows the elevations of three locations identified in the field as being on the same contact. These points could just as easily be the elevations of a formation boundary identified in three wells. What is the attitude of the contact? Draw a line between the highest and lowest points (Fig. 2.16b) and measure its length (3095 ft). Use Eq. (2.4) to find the distance along the line from the high point to the level of the intermediate elevation (ab = 1719 ft). Connect the two intermediate elevations to find the strike line (Fig. 2.16b). Draw a perpendicular from the strike line to the lowest point (Fig. 2.16b) and measure its length (gc = 1179 ft) and its azimuth in the down-dip direction (120°). Determine the dip from Eq. (2.5) (d = 04°).

A dip can be converted from degrees into feet/mile or meters/kilometer by solving Eq. (2.5) for v and letting h be the reference length (5280 ft or 1000 m for ft/mile or m/km, respectively):

$$v = h \tan \delta, \tag{2.6}$$

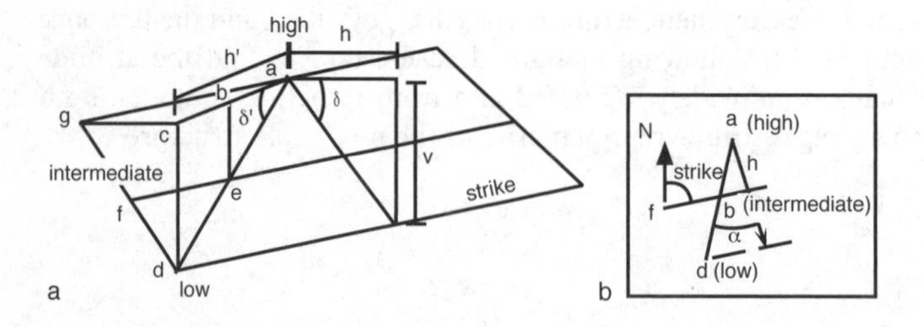

Fig. 2.15. True dip, δ, and apparent dip, δ′. **a** Perspective view. **b** Map view. *N* north. For explanation of *a–h* and v, see text (Sect. 2.4.1.1)

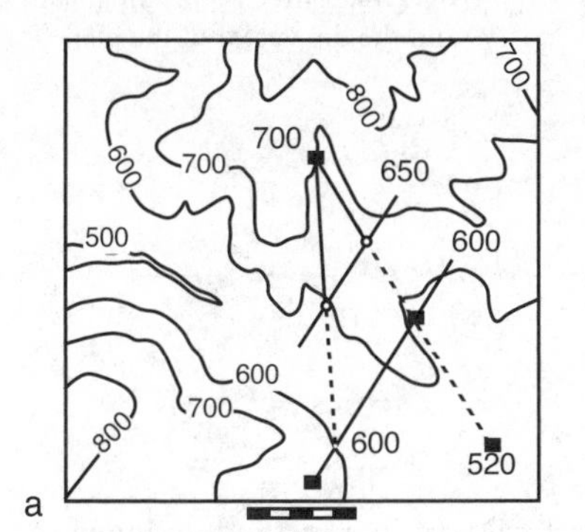

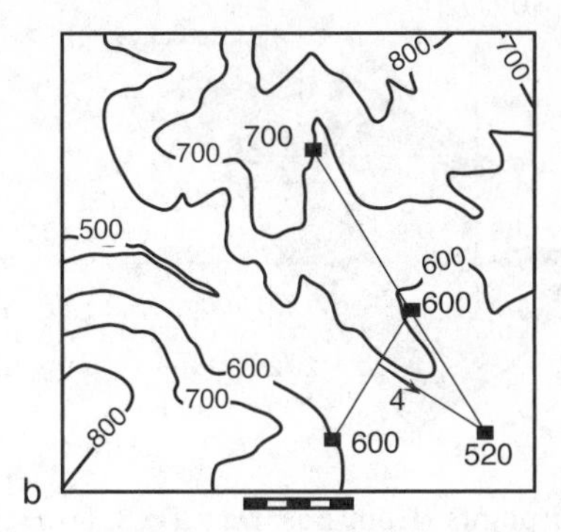

Fig. 2.16. Attitude determination from three points on a topographic map of a valley in a southeastern portion of the Blount Springs area. Elevations are in feet and the *scale bar* is 1000 ft. **a** Three points *(solid squares)* on the Mpm-Mh contact. **b** Results of 3-point attitude calculation in bold. Attitude of the contact is 04, 120

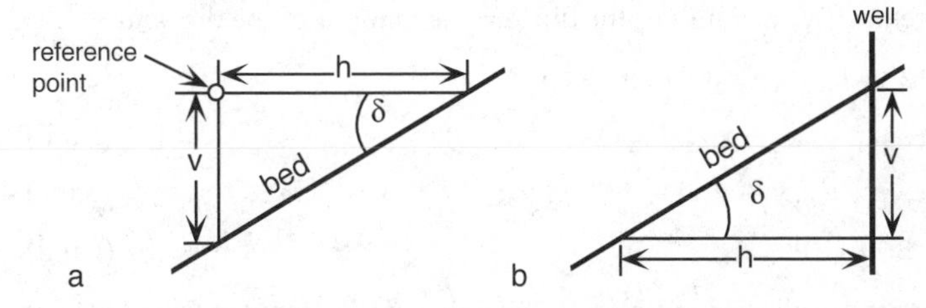

Fig. 2.17. Distance to a point on a dipping bed, in vertical cross sections in the dip direction. **a** Vertical distance from a reference point to a dipping bed. **b** Horizontal distance from a well to a dipping bed

where v = the vertical elevation change, h = reference length, and δ = dip. The same relationship can be used to determine the vertical distance from a reference point to a dipping horizon seen in a nearby outcrop (Fig. 2.17a).

Another useful application of Eq. (2.5) is to find the distance to the intersection between a horizontal plane (such as an oil-water contact) and a dipping plane (such as the top of the unit containing the contact) which are separated by a known distance in the well. Solve Eq. (2.5) for h (Fig. 2.17b):

$$h = v / \tan \delta, \tag{2.7}$$

where h = distance from the well bore to the intersection with a dipping bed and v = vertical distance in the well between the intersection of the dipping horizon and the horizontal horizon.

2.4.1.2
Analytical Three-Point Problem

The attitude of a plane is given by the trend and plunge of the dip vector (Fig. 2.18). The dip vector can be determined analytically from the xyz coordinates of three non-

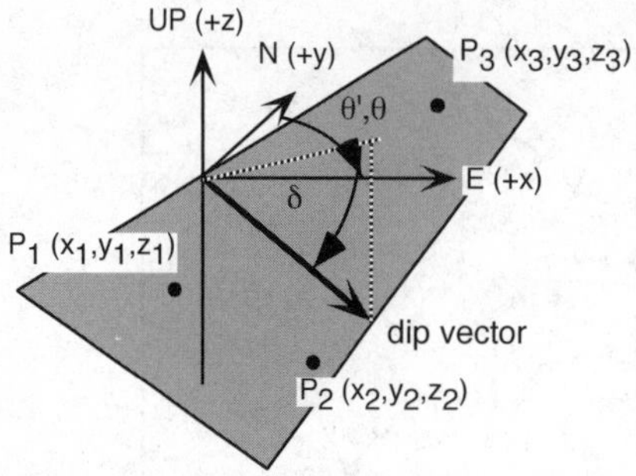

Fig. 2.18. Three points on a plane and the dip vector of the plane

colinear points (from Eqs. D2.7, D2.8, D2.18, D2.20, D2.21) in a method designed for spreadsheet calculation. The preliminary trend and plunge of the dip vector is

$$\theta' = \arctan (A / B); \quad (2.8)$$

$$\delta = \arcsin \{- \cos [90 + \arccos (C / E)]\}, \quad (2.9)$$

where θ' = the preliminary azimuth of the dip, δ = the amount of the dip, and

$$A = y_1z_2 + z_1y_3 + y_2z_3 - z_2y_3 - z_3y_1 - z_1y_2; \quad (2.10a)$$

$$B = z_2x_3 + z_3x_1 + z_1x_2 - x_1z_2 - z_1x_3 - x_2z_3; \quad (2.10b)$$

$$C = x_1y_2 + y_1x_3 + x_2y_3 - y_2x_3 - y_3x_1 - y_1x_2; \quad (2.10c)$$

$$E = (A^2 + B^2 + C^2)^{1/2}. \quad (2.10d)$$

Division by zero is not allowed in Eq. (2.8). The value of θ' computed from Eq. (2.8) is always between the values of 000 and 090° and is equal to 90° if B = 0. The true azimuth, θ, of the dip in the complete range from 000 to 360° can be determined from θ' and the signs of $\cos \alpha$ and $\cos \beta$ (Eqs. 2.11a, b, 2.12 and Table 2.1):

$$\cos \alpha = A / E; \quad (2.11a)$$

$$\cos \beta = B / E; \quad (2.11b)$$

$$D = z_1y_2x_3 + z_2y_3x_1 + z_3y_1x_2 - x_1y_2z_3 - y_1z_2x_3 - z_1x_2y_3, \quad (2.12)$$

where A, B, and E are given by Eqs. (2.10a,b,d) above. The sign of E = – sign D if D ≠ 0; = sign C if D = 0 and C ≠ 0; = sign B if C = D = 0. An Excel equation that returns the azimuth in the correct quadrant has the form:

=IF(cell B<=0,180+Cell E,IF(cell A>=0,cell E,360+cell E)), (2.13)

where cell A contains $\cos \alpha$, cell B contains $\cos \beta$ and cell E contains Eq. (2.8).

Table 2.1. True azimuth, θ, determined from preliminary azimuth, θ' based on the signs of $\cos \alpha$ and $\cos \beta$ (Eq. 2.11) or Δx and Δy (Eq. 2.19)

Azimuth	$\cos \alpha$; Δx	$\cos \beta$; Δy	θ
000+ to 090	+	+	θ'
090+ to 180	+	–	$180 + \theta'$
180+ to 270	–	–	$180 + \theta'$
270+ to 360	–	+	$360 + \theta'$

As an example, find the analytical solution to the 3-point problem in Fig. 2.16. Let the lower left data point be the origin of the x and y coordinates. The three points then have the coordinates, in xyz order, of 0, 0, 600; 1424, 82, 520; -193, 2643, 700. From Eqs. (2.8)–(2.12), the dip is 03.8° at an azimuth of 120°.

2.4.2 Structure Contours

Structure contours provide an efficient and effective representation of the attitude of a surface. The contours are strike lines and the dip directions are perpendicular to the contours (Fig. 2.19). The simplest way to construct structure contours from a geologic map is to connect points on a contact that lie at the same elevation. If two points on a contact have the same elevation, for example a and b in Fig. 2.19, then a line joining them is a strike line or structure contour. Finished maps normally show contours at even increments of elevation.

Structure contours are the primary method for illustrating the shape of structures in the subsurface. Generated from outcrop maps, structure contours provide an effective method for validating the outcrop pattern and for smoothing out local dip variations caused by bed roughness. If a map horizon is planar (unfolded) the contours on it are parallel and uniformly spaced. Folding produces curved contours that usually are not far from parallel at a local scale. Abrupt changes in direction indicate fold hinges, faults, or mistakes. Unless radical thickness changes occur, the structure contours on adjacent horizons are close to parallel.

The following sections describe the creation of straight-line structure contours from the elevations of three points and the interpretation of contours in the vicinity of a dip measurement. The generation of curved structure contours from multiple control points is discussed in Chapter 3.

2.4.2.1 *Structure Contours from Point Elevations*

Structure contours are usually generated from point elevations, especially in subsurface work. Except for the unlikely circumstance where all the points happen to fall on the selected contour elevations, contouring will require placing contours between control points that have different elevations. This means that contour values must be interpolated between the control points. Using linear interpolation (more complex forms of interpolation are considered in Chap. 3), an interpolated even-elevation con-

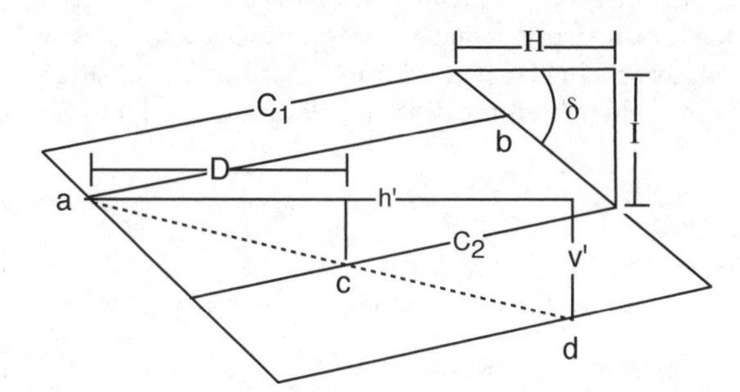

Fig. 2.19. Relationships between structure contours, point elevations, and dip. Points *a* and *b* are at the same elevation; *c* and *d* are at different elevations. C_1 and C_2 are structure contours. For explanation of other symbols, see text (Sect. 2.4.2.1)

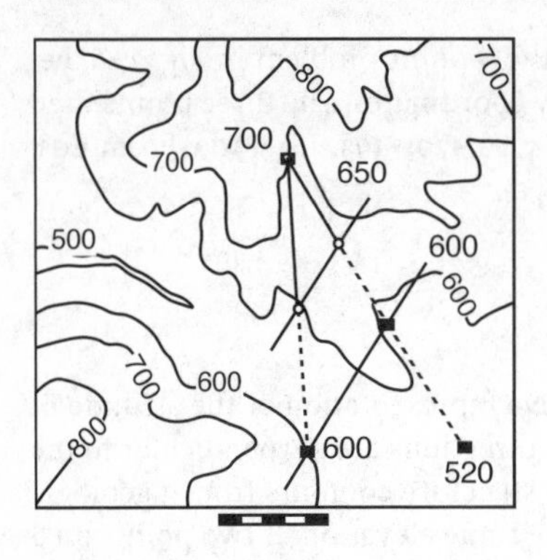

Fig. 2.20. Structure contours on the Mpm-Mh contact, Blount Springs map area (from Fig. 2.16). Three points *(solid squares)* are mapped as being on the contact. The 650-ft and 600-ft contours are interpolated between the points

tour lies at point c (Fig. 2.19) on the straight line between the two points at known elevations a and d. Following the method of Eq. (2.4), the elevation of the upper control point is a, the horizontal distance of the desired contour from the highest control point is D, the elevation of the desired contour is C_2, the map distance between the two control points is h′, and the vertical distance between the two control points is v′, giving:

$$D = h' (a - C_2) / v'. \tag{2.14}$$

As an example, find the position of the 650-ft contour between the three points of Fig. 2.20 using Eq. (2.14). Between the upper (700) and lower (520) points, h′ = 3064 ft, a = 700 ft, C_2 = 650, and v′ = 180, giving D = 851 ft from the 700-ft elevation point toward the lower point. Between the upper and the intermediate (600) point, h′ = 2674 ft, a = 700 ft, C_2 = 650, and v′ = 100 ft, giving D = 1337 ft from the 700-ft elevation point toward the intermediate point. Both points on the 650-ft contour have been plotted on Fig. 2.20 and connected to construct the contour. The 600-ft contour was constructed previously as part of a three-point problem (Fig. 2.16).

2.4.2.2
Structure Contours from Attitude

If the dip is known, then the spacing between structure contours is found from the method of Eq. (2.5) (Fig. 2.19) to be

$$H = I / \tan \delta, \tag{2.15}$$

where H = horizontal spacing between structure contours measured perpendicular to the contour strike, I = contour interval, and δ = dip. The trend of the structure contours is perpendicular to the dip direction. Equation (2.15) can be used to generate a ruler of the contour spacings needed to make or interpret a map at any given scale and contour interval (Fig. 2.21). Use of the contour-spacing ruler can greatly speed up the creation of a contour map from control points where both the dips and elevations are known.

2.4.2.3
Dip from Structure Contours

If the attitude is given by structure contours, the dip is found by solving Eq. (2.15):

$$\tan \delta = I / H, \tag{2.16}$$

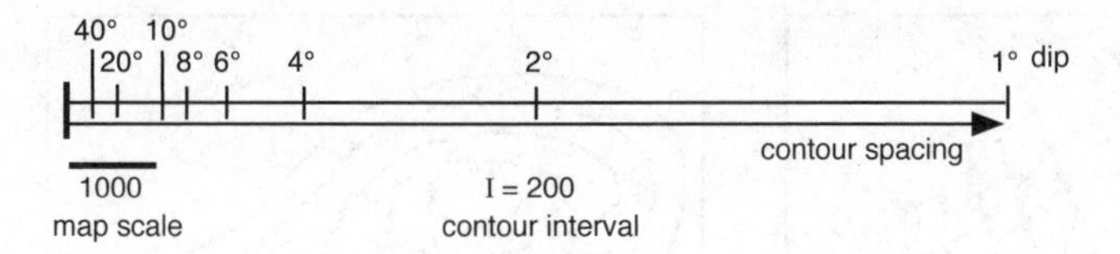

Fig. 2.21. Contour-spacing ruler for the map scale and contour interval shown, as a function of dip (Eq. 2.15). The distance from the left boundary to the labeled dip is the contour spacing

where δ = dip, I = contour interval, and H = spacing between contours. The dip direction is perpendicular to the contour, in the direction toward the lower contour.

As an example, find the dip of the Mpm-Mh contact between the 600- and 650-ft contours in Fig. 2.20 using Eq. (2.16). Measure the perpendicular distance between the two contours (H = 760 ft). The contour interval is I = 50 ft and so the dip is 04°. This result is equal to the dip used to construct the contours, as it should be.

2.4.2.4
Intersecting Contoured Surfaces

A geologic map represents the intersection of a topographic surface with the boundaries of the geologic units. Finding the intersection between two surfaces is a fundamental technique in three-dimensional interpretation. In Section 2.4.2.1, selected points along the outcrop trace of a unit boundary were used to generate a structure-contour map. Conversely, the outcrop trace of a contact can be determined from the structure contour map. The following procedure works for both plane and curved structure contour surfaces. The structure contour map is superimposed on the topographic map (Fig. 2.22a). All intersection points where both surfaces have identical elevations are marked by dots. Additional topographic and structural contours can be interpolated to add additional control. Marking the intersection points is most convenient if the topographic map is on a transparent overlay that can be placed above the structure contour map, something that is easily done in computer drafting programs. The intersection points are then connected to obtain the outcrop pattern of the horizon (Fig. 2.22b). The inferred outcrop pattern becomes a working hypothesis for the location of the horizon. Exactly the same procedure is used to find the intersection between a fault and a stratigraphic marker or the line of intersection between two faults.

2.4.2.5
Single-Unit Map Validation

Structure contours are a valuable aid in evaluating the validity of the contact locations on a geological map. As an example, compare the agreement between a 3-point dip determination and a geologic contact as mapped on a topographic base. In the Blount Springs map area (Fig. 2.23) the vegetative cover renders the mapped contact locations somewhat tentative. In Figs. 2.16b and 2.20, the calculated 600-ft elevation does not quite fall on the 600 ft topographic contour, suggesting that at least one of the points is not exactly on the contact, or that the dip changes between the three points. The computed dip is 04° (Sect. 2.4.2.3), slightly less than the 06° value measured nearby

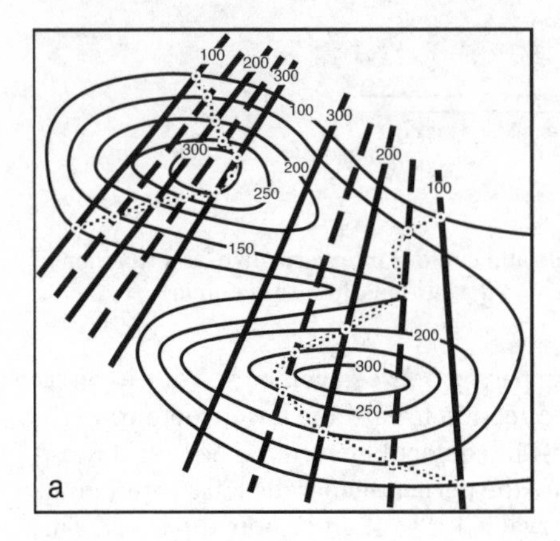

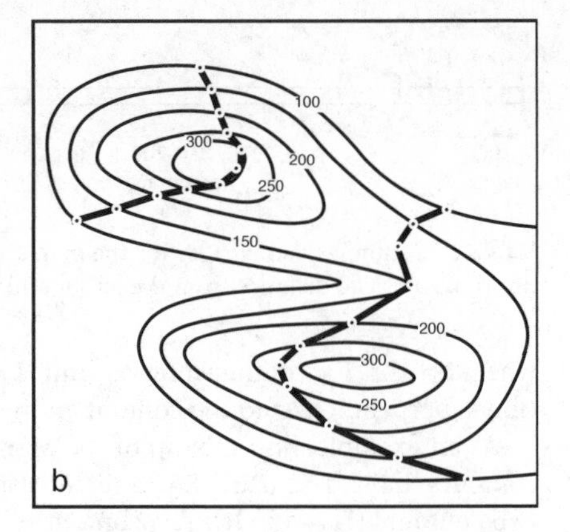

Fig. 2.22. Intersection between topographic contours *(thin lines)* and structure contours. **a** Structure contours on a geologic contact *(heavy lines)* and outcrop traces of the contact *(dotted lines)*. *Dashed* structure contours have been interpolated between selected contours to increase accuracy of the predicted outcrop trace. **b** Geologic map of the outcrop trace *(wide lines)* with the structure contours removed

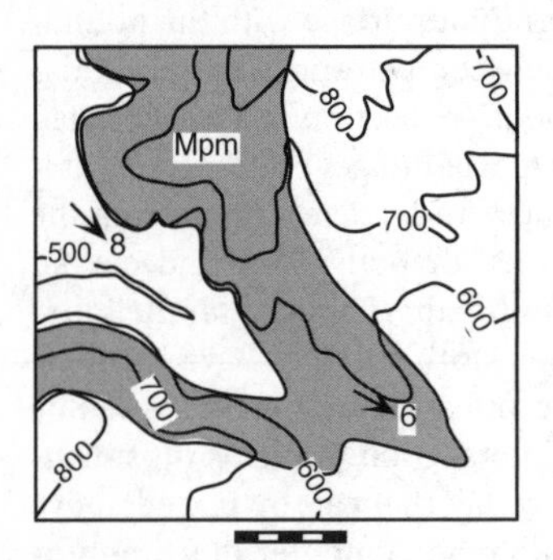

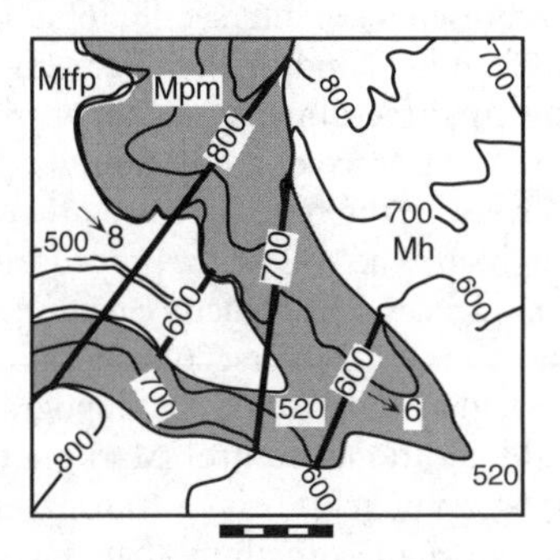

Fig. 2.23. Preliminary geologic map, Blount Springs area, southeastern detail. Topographic elevations are in feet and the scale bar is 1000 ft. **a** Geologic map. **b** Implied structure contours *(heavy lines)* determined from the intersection of mapped formation boundaries with the topographic contours

(Fig. 2.23a). This small difference in dip may or may not be significant. To further test the preliminary geologic map (Fig. 2.23a), find the orientations of the structure contours implied by both the top and bottom boundaries of the Mississippian Pride Mountain Formation (Mpm, Fig. 2.23b). The 800-ft Mpm-Mh contour (Fig. 2.23b) is parallel to the calculated contour (Fig. 2.23a), but the implied 700-ft contour is quite different, suggesting that the implied 700 ft contour could be wrong. The implied 600-ft contour on the Mtfp-Mpm is parallel to the 800-ft Mpm-Mh contour, as expected in this parallel-bedded sedimentary sequence.

Assuming that the structure contours in Fig. 2.23b should be essentially parallel to one another, revise the map accordingly. Assume, for the sake of argument, that the northeastern contact is better located and revise the southwestern Mpm-Mh contact

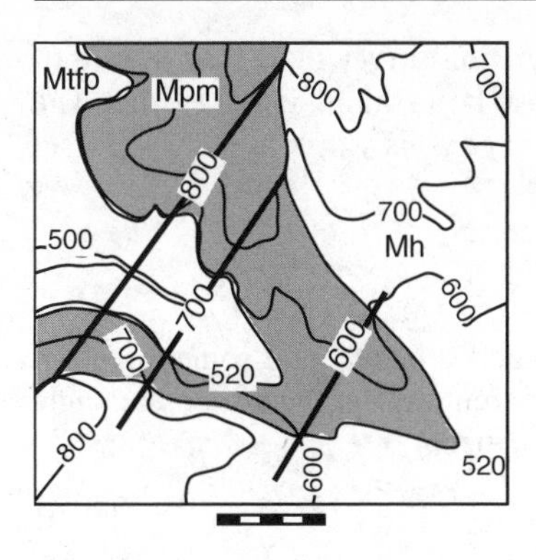

Fig. 2.24. Geologic and structure-contour map, southeastern Blount Springs area, revised from Fig. 2.23

to make the structure contours parallel (Fig. 2.24). The contact must be moved a relatively short distance. Check the new interpretation by finding the dips implied by the new structure contours (Eq. 2.16). The horizontal distance between the 700- and 600-ft contours is 1441 ft and the vertical distance is 100 ft, giving a dip of 04°, the same as obtained previously. The distance between the 800- and 700-ft contours is 683 ft, giving a dip of 08°, the same as found nearby on the map. We will tentatively accept the revised map as an improvement. The revised map becomes a new working hypothesis which should be field checked, if possible.

2.5 Finding the Orientations of Lines

2.5.1 From Two Points

Given the zyx coordinates of two points, it is possible to calculate the trend and plunge of the line between them (Fig. 2.7). If subscript 1 represents the higher of the two points, the plunge of the point determined from the following equations will be downward. The preliminary value of the azimuth θ' and plunge δ of the line between points 1 and 2 are (from Eqns. D2.13 and D2.14):

$$\theta' = \arctan (\Delta x / \Delta y); \quad (2.17)$$

$$\delta = \arcsin [(z_2 - z_1) / L], \quad (2.18)$$

where

$$\Delta x = (x_2 - x_1), \quad (2.19a)$$

$$\Delta y = (y_2 - y_1), \quad (2.19b)$$

and L is given by Eq. (2.2). Division by zero is not allowed in Eq. (2.17) (see Sect. 2.4.1.2). The preliminary azimuth, θ', calculated from Eq. (2.17), is always in the range of 000 to 090°. The angle must be converted to the true azimuth, θ, in the range of 000 to 360° using Table 2.1.

As an example of the calculation of the orientation of a line, find the trend and plunge of the well between points P_1 and P_2 from the example in Section 2.2.2.2. The locations of the two points are found from the deviation survey to be P_1: z = −540 ft, 50 ft northing, −1050 ft easting and P_2: z = −740 ft, 150 ft northing, -1150 ft easting. The plunge and trend of the segment from Eq. (2.17) and (2.18) is 55, 315.

2.5.2 Apparent Dip

Apparent dip, δ', is the angle in a plane between the horizontal and some direction other than the true dip (Fig. 2.15a). To find the apparent dip, let the horizontal angle between the true and apparent dip be α (Fig. 2.15b), then:

$$\delta' = \arctan(\tan\delta \cos\alpha). \tag{2.20}$$

On a completed structure-contour map, the apparent dip in a given direction is found from

$$\delta' = \arctan(I / h') \tag{2.21}$$

where I = the contour interval and h′ = the horizontal distance on the map between the contours in the direction of interest (Fig. 2.19). If the direction perpendicular to the contours is selected, then the apparent dip is the true dip. If the strike direction is selected, the apparent dip is zero.

2.6 Thickness of Plane Beds

The best thickness measurement is the most direct. In a completely exposed outcrop, bed thickness can be measured directly on the bed. In a well that is perpendicular to bedding, the thickness in the well is the true thickness. Commonly, however, thicknesses must be determined from poor exposures or from wells that are not perpendicular to the bed boundaries. The following sections give different methods for finding the thickness of beds that are uniform in dip between the top and base along the line of the measurement. The map-angle method (Sect. 2.6.1) applies to maps on a topographic base. Two equations are necessary with the choice between them depending on the direction of the topographic slope relative to the direction of the dip. If the topographic slope is nearly the same as but oblique to the dip, it is easy to accidentally choose the wrong equation. The pole-thickness method (Sect. 2.6.2) is based on a single equation that applies to both maps and wells. This method is suitable for automatic calculation in a spreadsheet. Where the top and base of a unit can be defined by structure contours, the best thickness measurement is derived from the contours (Sect. 2.6.3). The final section (2.6.4) considers the effect of measurement errors on calculated thicknesses.

2.6.1 Map-Angle Thickness Equations

The calculation of the thickness of a unit based on map distances and directions results in two equations, depending on whether the topographic slope is in the same direction as the dip of the unit (derived as Eqns. D2.24 and D2.27) or the opposite direction to it. For the bed dip and topographic slope in the same direction, the equation is:

$$t = |h \cos \alpha \sin \delta - v \cos \delta|, \tag{2.22}$$

where the notation $|\cdots|$ = the positive value of the expression between the bars. For a bed dip and topographic slope that are in opposite directions:

$$t = h \cos \alpha \sin \delta + v \cos \delta, \tag{2.23}$$

where h (Fig. 2.25) = the horizontal distance along a line between the upper and lower contacts on the map, α = the angle between the measurement line and the dip direction, δ = the true dip of the unit, and v = the elevation difference between the end points of the measurement line. If the measurement is in the dip direction, $\alpha = 0$ and $\cos \alpha = 1$; if the base and top of the unit are both at the same elevation, $v = 0$.

As an example, determine the thickness of the Pride Mountain Formation (Mpm) indicated by its outcrop width on the map of Fig. 2.26. Along line a on the map, the horizontal width of the outcrop is h = 1250 ft, the vertical drop from highest to lowest contact is v = 100 ft, the attitude of bedding is $\delta = 07°$ (the average of the 06 and 08° dips mapped) at an azimuth $\theta = 135°$, and the angle between the dip direction and the measurement line is $\alpha = 85°$. The bed and topography slope in the same direction (strike of bedding = 225° and the thickness traverse is along azimuth 220°), making Eq. (2.22) appropriate. The resulting thickness is 86 ft.

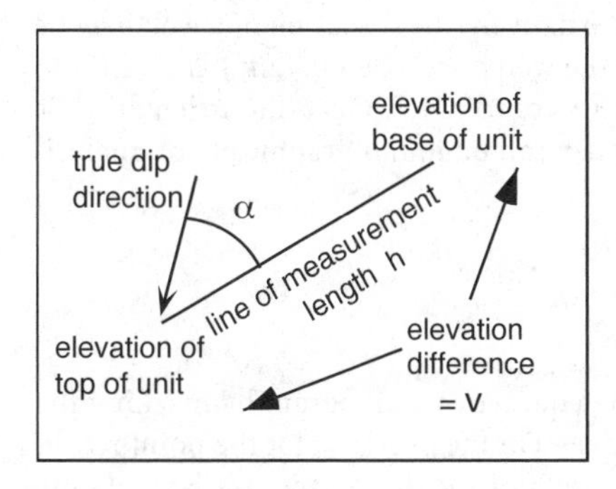

Fig. 2.25. Map data needed to determine the thickness of a unit from outcrop observations using the map-angle thickness Eqs. (2.22) and (2.23).

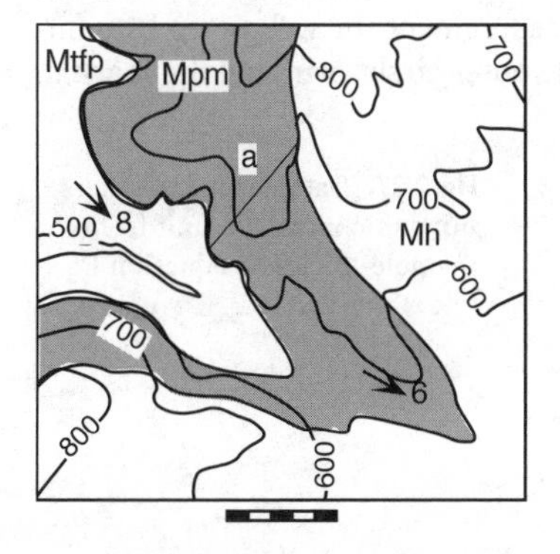

Fig. 2.26. Blount Springs map area from Fig. 2.24 showing the line of a thickness measurement (*a*)

2.6.2 Pole-Thickness Equation

The thickness can be determined from a single equation based on the angle between the direction of the thickness measurement and the pole to bedding (Fig. 2.27). The method is convenient for data from a well or a map. The advantage of this method is that the thickness is given by a single equation, eliminating the potential error of choosing the wrong equation. The disadvantage for hand calculation is the need to determine an angle in three dimensions, although this is easily done with a stereogram. The apparent thickness is measured along the direction L. In the plane defined by the line of measurement and the pole to bedding

$$t = L \cos \rho , \tag{2.24}$$

where t = the true thickness and L is the straight-line length between the top and base of the unit, measured along a well or between two points on a map. The angle ρ = the angle between L and the pole to the bed. If the acute angle is used in the equation, the thickness is positive. If the obtuse angle is used the thickness will be correct in magnitude but negative in sign; taking the absolute value gives the correct result for either possibility.

If L is vertical, then the angle $\rho = \delta$, the dip of the bed. If L is not vertical, its inclination must be determined. For a well, a directional survey will give the azimuth of the deviation direction and the amount of the deviation; the latter may be reported as the hade, which is the angle from the vertical. Alternatively, the deviation of a well may be given as the xyz coordinates of points along the well bore. Use Eq. (2.17) and (2.18) to determine the orientation (θ, δ) from the xyz coordinates. Once the orientation is known, then the pole to the bed and the angle r can be found graphically or analytically, as explained in the next two sections.

2.6.2.1 *Angle Between Two Lines, Stereogram*

To find the angle r, prepare an overlay on the equal-area stereogram by marking the positions of the north, south, east, and west axes. On the overlay, plot the point representing the pole to bedding by marking the trend of the dip on the overlay, rotating the overlay to bring this mark to the east-west axis, and counting inward from the outer circle (the zero-dip circle) the amount of the dip, and mark the point. Return the overlay to its original position. Plot the line of measurement (or well bore) by similarly marking the trend of the measurement on the outer circle, bringing the mark to

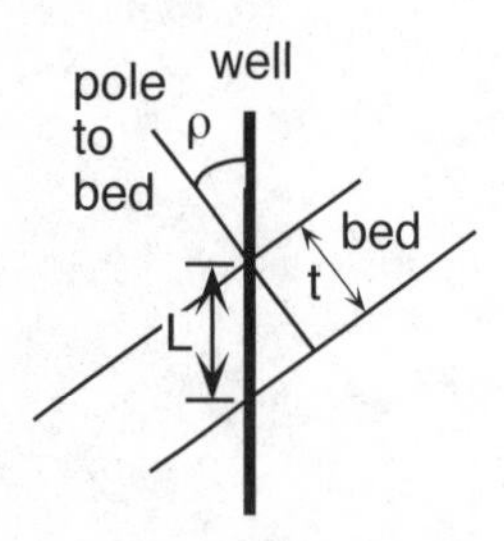

Fig. 2.27. Data needed to determine thickness of a unit from the pole-thickness equation Eq. (2.24)

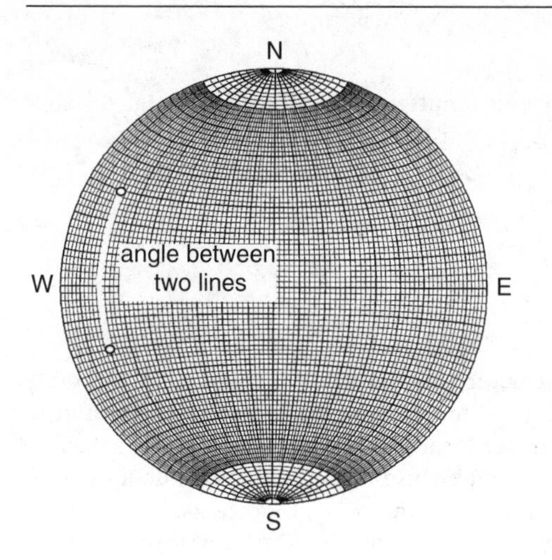

Fig. 2.28. Equal-area stereogram showing the angle between two lines. Angle ρ is the great circle distance between the points giving the pole to bedding and the orientation of the apparent thickness measurement. Lower-hemisphere projection

the east-west axis and measuring the dip inward from the outer circle if given as a plunge, or outward from the center of the graph if given as a hade. Rotate the overlay until the two points fall on the same great circle (Fig. 2.28). The angle ρ is measured along the great circle between the two points.

As an example of the thickness calculation based on the pole-thickness equation (Eq. 2.24), find the true thickness of a bed that is L = 10 m thick in a well. The well hades 10° to 310° and the bed dip vector is 20, 015. Plot the bed on the stereogram and find its pole. Then plot the well and measure the angle between the two lines (ρ = 27°). Equation (2.24) gives t = 8.9 m.

2.6.2.2
Angle Between Two Lines, Analytical

If the locations of the unit boundaries are given by their xyz coordinates, and the orientation of bedding by its azimuth and dip, the angle ρ required in Eq. (2.24) may be found from Eq. (D2.31) as:

$$\rho = \arccos\{[(x_1 - x_2) / L] \sin\theta_b \sin\delta_b + [(y_1 - y_2) / L] \cos\theta_b \sin\delta_b + [(z_2 - z_1) / L] \cos d_b\}, \quad (2.25)$$

where θ_b and δ_b =, respectively, the azimuth and dip of the bed dip vector, L, the apparent length, is given by Eq. (D2.2) as:

$$L = [(x_2 - x_1)^2 + (y_2 - y_1)^2 + (z_2 - z_1)^2]^{1/2}. \quad (2.26)$$

If the line of the thickness measurement is defined by its map length, h, change in elevation, v, and orientation given by the azimuth, θ, to the lower end of the line, then the angle ρ required in Eq. (2.24) may be found from Eq. (D2.30) as:

$$\rho = \arccos\{\cos\delta_b \sin[\arctan(v / h)] - \sin\delta_b \cos[\arctan(v / h)] (\cos\theta_b \cos\theta + \sin\theta_b \sin\theta)\}, \quad (2.27)$$

where θ_b and δ_b =, respectively, the azimuth and dip of the bed dip vector. L, the apparent length, is:

$$L = (v^2 + h^2)^{1/2}, \tag{2.28}$$

where v = vertical distance between end points and h = the horizontal distance between end points. The angle in Eq. (2.27) must be acute and must be changed to = 180 – ρ if it is obtuse. The results of Eq. (2.25) or (2.27) are substituted into (2.24) to find the thickness.

2.6.3 Thickness Between Structure Contours

The thickness determined between structure contours shows much less variability than that determined between individual points and provides a more reliable value in situations where the attitudes and contact locations are uncertain, as on a surface map. Determining the best-fit structure contours uses a large amount of data simultaneously to improve the attitude of bedding and the contact locations. This method requires a structure contour at the top and base of the bed (Fig. 2.29a). The width of the unit is always measured in the dip direction, perpendicular to the structure contours. The simplest situation arises if contours at the same elevation can be constructed on the top and base, because then, from the geometry of Fig. 2.29b, the thickness of the unit is

$$t = h_c \sin \delta, \tag{2.29}$$

where h_c = horizontal distance between contours at equal elevations on the top and base of the unit, t = true thickness, and δ = true dip. If the contours on the top and base of the unit are at different elevations (Fig. 2.29c), then the line on the map that connects the upper and lower contours has the length L, and the thickness can be calculated from Eq. (2.24). Equation (2.24) gives the same result as Eq. (2.29) for the special case where L is horizontal.

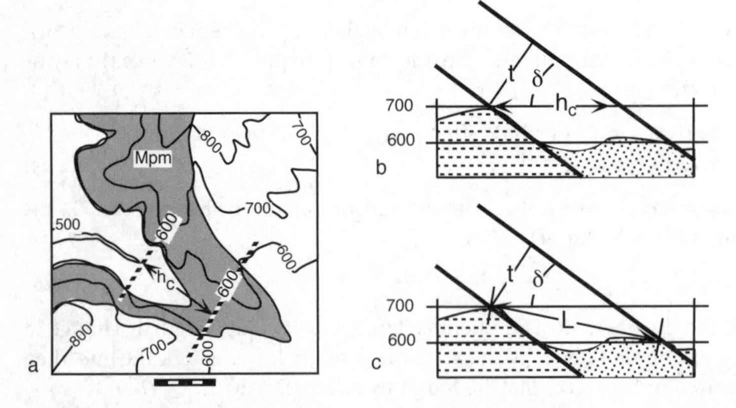

Fig. 2.29. Thickness measured between structure contours. **a** Structure contours at 600-ft elevation on the top and base of a formation; h_c is perpendicular to structure contours. **b** Measurement along a constant elevation on a vertical cross section in the dip direction. **c** Measurement between points of different elevations on a vertical cross section in the dip direction. For explanation of *symbols,* see text (Sect. 2.6.3)

As an example, in Fig. 2.29a, the 600-ft contour has been located on both the top and base of the Mpm, and so the simplest means of thickness determination is with Eq. (2.29). The value of h_c measured from the map is 1387 ft, and the dip, δ, is 04°, giving a thickness of 97 ft.

2.6.4 Effect of Measurement Errors

In the determination of thickness from measurements on a map, there are two sources of error: the length measurement and the direction measurement. Errors in length will result from the finite widths of the lines at the scale of the map and from errors in the positions of the unit boundaries. The minimum possible length error is approximately equal to the thickness of the geologic contacts as drawn on the map. For example, the geologic contacts in Fig. 2.29a are about 20 ft wide. Thus a thickness measurement of about ± 20 ft is about the best that could be done on this map. Small errors in contact locations will lead to small errors in the calculated thickness. Direction errors will arise mainly from uncertainties in the dip amount and dip direction. The effects of error in the dip range from very small if the thickness measurement is at a high angle to the bed boundary, to very large if the measurement is at a low angle to the boundary.

The effect of measurement errors in directions on a thickness determination can be estimated using Eq. (2.24) which puts the variables in their simplest form. The true thickness (t) in Eq. (2.24) is directly proportional to the apparent thickness (L) and so the erroneous thickness (t_e) is directly proportional to the error in the length of the apparent thickness measurement. The error measure (t_e) is normalized in Fig. 2.30 to remove the effect of the length scale. The true thickness in Eq. (2.24) is proportional to the cosine of the angle between the pole to the bed and the line of the thickness

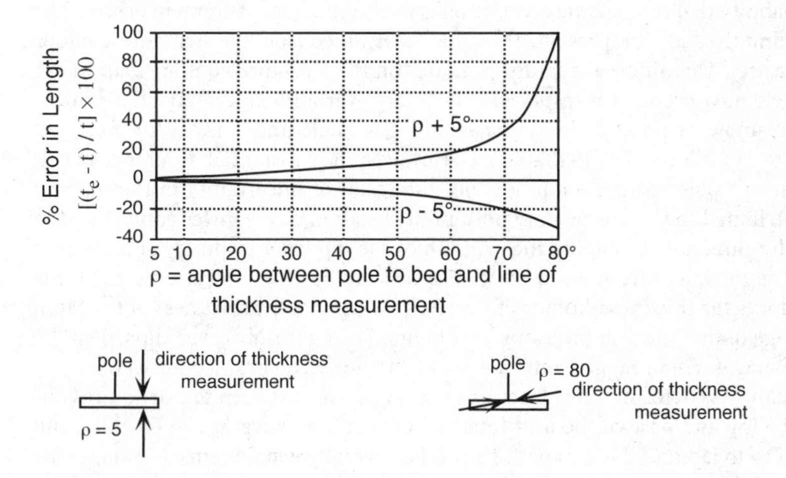

Fig. 2.30. Effect of attitude and angle measurement errors on thickness calculation. Normalized error in thickness measurement, in percent as a function of the angle ρ (Eq. 2.24) for angle errors of ρ = ± 5°. *t* True thickness of bed; t_e erroneous thickness given the angle error; ρ angle between the pole to the bed and direction along which thickness is measured

measurement (ρ), leading to a non-linear dependence of the error on the size of the angle (Fig. 2.30). The angle error could be in the orientation of the bed, in the orientation of the apparent thickness, or a combination of both. A combined error in ρ of $\pm 5°$ from the correct value seems like the upper limit for careful field or well measurements. An angle of $\rho = 0°$ means that the thickness measurement is perpendicular to bedding; this is the most accurate measurement direction. An angle of $\rho = 90°$ means that the measurement direction is parallel to bedding, an impossibility. The sensitivity of a thickness measurement to error in the angle goes up rapidly as ρ increases (Fig. 2.30). The thickness error exceeds ±10% at an angle of 40° and is about 100% at 80+5° and −35% at 80−5°. This means that thickness measurements made at a low angle to the plane of the bed may produce very large errors, with a strong bias toward overestimation. At angles between the measurement direction and bedding of 10° or less ($\rho = 80°$ or greater), very large thickness errors will occur with very small orientation errors and thickness determinations made from maps should be considered suspect.

From Fig. 2.30, a bed having a true thickness of 100 m measured at an angle of 10° to bedding ($\rho = 80°$) for which the angle is overestimated by 5° (giving $\rho = 85°$) will yield a thickness of nearly 200 m. The same measurement with a 5° underestimate in the angle will give a thickness of 65 m. The average thickness for these two measurements is 132.5 m, still an overestimate. Thickness measurements made nearly perpendicular to bedding ($\rho \approx 0°$) are rather insensitive to errors in the angle. At $\rho = 20°$ the error is about ±3%; a 100 m thick bed would be measured as being between 97 and 103 m thick.

Thickness calculations between two points on a map are very dependent on the exact points chosen and the attitude of bedding, as illustrated by the data obtained from Fig. 2.31a. The thickness of the Mpm from the seven locations a–g (Table 2.2) ranges from 84 to 230 ft, as calculated from Eq. (2.24). This is an unreasonably large variation in thickness at what is nearly a single location at the scale of the map. What is the probability that the thickness variation is due to small measurement errors? The bedding azimuth of 4, 125 represents the value determined from the structure-contour map (Fig. 2.31a). The difference in dip from 04° on the structure contour map to 06° from the field measurement is responsible for a large variation in the calculated thickness. For example, at point c, the thickness along a single line is 127 ft for a 04° dip compared to 230 ft for a 06° dip (Table 2.2). However, the substantial difference in calculated thickness along lines a and g is not the result of uncertainty in the dip, but must be attributed to the uncertainty both in the location of the lower contact and in the exact dip direction. Changing the azimuth of the dip from 135 to 127 at location a increases the thickness from 84 to 108 ft at a bed dip of 08°; at location g the same change reduces the thickness from 173 to 159 ft. If we say that the thickness of the Mpm is the average of the values at locations a-g obtained using the observed dips, then the thickness is 181 ft with a range from 84 to 230 ft, a poorly constrained answer.

The thicknesses determined at locations 1–4 (Fig. 2.31b) between the structure contours on the top and base of the unit (method of Sect. 2.6.3) average 108 ft thick and range from 97 to 120 ft (Table 2.2) or 108 ±12 ft based on the whole range of values. The SE lengths (Table 2.2) are measured to a southeasterly position on the base of the Mpm, at the 600-ft contour, which lies directly beneath the 700-ft contour on the top of the unit; the NW measurements are from the more northwesterly position of the lower contact. Changing the location of the structure contour of the base has only a

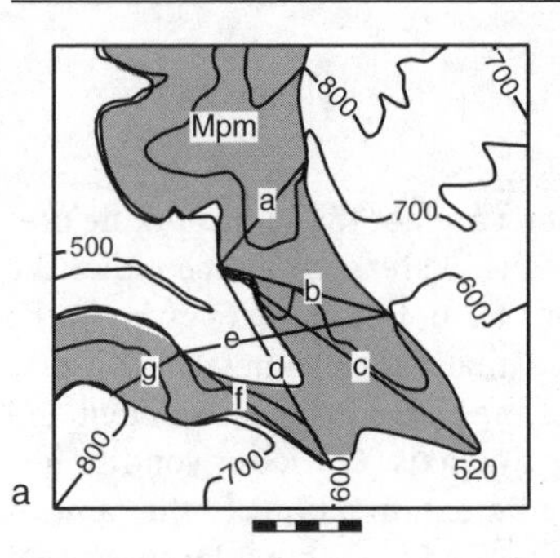

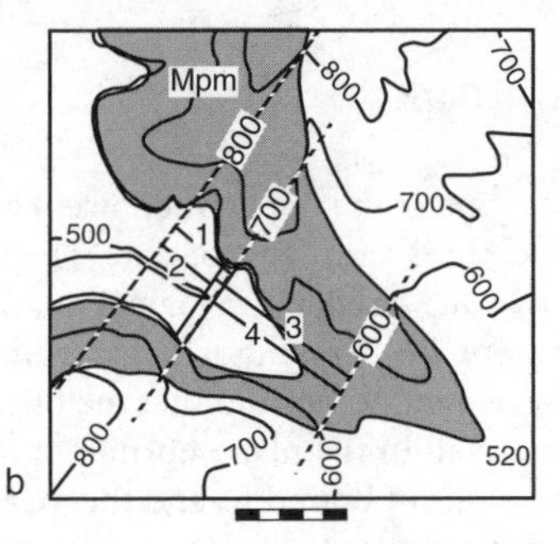

Fig. 2.31. Alternative thickness measurements of Mpm on the map of the Blount Springs area (from Fig. 2.24). **a** Point-to-point measurement lines *a–g*. **b** Measurements *1–4* between structure contours

Table 2.2. Data for thickness calculation in Sequatchie anticline map area of Fig. 2.31. All thicknesses calculated with Eq. (2.24)

Location	Bed Dip, azimuth	Apparent t azimuth	h (ft)	v (ft)	α	Calculated t (ft)
a	7, 135	220	1242	100	85	86
	8, 135	220	1242	100	85	84
b	7, 125	105	1658	0	20	190
	4, 125	105	1658	0	20	109
c	6, 125	126	2960	80	1	230
	4, 125	126	2960	80	1	127
d	6, 125	150	2104	0	25	199
	4, 125	150	2104	0	25	133
e	7, 125	080	1030	0	45	174
	4, 125	080	1030	0	45	99
f	7, 125	127	1770	0	2	216
	4, 125	127	1770	0	2	123
g	7, 135	059	741	200	76	177
	8, 135	059	741	200	76	173
1 SE	8, 127	127	631	200	0	110
2 NW	8, 127	127	564	200	0	120
3 SE	4, 124	124	1503	0	0	105
4 NW	4, 124	124	1387	0	0	97

small effect on the thickness. The average thickness determined from the structure-contour-based measurements falls within the range of the point-to-point measurements, but is much smaller than the average of the point-to point measurements, as expected from the behavior of the thickness equation (Fig. 2.30). Thickness measurements between two points (Eq. 2.22 and 2.23, or 2.24) exhibit a non-linear sensitivity to error at low angles between the dip vector and the measurement orientation, leading to a high probability of an artificially high average from multiple measurements. Smoothing of the attitude errors by structure contouring leads to a better average thickness.

Where the thickness is known accurately from a complete exposure or from well-defined contacts in a borehole, the structure contours or bedding attitudes might be adjusted to conform to the thicknesses. The thickness measured between structure contours is the best approach at the map scale where there is uncertainty in the data.

2.7 Thickness of Folded Beds

In a folded bed, the dips of the upper and lower contact are not the same and the previous thickness equations are inappropriate. The fold is likely to approach either the planar dip domain or the circular arc form. Equations for both forms are given in the next two sections. For both methods it is assumed that the thickness is constant between the measurement points and that the line of the thickness measurement and the bedding poles are all in the plane normal to the fold axis. The latter condition is satisfied if the directions of both dips and the measurement direction are the same. If the geometry is more complex than this, then a cross section perpendicular to the fold axis should be constructed to find the thickness and projection may be required, as discussed in Chapter 7.

2.7.1 Circular-Arc Fold

If the bed forms a portion of a circular arc (Fig. 2.32), then its thickness is (from Eq. D2.36):

$$t = (L / \sin \gamma)(\sin \rho_2 - \sin \rho_1), \tag{2.30}$$

where t = true thickness, L = apparent thickness along the well or traverse (Eqs. 2.26 or 2.28), ρ_1 = the smaller angle between the well and the pole to bedding, ρ_2 = the larger angle between the well and the pole to bedding, and $\gamma = \rho_2 - \rho_1$.

A typical data set is shown in Fig. 2.33a. The cross section is in the dip direction. The angles between the line of measurement and the poles to bedding are determined as well as the acute angle between the poles (Fig. 2.33b). From Eq. (2.30), the true thickness of the bed is 234 ft.

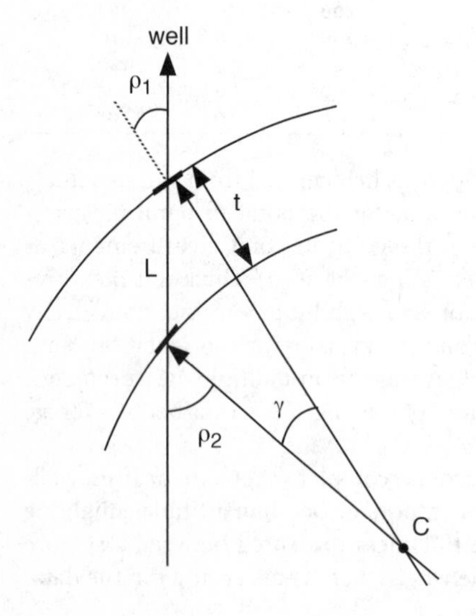

Fig. 2.32. Thickness of a bed folded into a circular arc, in the plane normal to the fold axis

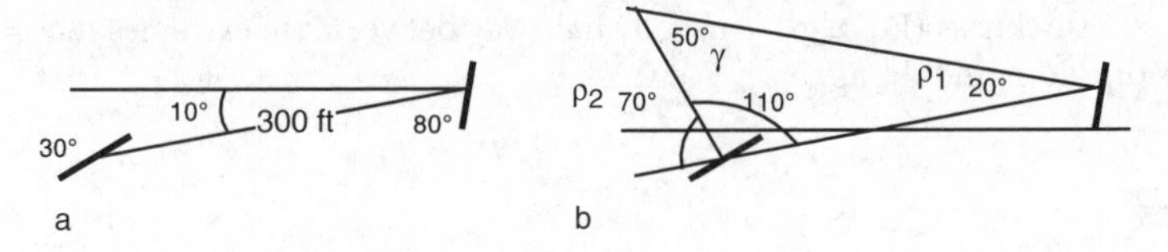

Fig. 2.33. Example of thickness determination of a circularly folded bed. **a** Field data: bedding dips are shown by *heavy lines*, distance between exposures of upper and lower contacts is 300 ft on a line that plunges 10°. **b** Angles required for the thickness calculation

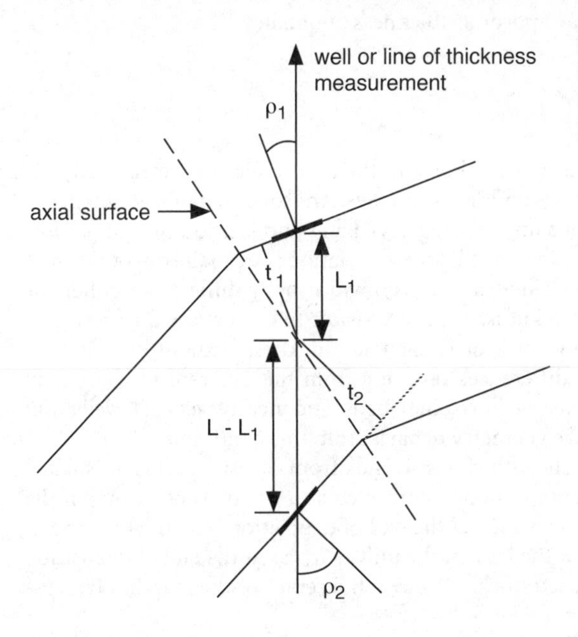

Fig. 2.34. Thickness of a dip-domain bed that changes dip in the measured interval, in the plane normal to the fold axis

2.7.2 Dip-Domain Fold

The dip-domain method can be used to find or place bounds on the thickness of a unit that changes dip from its upper to lower contact (Fig. 2.34). For constant bedding thickness, the axial surface bisects the angle of the bend. The total thickness of the bed along the measurement direction, t, is the sum of the thickness in each domain, found from Eq. (2.24) as

$$t = L_1 \cos \rho_1 + (L - L_1) \cos \rho_2, \tag{2.31}$$

where ρ_1 = the angle between the well and the pole to the upper bedding plane, ρ_2 = the angle between the well and the pole to bedding of the lower bedding plane, L = total apparent thickness, and L_1 = the apparent thickness of the upper domain. If the position of the dip change can be located, for example with a dipmeter, then it is possible to specify L_1 and find the true thickness. If the location of the axial surface is unknown, the range of possible thicknesses is between the values given by setting $L_1 = 0$ and $L_1 = L$ in Eq. (2.31).

The circular-arc thickness (Eq. 2.30) is usually half-way between the extremes that are possible for dip-domain folding.

2.8 Thickness Maps

Thickness maps are valuable for both structural and stratigraphic interpretation purposes. Multiple measures of thicknesses can be mapped, requiring care in measurement and interpretation. The calculated thickness is related to the dip and so uncertainties or errors in the dip may appear as thickness anomalies.

2.8.1 Construction and Interpretation

An isopach map is a map of the true thickness of the unit (t, Fig. 2.35) measured normal to the unit boundaries (Bates and Jackson 1987). An isocore map is defined as a map of the vertical thickness of a unit (t'_V; Fig. 2.35; Bates and Jackson 1987). Both isocore and drilled thickness maps include thickness variations due to the dip of the unit. The drilled thickness in a deviated well (t'_S, 2.35) will usually differ from either the true thickness or the vertical thickness. It is not possible to correct the thickness in a deviated well to the vertical thickness or to the true thickness without knowing the dip of the bed. The thickness differences resulting from the different measurement directions are not large for nearly horizontal beds and nearly vertical wells, but increase significantly as the true geometry departs from this condition.

An isopach map is used to show thickness trends from measurements at isolated points (Fig. 2.36a). An isopach map can be interpreted as a paleostructure map if the upper surface of the unit was horizontal at the end of deposition. The thickness variations represent the structure at the base of the unit as it was at the end of deposition of the unit. The trend of increased thickness down the center of the map in Fig. 2.36a could imply a filled paleovalley.

The slope of the base of the paleovalley can be determined from the thickness difference and the spacing between the contours according to the geometry of Fig. 2.36b:

$$\delta = \arctan(\Delta t / h), \qquad (2.32)$$

where δ = the slope, Δt = the difference in thickness between two contours, and h = the horizontal (map) distance between the contours, measured perpendicular to the contours. For the map in Fig. 2.36a, the slope implied for the western side of the pale-

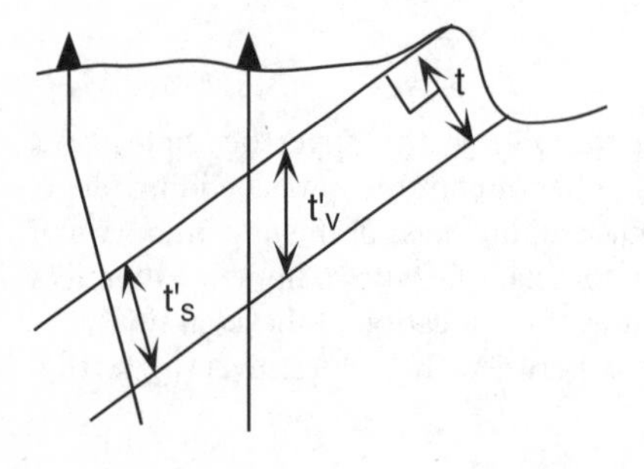

Fig. 2.35. Vertical cross section in the dip direction showing true and apparent thicknesses: t true thickness; t'_V vertical thickness; t'_S slant thickness

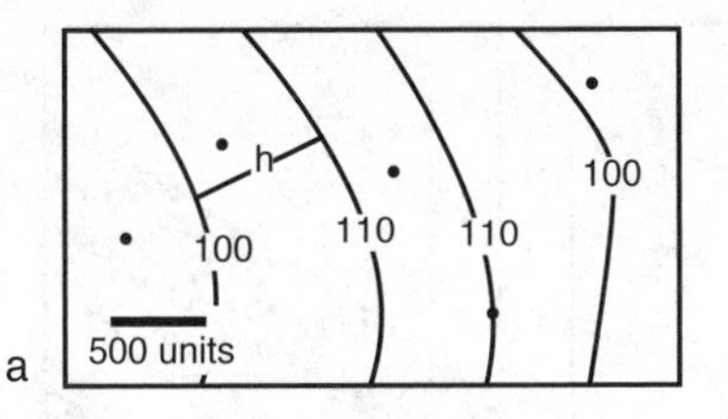

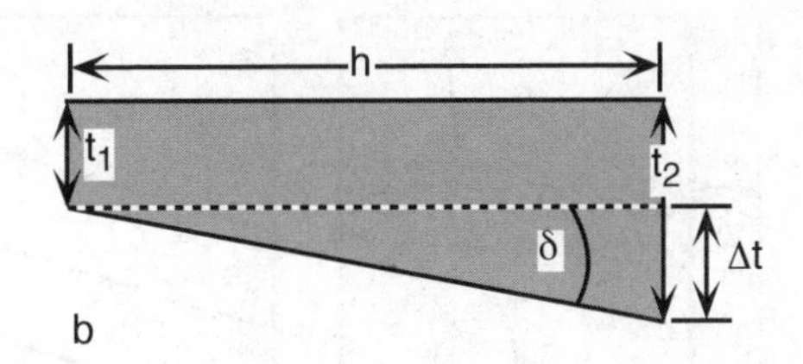

Fig. 2.36. Paleoslope from thickness change. **a** Isopach map. Dots are measurement points; h is the location of a cross section. **b** Cross section *(shaded)* perpendicular to the trend of the thickness contours, interpreted as if the upper surface of the unit were horizontal

ovalley is about 0.5° ($\Delta t = 10$, $h \sim 900$). Stratigraphic thickness variations could be caused by growing structures. The dip calculated from an isopach map using Eq. (2.35) could represent the structural dip that developed during deposition. According to this interpretation, Fig. 2.36a would show a depositional syncline.

2.8.2 Dip vs. Thickness

True dip is not known in a well unless the interval of interest has been cored or a dipmeter log is available. Inferred thickness variations might be due to structural dip. The exaggerated thickness due to the dip in a vertical well has the same form as Eq. (2.24) (replacing L by t′) and can be rewritten as:

$$t' = t / \cos \delta, \tag{2.33}$$

where t = true thickness, t′ = exaggerated thickness, and δ = true dip of the unit. The same relationship applies to a well deviated from the vertical an amount δ and intersecting horizontal beds. If a unit has a true thickness of 100 m, a dip of 10° gives an exaggerated thickness of 102 m, 20° gives 106 m, 30° gives 115 m, 40° gives 131 m, 50° gives 156 m. The importance of this effect will clearly depend on the level of detail being interpreted, but will become significant for nearly any purpose at dips over 20–30°.

Apparent thickness variations can provide a very sensitive tool for structural analysis. Although the thickness exaggeration at low dips may not be significant for the purpose of stratigraphic interpretation, the implied dips are critical in structural interpretation. If the true thickness is known, the dip (or well deviation in horizontal beds) can be determined by solving Eq. (2.24) to obtain:

$$\delta = \arccos (t / t'). \tag{2.34}$$

For example, a measured thickness of 103 ft for a unit having a true thickness of 100 ft is stratigraphically hardly significant, yet implies a dip of 14° which is steeper than the dip that produces the closure in many oil fields. If the unit mapped in Fig. 2.36a actually has a constant thickness of 100 units, then the dips in the center of the map where the isocore thickness is 110 units must be 25°. Alternatively, the unit could be horizontal and the wells in which the thicknesses were observed could deviate 25° from the vertical.

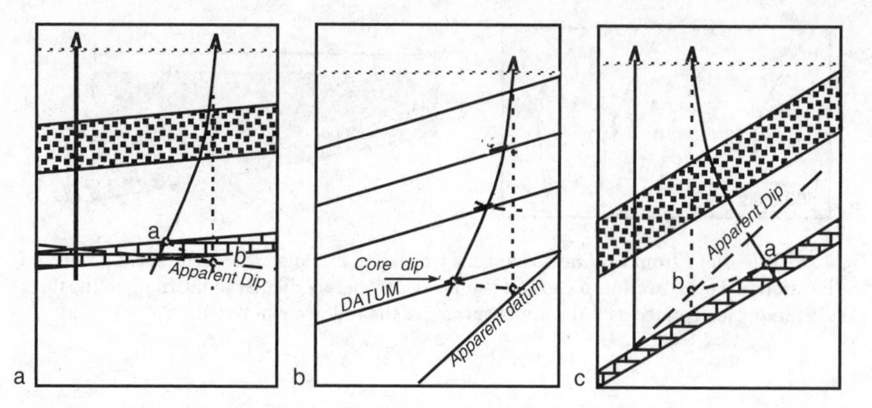

Fig. 2.37. False apparent dips caused by unrecognized well deviation. **a** Apparent dip determined between two wells, one unknowingly deviated down dip. **b** Dip determined from core dip in the unknowingly deviated well. **c** Apparent dip determined between two wells, one unknowingly deviated up dip. (After Low 1951)

2.8.3 Location Anomalies, Dip and Thickness

Deviated wells that are mistakenly interpreted as being vertical will cause dip and thickness anomalies. The log depth to a formation boundary is larger in a well that deviates down dip than in a vertical well at the same surface location (Fig. 2.37a). If the formation boundary is plotted vertically beneath the well location, its depth will be too great. If an apparent dip is then determined between this well and a correctly located formation top in another well, the apparent dip will be wrong, perhaps even in the wrong direction (Fig. 2.37a). If the dip is determined from a core in a deviated well (Fig. 2.37b) and the well is mistakenly thought to be vertical, then the inferred dip will be too large. The apparent thicknesses are too large in both situations. A well that deviates up dip will result in an apparent steepening of the dip between two wells and thicknesses (if mistakenly corrected for dip) that will be too small (Fig. 2.37c).

2.9 Derivations

2.9.1 Location

In a right-handed xyz coordinate system (Fig. 2.4) with z positive upward, positive x = east and positive y = north, the point P (xyz) dividing P1 ($x_1y_1z_1$) and P2 ($x_2y_2z_2$) by the ratio r/s has the coordinates (Eves, 1984):

$$x = (rx_2 + sx_1)/(r + s); \quad \text{(D2.1a)}$$

$$y = (ry_2 + sy_1)/(r + s); \quad \text{(D2.1b)}$$

$$z = (rz_2 + sz_1)/(r + s), \quad \text{(D2.1c)}$$

where r + s = L is found from the Pythagorean theorem:

$$L = [(x_2 - x_1)^2 + (y_2 - y_1)^2 + (z_2 - z_1)^2]^{1/2}. \quad \text{(D2.2)}$$

Substitute s = L – r into Eq. (D2.1) to obtain

$$x' = (rx_2 - rx_1 + Lx_1) / L; \quad \text{(D2.3a)}$$

$$y' = (ry_2 - ry_1 + Ly_1) / L; \quad \text{(D2.3b)}$$

$$z' = (rz_2 - rz_1 + Lz_1) / L. \quad \text{(D2.3c)}$$

2.9.2 Solid Geometry

The orientations of structural elements are most readily obtained analytically by vector geometry. A line is represented by a vector of unit length and a plane by its pole vector. The direction cosines of a line describe the orientation of the unit vector parallel to the line. Structural information such as bearing and plunge is converted into direction cosine form, the necessary operations performed, and then the values converted back to standard geological format.

2.9.2.1 *Direction Cosines*

Direction cosines define the orientation of a vector in three dimensions (Fig. D2.1). The direction angles between the line OC and the positive coordinate axes x, y, z are α, β, γ, respectively. The direction cosines of the line OC are:

$$\cos \alpha = \cos EOC; \quad \text{(D2.4a)}$$

$$\cos \beta = \cos GOC; \quad \text{(D2.4b)}$$

$$\cos \gamma = \cos (90 + AOC). \quad \text{(D2.4c)}$$

In the geological sign convention, an azimuth of 0 and 360° represents north, 90° east, 180° south, and 270° west. Dip (of a plane) or plunge (of a line) is an angle from the horizontal between 0 and 90°, positive downward. From the geometry of Fig. D2.1:

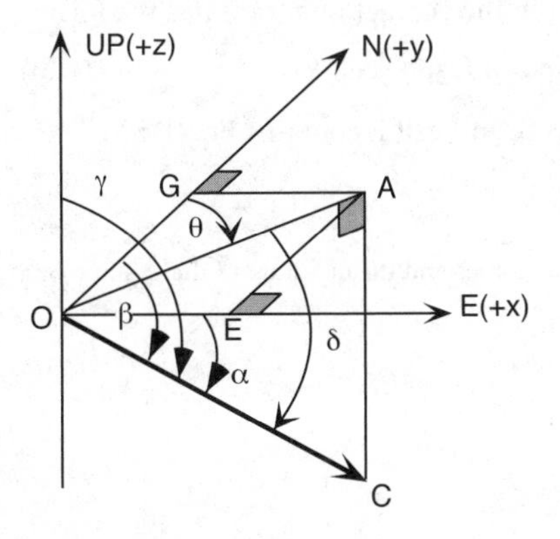

Fig. D2.1. Orientation of vector *OC* in right-handed xyz space; θ is the azimuth of the vector and δ is the plunge. The direction angles are α, β, γ

$$\cos \delta = OA / OC; \quad \text{(D2.5a)}$$

$$\cos \theta = OG / OA; \quad \text{(D2.5b)}$$

$$\sin \theta = OE / OA; \quad \text{(D2.5c)}$$

$$\cos \alpha = OE / OC; \quad \text{(D2.5d)}$$

$$\cos \beta = OG / OC; \quad \text{(D2.5e)}$$

$$\cos \gamma = \cos (90 + \delta) = - \sin (\delta). \quad \text{(D2.5f)}$$

2.9.2.2
Direction Cosines of a Line from Azimuth and Plunge

The direction cosines of a line in terms of its azimuth and plunge are obtained from the relationships in Eq. (D2.5):

$$\cos \delta \sin \theta = (OA / OC)\ (OE / OA) = OE / OC = \cos \alpha; \quad \text{(D2.6a)}$$

$$\cos \delta \cos \theta = (OA / OC)\ (OG / OA) = OG / OC = \cos \beta; \quad \text{(D2.6b)}$$

$$- \sin \delta = \cos \gamma. \quad \text{(D2.6c)}$$

2.9.2.3
Azimuth and Plunge from Direction Cosines

To reverse the procedure and find the azimuth and plunge of a line from its direction cosines, divide Eq. (D2.6a) by (D2.6b) and solve for θ, then solve Eq. (D2.6c) for δ:

$$\theta' = \arctan (\cos \alpha / \cos \beta); \quad \text{(D2.7)}$$

$$\delta = \arcsin (- \cos \gamma). \quad \text{(D2.8)}$$

The value θ' given by Eq. (D2.7) will be in the range of ±90° and must be corrected to give the true azimuth over the range of 0 to 360°. The true azimuth, θ, of the line can be determined from the signs of $\cos \alpha$ and $\cos \beta$ (Table D2.1). The direction cosines give a directed vector. The vector so determined might point upward. If it is necessary to reverse its sense of direction, reverse the sign of all the direction cosines. Note that division by zero in Eq. (D2.7) must be prevented.

An Excel equation that returns the azimuth in the correct quadrant has the form:

= IF(cell B<=0,180+Cell E,IF(cell A>=0,cell E,360+cell E)) (D2.9)

where cell A contains $\cos \alpha$, cell B contains $\cos \beta$, and cell E contains Eq. (D2.7).

Table D2.1. Relationship between signs of the direction cosines and the quadrant of the azimuth of a line

Azimuth	$\cos \alpha$	$\cos \beta$	θ
000+ to 090	+	+	θ'
090+ to 180	+	–	$180 + \theta'$
180+ to 270	–	–	$180 + \theta'$
270+ to 360	-	+	$360 + \theta'$

2.9.2.4
Direction Cosines of a Line on a Map

The direction cosines of a line defined by its horizontal, h, and vertical, v, dimensions on a map can be found by letting $\delta = \arctan(v / h)$ in Eqs. (D2.6):

$$\cos \alpha = \cos(\arctan(v / h)) \sin \theta; \quad \text{(D2.10a)}$$

$$\cos \beta = \cos(\arctan(v / h)) \cos \theta; \quad \text{(D2.10b)}$$

$$\cos \gamma = -\sin(\arctan(v / h)). \quad \text{(D2.10c)}$$

Both v and h are taken as positive numbers in Eq. (D2.10) and the resulting dip is positive downward.

The direction cosines of a line defined by the x, y, z coordinates of its two end points are obtained by letting point 1 be at O and point 2 be at C in Fig. D2.1. Then

$$x_2 - x_1 = OE; \quad \text{(D2.11a)}$$

$$y_2 - y_1 = OG; \quad \text{(D2.11b)}$$

$$z_2 - z_1 = AC. \quad \text{(D2.11c)}$$

Substitute Eqs. (D2.11) into (D2.6) to obtain:

$$\cos \alpha = (x_2 - x_1) / OC; \quad \text{(D2.12a)}$$

$$\cos \beta = (y_2 - y_1) / OC; \quad \text{(D2.12b)}$$

$$\cos \gamma = -\sin \delta = -AC / OC = -(z_2 - z_1) / OC = (z_1 - z_2) / OC, \quad \text{(D2.12c)}$$

where OC = L, given by Eq. (D2.2). Using the convention that point 1 is higher and point 2 is lower, a downward-directed bearing is positive in sign.

2.9.2.5
Azimuth and Plunge of a Line from the End Points

To find the azimuth and plunge of a line from the coordinates of two points, substitute Eq. (D2.12) into (D2.7) and (D2.8):

$$\theta' = \arctan((x_2 - x_1) / (y_2 - y_1)); \quad \text{(D2.13)}$$

$$\delta = +\arcsin((z_2 - z_1) / L). \quad \text{(D2.14)}$$

The value of L is given by Eq. (D2.2). If $y_2 = y_1$, there is a division by zero in Eq. (D2.13) which must be prevented. The value of θ is obtained from Table D2.1.

2.9.2.6
Pole to a Plane

The pole to a plane defined by its dip vector has the same azimuth as the dip vector, and a plunge of $\delta_p = \delta + 90$. Substitute this into Eq. (D2.6) to obtain the direction cosines of the pole, p, in terms of the azimuth, θ, and plunge, δ, of the bedding dip:

$$\cos \alpha_p = \cos(\delta + 90) \sin \theta = -\sin \delta \sin \theta; \quad \text{(D2.15a)}$$

$$\cos \beta_p = \cos(\delta + 90) \cos\theta = -\sin\delta \cos\theta; \quad \text{(D2.15b)}$$

$$\cos \gamma_p = -\sin(\delta + 90) = -\cos\delta. \quad \text{(D2.15c)}$$

2.9.2.7
Attitude of a Plane from Three Points

The general equation of a plane (Foley and Van Dam 1983) is:

$$Ax + By + Cz + D = 0. \quad \text{(D2.16)}$$

The equation of the plane from the xyz coordinates of three points is

$$\begin{vmatrix} x & y & z & 1 \\ x_1 & y_1 & z_1 & 1 \\ x_2 & y_2 & z_2 & 1 \\ x_3 & y_3 & z_3 & 1 \end{vmatrix} = 0. \quad \text{(D2.17)}$$

This expression is expanded by cofactors to find the coefficients A, B, C, and D:

$$A = y_1z_2 + z_1y_3 + y_2z_3 - z_2y_3 - z_3y_1 - z_1y_2; \quad \text{(D2.18a)}$$

$$B = z_2x_3 + z_3x_1 + z_1x_2 - x_1z_2 - z_1x_3 - x_2z_3; \quad \text{(D2.18b)}$$

$$C = x_1y_2 + y_1x_3 + x_2y_3 - y_2x_3 - y_3x_1 - y_1x_2; \quad \text{(D2.18c)}$$

$$D = z_1y_2x_3 + z_2y_3x_1 + z_3y_1x_2 - x_1y_2z_3 - y_1z_2x_3 - z_1x_2y_3. \quad \text{(D2.18d)}$$

The direction cosines of the pole to this plane are (Eves 1984):

$$\cos \alpha_p = A / E; \quad \text{(D2.19a)}$$

$$\cos \beta_p = B / E; \quad \text{(D2.19b)}$$

$$\cos \gamma_p = C / E, \quad \text{(D2.19c)}$$

where

$$E = (A^2 + B^2 + C^2)^{1/2} \quad \text{(D2.20)}$$

and the sign of E = – sign D if $D \neq 0$; = sign C if $D = 0$ and $C \neq 0$; = sign B if $C = D = 0$.

The direction cosines of the pole must be converted to the direction cosines of the dip vector. If $\cos \gamma_p$ is negative, the pole points downward. Reverse the signs on all three direction cosines to obtain the upward direction. If $\cos \gamma_p$ is positive, the pole points upward, and the dip azimuth is the same as that of the pole and the dip amount is equal to 90° plus the angle between the pole and the z axis:

$$\cos \alpha = A / E; \quad \text{(D2.21a)}$$

$$\cos \beta = B / E; \quad \text{(D2.21b)}$$

$$\cos \gamma = \cos(90 + \arccos(C / E)). \quad \text{(D2.21c)}$$

To find the azimuth and plunge of the dip direction, use Eqs. (D2.7) and (D2.8) and Table D2.1.

2.9.3
Thickness

The trigonometric method and the bed-normal form of thickness calculation are based on the assumption that the dip is constant between the end points of the measurement. The circular arc-calculation is for a bed segment folded into the shape of a circular arc.

2.9.3.1
Trigonometric Method

The trigonometric method for determining thickness results in two equations, depending on the relative dip of topography and bedding. The following derivations are after Dennison (1968). If the ground slope and the dip are in the same general direction Fig. D2.2 shows that $t = ah$ = the true thickness, $bc = fe = v$ = the vertical elevation change, $ac = h$ = the horizontal distance from the upper to the lower contact, angle $cae = \alpha$ = the angle between the measurement direction and the true dip, and angle $aej = feg = \delta$ = the true dip. The thickness is:

$$t = aj - hj = aj - eg, \tag{D2.22}$$

where

$$eg = v \cos \delta; \tag{D2.23a}$$

$$aj = ea \sin \delta; \tag{D2.23b}$$

$$ea = h \cos \alpha. \tag{D2.23c}$$

Substitute Eqs. (D2.23) into (D2.22) to obtain:

$$t = |h \cos \alpha \sin \delta - v \cos \delta|. \tag{D2.24}$$

If the dip of bedding is less than the dip of the topography, the second term in Eq. (D2.24) is larger than the first, giving the correct, but negative, thickness. Taking the absolute value corrects this problem.

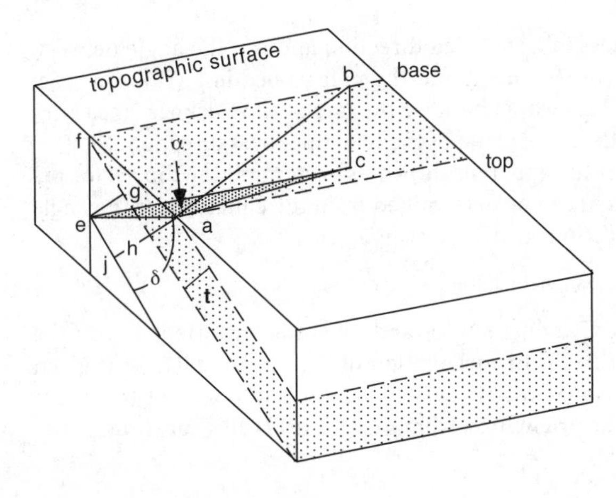

Fig. D2.2. Thickness parameters for a bed and topographic surface dipping in the same general direction

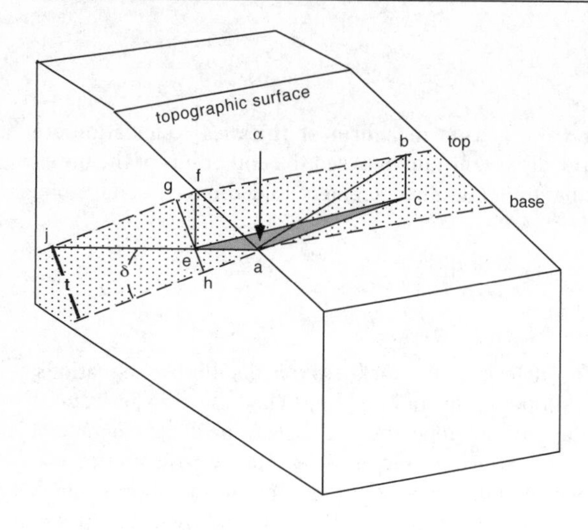

Fig. D2.3. Thickness parameters for a bed and topographic surface dipping in opposite general directions

The thickness of a unit which dips opposite to the slope of topography (Fig. D2.3) is:

$$t = eg + eh, \tag{D2.25}$$

where

$$eh = ea \sin \delta. \tag{D2.26}$$

Substituting Eqs. (D2.23a), (D2.23c), and (D2.26) into (D2.25):

$$t = h \cos \alpha \sin \delta + v \cos \delta. \tag{D2.27}$$

2.9.3.2 *Bed-Normal Form*

The true thickness, t, can be calculated from (Fig. 2.27)

$$t = L \cos \rho, \tag{D2.28}$$

where L = the apparent thickness in a specified direction and ρ = the angle between the direction of the thickness measurement and the pole to bedding (Hobson 1942; Charlesworth and Kilby 1981). This works because the line of the thickness measurement and the pole to bedding lie in the plane of the true thickness direction.

If the length and direction of the apparent thickness vector and the dip vector are known, the angle ρ between them can be determined from the equation for the angle between two lines, given as direction cosines (Eves 1984):

$$\cos \rho = \cos \alpha_1 \cos \alpha_2 + \cos \beta_1 \cos \beta_2 + \cos \gamma_1 \cos \gamma_2. \tag{D2.29}$$

Let the orientation of the bedding dip be δ_b, θ_b and substitute the orientation of the pole from Eqs. (D2.15) into (D2.29). If the orientation of the apparent length is given by the horizontal (h) and vertical (v) distances between the end points of the measurement and the azimuth (θ), the orientation from Eq. (D2.10) is also substituted into (D2.29) to give:

$$\cos \rho = \cos \delta_b \sin [\arctan (v / h)] - \sin \delta_b \cos [\arctan (v / h)] (\cos \theta_b \cos \theta + \sin \theta_b \sin \theta). \quad (D2.30)$$

The direction of this vector is downward, that is, from the higher contact location to the lower. If the orientation of the apparent length is given by the coordinates of the end points of the measurement, the orientation from Eqs. (D2.12) is substituted into (D2.29) to give

$$\cos \rho = ((x_1 - x_2) / L) \sin \theta_b \sin \delta_b + ((y_1 - y_2) / L) \cos \theta_b \sin \delta_b + ((z_2 - z_1) / L) \cos \delta_b, \quad (D2.31)$$

where L is given by Eq. (D2.2). Equations (D2.30) and (D2.31) will give an obtuse angle if the vectors point away from each other. The thickness equation gives a negative thickness if the obtuse angle is used. To convert an obtuse angle to the acute angle, take $\rho = 180 - \rho$. An Excel equation to do this is:

=IF(cell R – 90 < 0, cell R, 180 – cell R), (D2.32)

where cell R = the location of the angle ρ.

2.9.3.3
Circular Arc, in Dip Direction

The thickness of a bed that is folded into a circular arc (Fig. D2.4) can be found if the dip direction of the bed and the well or traverse line are coplanar. In this situation the bedding poles intersect at a point. Let ρ_1 be the smaller angle between the well and the pole to bedding, thus always associated with the longer radius, r_1. The thickness, t, is:

$$t = r_1 - r_2. \quad (D2.33)$$

From the law of sines:

$$r_2 = (L \sin \rho_1) / \sin \gamma; \quad (D2.34)$$

$$r_1 = (L \sin (180 - \rho_2)) / \sin \gamma, \quad (D2.35)$$

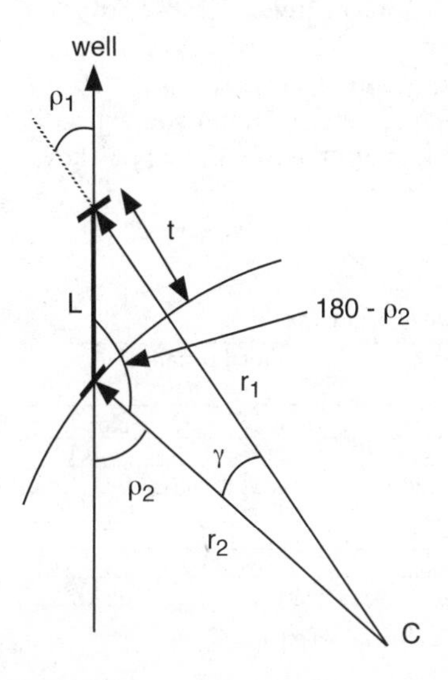

Fig. D2.4. Thickness of a bed folded into a circular arc. The dip of the bed and the well are co-planar. C is the center of curvature where the poles to bedding intersect where ρ_2 and ρ_2 =, respectively, the angle between the bed pole and the well and the radius associated with the larger angle, and $\gamma = \rho_2 - \rho_1$. Substitute Eqs. (D2.34) and (D2.35) into (D2.33) and replace sin $(180 - \rho_2)$ with sin ρ_2 to obtain the thickness:

$$t = (L / \sin \gamma)\,(\sin \rho_2 - \sin \rho_1). \qquad \text{(D2.36)}$$

Hewett (1920) gives a calculation for the thickness of a bed folded into a circular arc that is based on the elevations of the outcrop of the top and base of the bed and the distance between these points. It is simpler for computer calculation to find the length and direction of the line between the points using Eqs. (D2.2), (D2.18), and (D2.19), and then find the thickness from Eqs. (D2.35).

2.10 Problems

2.10.1 Interpretation of Data from an Oil Well

The Appleton oil field is located near the northern rim of the Gulf of Mexico basin. Use the data from well 4835-B in this field (Table 2.3) to solve the following problems:

1. Find the subsea depths of the stratigraphic markers.
2. Write a spreadsheet program using Eqs. (2.1) and (2.2) to find the coordinates of a point between two known points. Find the true vertical depths and the total rectangular coordinates of the formation contacts in the well.
3. What is the orientation of the well where it penetrates the Smackover? Write a spreadsheet program to solve this problem based on Eq. (2.3).
4. Find the true vertical depths and the total rectangular coordinates of the oil–water contact (OWC) and the base of the Smackover.
5. Plot the locations of the stratigraphic tops relative to the surface location of the well on the map (Fig. 2.38). Connect the points to show the shape of the well in map view.
6. What is the isocore thickness of the Smackover?
7. What is the true thickness of the Smackover given its attitude of 12, 056 from the dipmeter log and the orientation of the well from problem 3? Discuss the significance of the difference between the isopach and isocore thickness.
8. How far from the well (horizontally and in which direction) would you expect to first find the intersection of the oil – water contact with the top of the Smackover Formation, assuming constant dip for the Smackover?

Table 2.3. Data from well 4835-B, Appleton field, Alabama. Kelly Bushing: 244 ft.

Measured depth = log depth (ft)	True vertical depth (ft)	Depth subsea (ft)	Total rectangular coordinates (ft)	
0	0		0.00 N	0.00 E
928	928		1.72 N	1.93 E
2308	2308		4.53 N	9.51 W
3282	3282		6.37 N	6.29 W
4150 Eutaw				
4400 U. Tuscaloosa				
4811	4811		13.38 N	5.99 W

Table 2.3. countinued

Measured depth = log depth (ft)	True vertical depth (ft)	Depth subsea (ft)	Total rectangular coordinates (ft)	
4890 M. Tuscaloosa				
5150 L. Tuscaloosa				
5525 L. Cretaceous				
6198	6198		30.82 N	17.30 W
7696	7695		61.48 N	7.50 W
8328	8327		74.90 N	1.25 E
8482	8480		74.59 N	18.82 E
8661	8655		71.22 N	53.19 E
8935	8922		66.16 N	117.74 E
9186	9161		62.25 N	194.12 E
9556	9512		54.50 N	310.78 E
10006	9940		34.61 N	447.09 E
10266 Cotton Valley				
10431	10346		7.63 N	568.29 E
11103	10982		55.68 S	777.65 E
11621	11470		116.46 S	938.37 E
11960	11791		142.28 S	1045.58 E
12370 Haynesville				
12450	12253		173.48 S	1206.13 E
12641	12432		185.03 S	1271.62 E
13020 Buckner				
13072 Smackover*				
13086	12851		210.98 S	1418.79 E
13190 OWC**				
13286 Base Smack.				

* Attitude from dipmeter 12, 056.
** Oil–water contact

2.10.2 Attitude

1. Given the dip vector of the plane 35, 240, show on an overlay of a stereogram the trace of the plane, its dip vector, and its pole.
2. Show the dip vector 25, 240 on a tangent diagram. What is the apparent dip along a line that trends 260?
3. Use a stereogram to find the acute angle between the two lines 45, 300 and 16, 080.
4. Use a stereogram to find the angle between a vertical well and the pole to a bed with dip vector 10, 290.
5. Write a spreadsheet program to find the angle between two lines based on Eqs. (2.25)–(2.28). Use the program to solve problems 3 and 4.

2.10.3 Thickness

1. Given a bed with dip vector 10, 290, and a thickness of 75 m in a vertical well, use the pole-thickness equation to determine its true thickness.

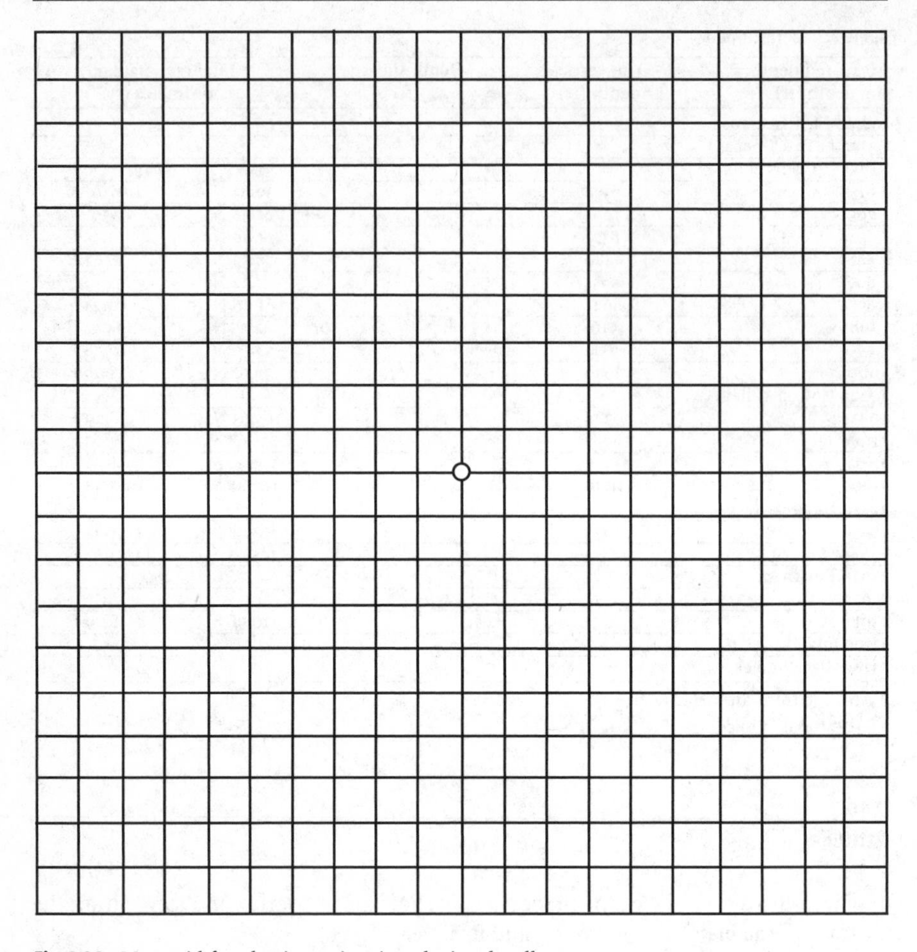

Fig. 2.38. Map grid for plotting points in a deviated well

2.10.4
Attitude and Thickness from Map

Use the map of the Blount Springs area (Fig. 2.39) to answer the following questions:

1. Find the attitude of the Mpm in its southeastern outcrop belt using the 3-point method.
2. Draw structure contours for the Mtfp, Mpm, and Mh on the eastern side of the map. Are the upper and lower contacts of each unit parallel to each other? Do you think the contacts are mapped correctly?

3. Determine the attitude of the eastern contact between the Mpm and the Mtfp by the 3-point method and from the structure contours. Are they the same? If they are different, discuss which answer is better.
4. Determine the thickness of the Mpm between the structure contours using the map-angle equations and the pole-thickness equation. Are the results the same? If they are different, discuss which answer is better.
5. What is the difference between the true thickness (from question 4) and the vertical thickness of the Mpm?
6. What is the thickness of the Mpm in its northeastern outcrop belt, assuming that the dip is 28° at its northwestern contact and the value determined in question 3 occurs at its southeastern contact? Use the concentric fold model and the dip-domain model. Discuss the effect of changing the location of the axial surface on the thickness computed with the dip-domain model.
7. Measure the thickness of the Mpm at 5–10 locations evenly distributed across the map. Measure thicknesses between structure contours where possible. Construct an isopach map from your thickness measurements. Is the unit constant in thickness?
8. What would be the apparent dip of the Mpm in a north-south roadcut through the northwestern limb of the anticline?

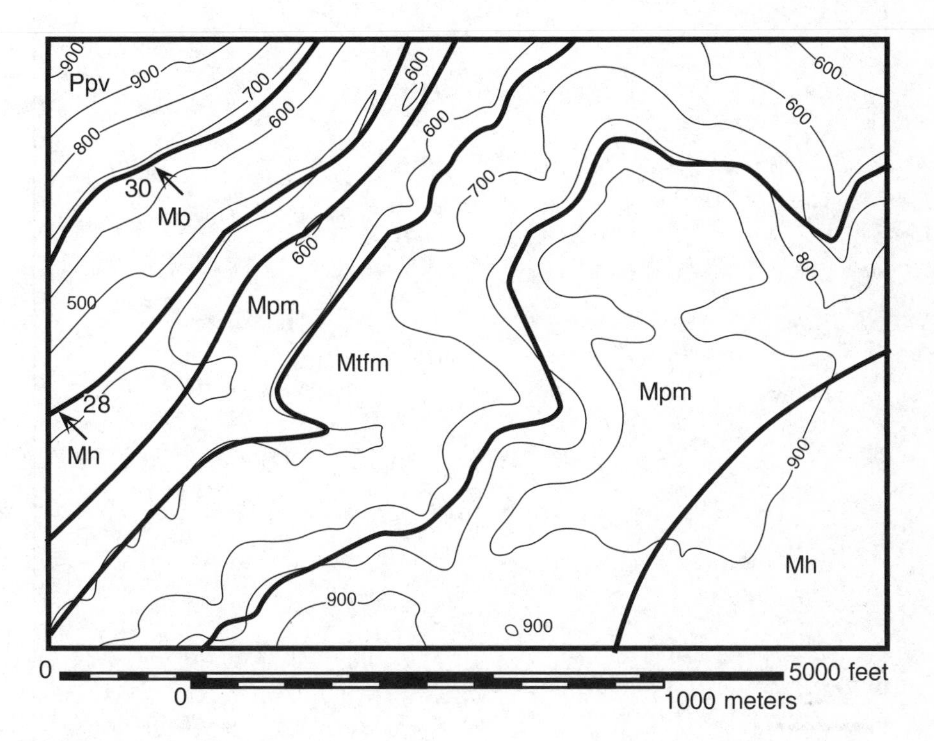

Fig. 2.39. Geological map of the northeast corner of the Blount Springs area, southern Appalachian fold-thrust belt. *Thin lines* are topographic contours (elevations in feet). *Thick lines* are geologic contacts. *Arrows* are dip directions; *numbers* give the amount of dip

9. What would be the apparent thickness of the Mpm in a north-south, vertical-sided roadcut through the northwestern limb of the anticline?

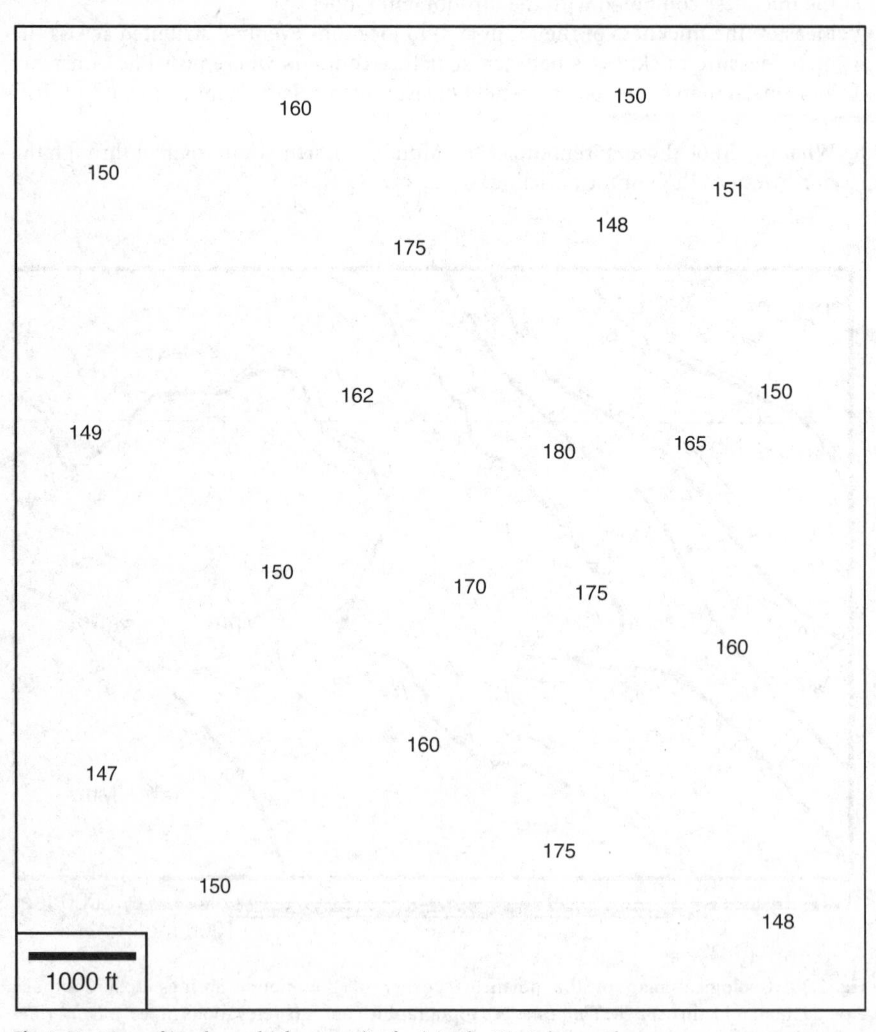

Fig. 2.40. Map of isochore thicknesses (in feet) in the Big John sandstone

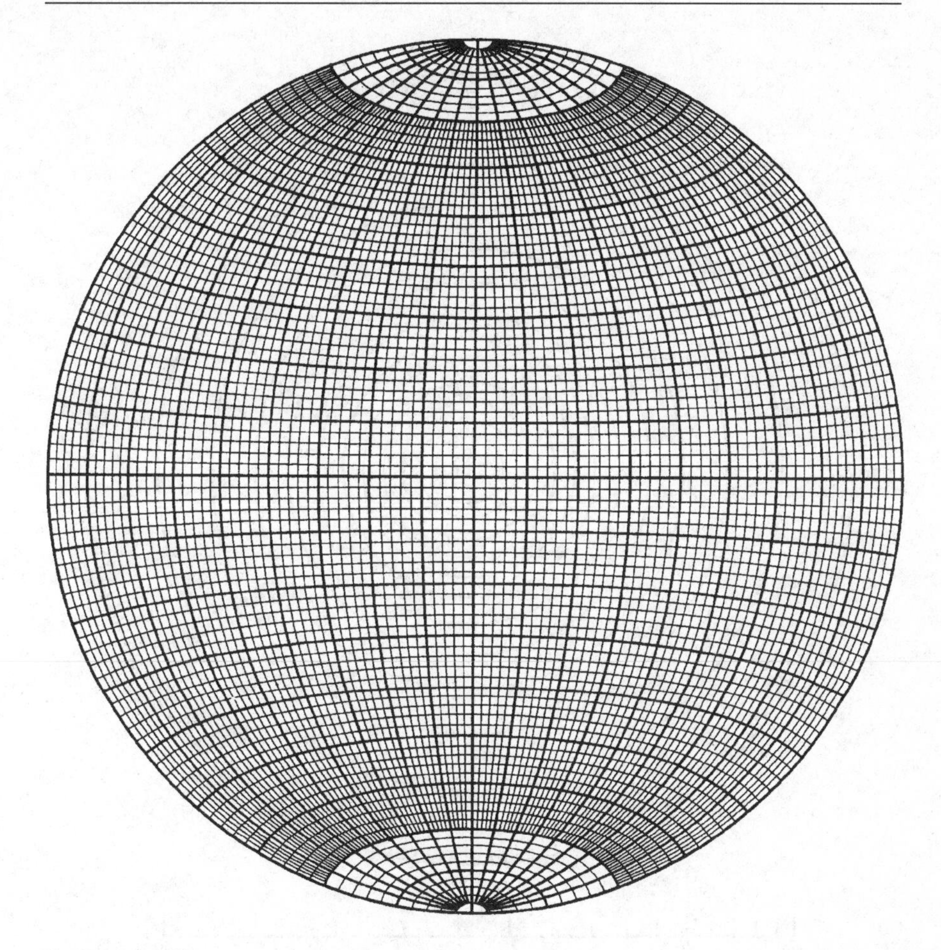

Fig. 2.41. Equal-area stereogram

2.10.5 Isopach Map.

Make an isopach map of the sandstone thicknesses on the map of Fig. 2.40. The thickest measurements form a trend that could be a channel or the limb of a monocline.

1. If the thickness anomaly is due to a dip change, what is the amount?
2. If the thickness anomaly is due to paleotopography, what is the maximum topographic slope?

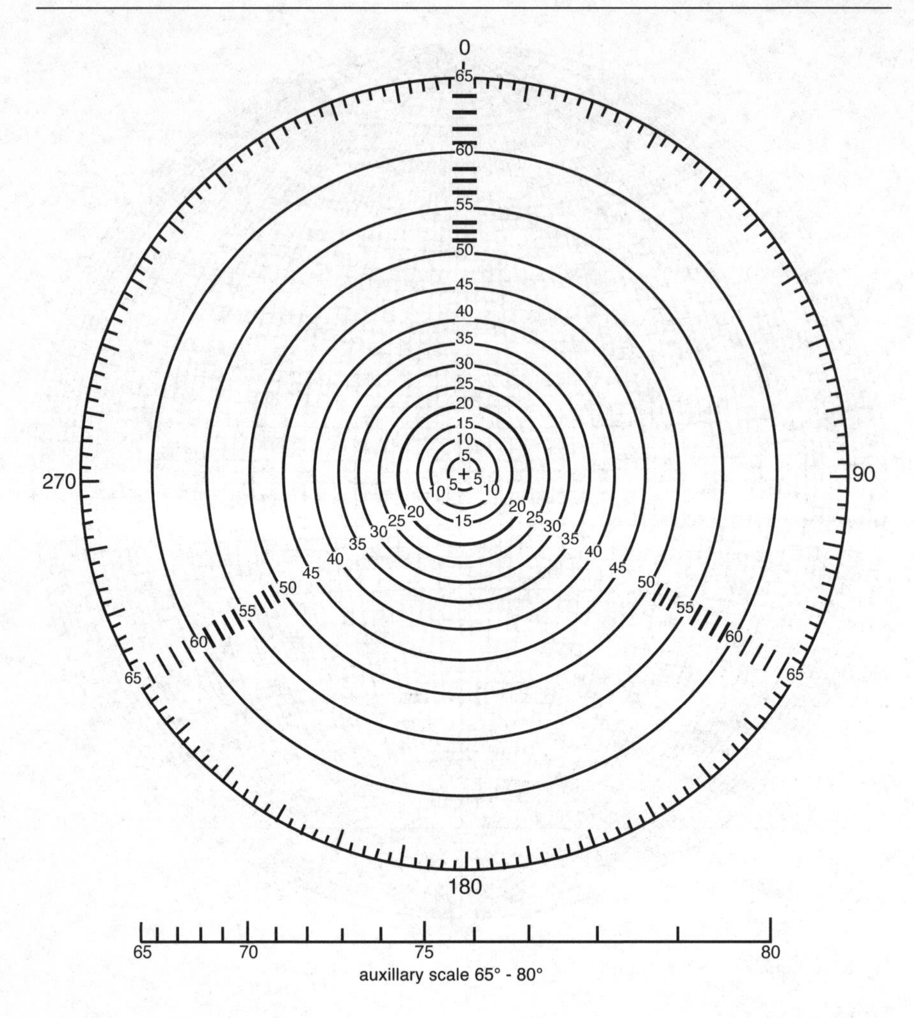

Fig. 2.42. Tangent diagram

References

Adams MG, Mallard LD, Trupe CH, Stewart KG (1996) Computerized geologic map compilation. In: DePaor DG (ed) Structural geology and personal computers: Pergamon Elsevier Science, Tarrytown, New York, pp 457–470
Aitken FK (1994) Well locations and mapping considerations. Houston Geol. Soc. Bull. Sept.: 15–16
Bates RL, Jackson JA (1987) Glossary of geology, 3 rd edn. American Geological Institute, Alexandria, Virginia, 788 pp
Bengtson CA (1980) Structural uses of tangent diagrams. Geology 8: 599–602
Bengtson CA (1981) Statistical curvature analysis techniques for structural interpretation of dipmeter data. Am. Assoc. Pet. Geol. Bull. 65: 312–332
Bishop MS (1960) Subsurface mapping. John Wiley, New York, 198 pp
Charlesworth HAK, Kilby WE (1981) Calculating thickness from outcrop and drill-hole data. Bull. Can. Pet. Geol. 29: 277–292
Cruden DM, Charlesworth HAK (1976) Errors in strike and dip measurements. Geol. Soc. Am. Bull. 87: 977–980
Dennis JG (1967) International tectonic dictionary. American Association of Petroleum Geologists, Mem. 7, 196 pp
Dennison JM (1968) Analysis of geologic structures. WW Norton, New York, 209 pp
Eves H (1984) Analytic geometry. In: Beyer WH (ed) CRC standard mathematical tables. CRC Press, Boca Raton, Florida, pp 193–226
Foley JD, Van Dam A (1983) Fundamentals of computer graphics. Addison-Wesley, London, 664 pp
Fowler RA (1997) Map accuracy specifications. EOM 6: 33–35
Greenhood D (1964) Mapping. University of Chicago Press, Chicago, 289 pp
Hewett DF (1920) Measurements of folded beds. Econ. Geol. 15: 367–385
Hobson GD (1942) Calculating the true thickness of a folded bed. Am. Assoc. Pet. Geol. Bull. 26: 1827–1842
Hubbert MK (1931) Graphic solution of strike and dip from two angular components. Am. Assoc. Pet. Geol. Bull. 15: 283–286
Low JW (1951) Subsurface maps and illustrations. In: LeRoy LW (ed) Subsurface geological methods. Colorado School of Mines, Golden Colorado, pp 894–968
Maltman A (1990) Geological maps: an introduction. Van Nostrand, Reinhold, New York, 184 pp
Ragan DM (1985) Structural geology, 3rd edn. John Wiley, New York, 393 pp
Robinson AH, Sale RD (1969) Elements of cartography, 3rd edn. John Wiley, New York, 415 pp
Rowland SM, Duebendorfer EM (1994) Structural analysis and synthesis, 2nd edn. Blackwell, Oxford, 279 pp
Snyder JP (1987) Map projections – a working manual. US Geol. Surv.Prof. Pap. 1395, 383 pp
Woodcock NH (1976) The accuracy of structural field measurements. J. Geol. 84: 350–355

Chapter 3
Structure Contouring

3.1
Introduction

This chapter covers the basic techniques of contouring continuous surfaces, the construction of composite surface maps, and map validation using structure contours. The contouring of faults and faulted surfaces is treated in Chapter 6.

3.2
Structure Contour Maps

A structure contour map is one of the most important tools for three-dimensional structural interpretation because it represents the full three-dimensional form of a map horizon. The mapping techniques to be discussed are equally applicable in surface and subsurface interpretation. The usual steps required to produce a structure contour map are:

1. Plot the points to be mapped.
2. Determine an appropriate contour interval.
3. Interpolate the locations of the contour elevations between the control points. There are several techniques for doing this and they may give very different results when only a small amount of data is available.
4. Project data from multiple horizons to the map horizon in order to increase the number of control points or
5. Map multiple surfaces and test the maps for compatibility.
6. The final step is to draw one or more cross sections to test the map or maps for geological reasonableness and internal consistency (Chapter 7).

A given set of points can be contoured into a nearly infinite number of shapes, depending on the methodology followed. The data points may either be contoured directly, as done using a triangulated irregular network (TIN), or can be interpolated into the elevations at the nodes of a grid (gridding) and then contoured (Jones and Hamilton 1992). Different styles of contouring can be applied to either format. The standard contouring styles are triangulation, equal spacing, parallel, and interpretive. These basic techniques are discussed next. Later sections give methods for increasing the amount of information used to derive the contours. The best maps are based on the structural styles and trends known in the area.

A geologic map of the Blount Springs area (Fig. 3.1) will provide the data for an ongoing example of the process of creating and validating a structure contour map. The map area is located along the Sequatchie anticline at the southern end of the Appalachian fold-thrust belt and is the frontal anticline of the fold-thrust belt.

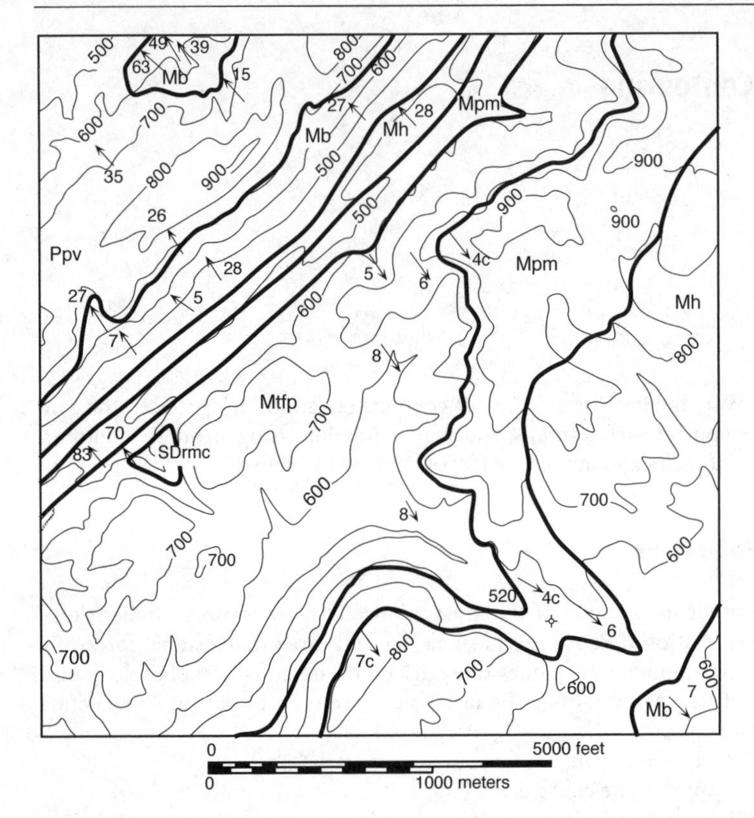

Fig. 3.1. Geologic map of a portion of the southern Sequatchie anticline in the vicinity of Blount Springs, Alabama. The units are, from oldest to youngest, Silurian Red Mountain Formation and Devonian Chattanooga Shale (*SDrmc*), Mississippian Tuscumbia Limestone and Fort Payne Formation (*Mtfp*), Mississippian Pride Mountain Formation (*Mpm*), Mississippian Hartselle Sandstone (*Mh*), Mississippian Bangor Limestone (*Mb*), Pennsylvanian Pottsville Formation (*Ppv*). *Thick lines* are geologic contacts, *thin lines* are topographic contours; elevations in feet. *Arrows* are bedding attitudes. Dip amounts are indicated next to the arrows; if followed by *c*, the dip is computed from three points. (After Cherry 1990; Thomas 1986)

3.2.1 Control Points

A structure contour map is constructed from the information at a number of observation points. The observations may be either the xyz positions of points on the surface, the attitude of the surface, or both. A relatively even distribution of points is desirable and which, in addition, includes the local maximum and minimum values of the elevation. If the data are from a geologic map or from 2-D seismic-reflection profiles, a very large number of closely spaced points may be available along widely spaced lines that represent the traces of outcrops or seismic lines, with little or no data between the lines. The number of points in a data set of this type will probably need to be reduced to make it more interpretable. Even if the contouring is to be done by

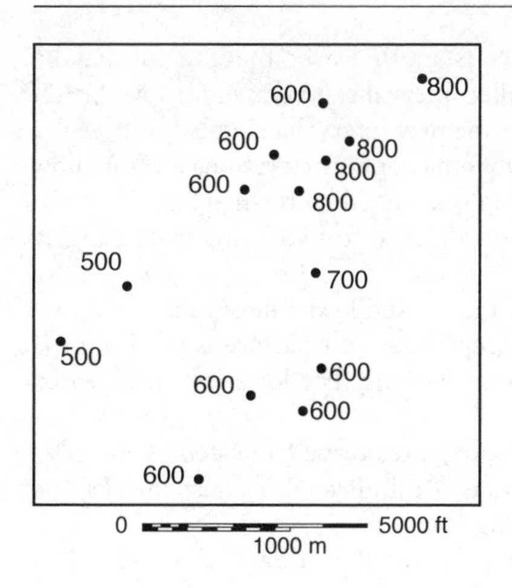

Fig. 3.2. Elevations (in feet) of the top of the Mtfp from the Blount Springs map area (Fig. 3.1)

computer, it is possible to have too much information (Jones et al. 1986). This is because the first step in contouring is always the identification of the neighboring points in all directions from a given data point and it is difficult and usually ambiguous to choose the neighbors between widely spaced lines of closely spaced points. Before contouring, the lines of data points should be resampled on a larger interval that is still small enough to preserve the form of the surface.

To create a structure contour map on the top of the Mtfp (Fig. 3.1), the geologic map has been sampled at representative locations where the top of the Mtfp crosses a topographic contour. This results in about as even a distribution of points as possible, given the linear outcrop pattern (Fig. 3.2). In order to illustrate the process of constructing a subsurface contour map, the points in Fig. 3.2 will be treated as if they come from wells where there is no knowledge of the shape of the surface between the wells. The alternative interpretations so obtained will then be tested and refined with the additional knowledge that can be obtained from the complete map.

3.2.2 Rules of Contouring

A series of rules for contouring has been developed over the years to produce visually acceptable maps that reasonably represent the surface geometry. The reference frame for a contour map is defined by specifying a datum plane, such as sea level. Elevations are customarily positive above sea level and negative below sea level. The surface to be contoured first should be represented by the largest amount of data. The contour interval selected depends on the range of elevations to be depicted and the number of control points. There should be more contours where more data are available. The contour interval should be greater than the limits of error involved. Errors of 20 ft (7 m) are typical in correlating well logs and as a result of minor deviations of a well from vertical. Uncertainties in locations in surface mapping are likely to produce errors of similar magnitude. The contour interval should be small enough to show the structures of interest. The map can be read more easily if every fifth or tenth contour is heavier. The map should always have a scale. A bar scale is best because if the map is enlarged or reduced, the scale will remain correct.

The contour interval on a map is usually constant; however, the interval may be changed with the steepness of the dips. A smaller interval can be used for low dips. If the interval is changed in a specific area, make the new interval a simple multiple (or fraction) of the original interval and clearly label the contour elevations and/or show the boundaries of the regions where the spacing is changed on the map.

The contours must obey the following rules (modified from Sebring 1958; Badgley 1959; Bishop 1960):

1. Every contour must pass between points of higher and lower elevation.
2. Contour lines should not merge or cross except where the surface is vertical or is repeated due to overturned folding or reverse faulting. The lower set of repeated contours should be dashed.
3. Contour lines should either close within the map area or be truncated by the edge of the map or by a fault. Closed depressions are indicated by hash marks (tic marks) on the low side of the inner bounding contour.
4. Contour lines are repeated to indicate reversals in the slope direction. Rarely will a contour ever fall exactly on the crest or trough of a structure.
5. Faults cause breaks in a continuous map surface. Normal separation faults cause gaps in the contoured horizon, reverse separation faults cause overlapping contours and vertical faults cause linear discontinuities in elevation. Where beds are repeated by reverse faulting, it will usually be clearer to prepare separate maps for the hangingwall and footwall.
6. The map should honor the trend or trends present in the area. Crestal traces, trough traces, fold hinges and inflection lines usually form straight lines or smooth curves as appropriate for the structural style.

Mapping should begin in regions of tightest control and move outward into areas of lesser control. It is usually best to map two or three contours simultaneously in order to obtain a feel for the slope of the surface. The contours will almost certainly be changed as the interpretation is developed; therefore, if drafting by hand, do the original interpretation in pencil. The map should be done on tracing paper or clear film so that it can be overlaid on other maps. In computer drafting, make the contour map a separate layer in the program so that it can be easily changed and superimposed on other layers.

3.2.3
Choosing the Neighboring Points: TIN or Grid

Drawing a contour between two control points requires first deciding which two points from the complete data set are to be used. This decision is not trivial or simple. The choice of neighboring points between which the contours are to be drawn has a major impact on the shape of the final surface. Two procedures are in wide use, triangulation and gridding. Triangulation involves finding the TIN network (Sect. 1.2.2) of nearest neighbors in which the data points form the nodes of the network (Fig. 3.3a). Gridding involves superimposing a grid on the data (Fig. 3.3b) and interpolating to find the values at the nodes (intersection points) of the grid. Many different interpolation methods are used in gridding. All involve some form of weighted average of points within a specified distance from each grid node (Hamilton and Jones 1992). Contours developed from either type of network are usually smoothed, either as part of the contouring procedure or afterward.

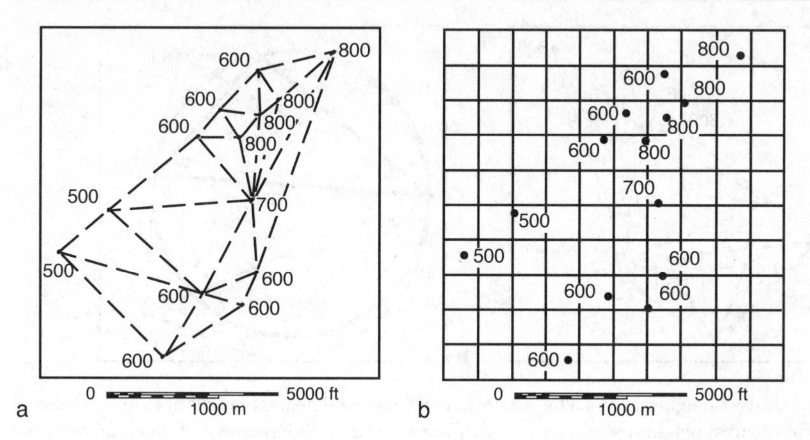

Fig. 3.3. Methods for relating point elevations of the top Mtfp to each other. **a** Triangulated irregular network (TIN) with data points at the vertices. **b** Grid superimposed on the data. Elevations will be interpolated at the grid nodes

The first decision is whether the contouring will be based on a TIN or on a grid. The most direct relationship is to connect adjacent points with straight lines, producing a TIN. This has long been a preferred approach in hand contouring and is increasingly popular in computer contouring (Banks 1991; Jones and Nelson 1992). The primary advantages of the method are that it is very fast, the contoured surface precisely fits the data, and it is easy to do by hand. For structural interpretation, fitting the data exactly, including the extreme values, is a valuable property, because the extreme values may provide the most important information. Plotted in three dimensions, the TIN network alone will show the approximate shape of the surface. The advantage of gridding is that once the data are on a regular grid, other operations, such as the calculation of the distance between two gridded surfaces, are relatively easy. The major disadvantage is that the contoured surface does not necessarily go through the data points and it may be difficult to make the surface fit the data. Gridding is extremely calculation intensive and requires specialized computer programs for practical implementation. For discussions of how to work with gridding-based computer contouring, see Walters (1969), Jones et al. (1986), and Hamilton and Jones (1992). Approaches based on TINs are used here because the results are appropriate for structural problems and because they are practical to apply by hand or with a hand calculator or computer spreadsheet.

3.2.4 Triangulated Irregular Networks

Creating a TIN requires determining the nearest neighbor points, between which the contours will be located. The possible choices of nearest neighbors are seen by connecting the data points with a series of lines to form triangles (Fig. 3.4a). The points could be connected differently to form different networks. Delauney triangles and greedy contouring are two unbiased approaches to choosing the nearest neighbors. An interpreter may wish to introduce a bias in the choice of nearest neighbors and a com-

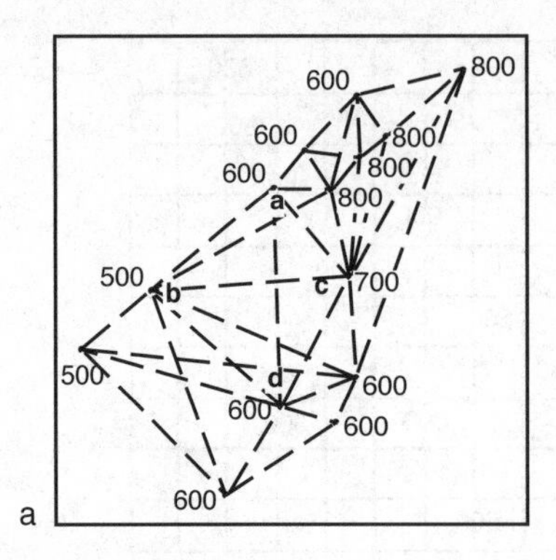

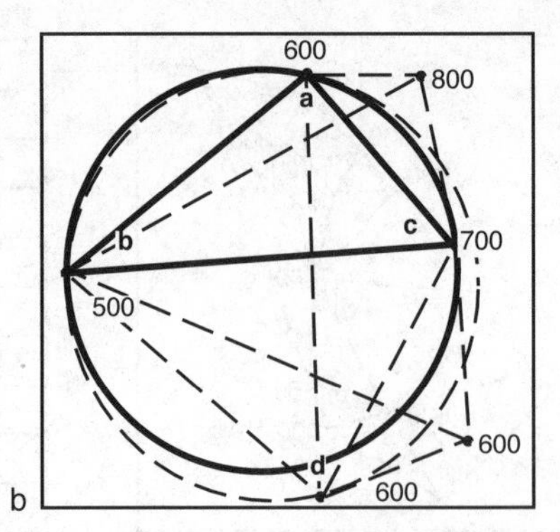

Fig. 3.4. Nearest neighbors in a TIN network. Four potential nearest neighbors are labeled *a–d*. **a** Possible nearest neighbors are connected by *dashed lines*. **b** Enlargement of potential neighbor points *a–d*. The *solid circle* includes three points (*a–c*) to define a Delauney nearest-neighbor triangle (*heavy lines*). The *dashed circle* through *a*, *b*, and *d* includes four points

puter program should allow this option. A biased choice of neighbors is used to control the grain of the final contours or to overcome a poor choice of neighbors that results from inadequate sampling of the surface.

3.2.4.1
Delauney Triangles

A Delauney triangle is one for which a circle through the three vertices does not include any other points (Jones and Nelson 1992). In Fig. 3.4b, the solid circle through vertices a, b and c is a Delauney triangle because no other points occur within the circle. The vertices of the triangle are nearest neighbors. The dashed circle through vertices a, b and d (Fig. 3.4b) includes point c and therefore does not define a Delauney triangle. This method is not as practical for use by hand as the next technique.

3.2.4.2
Greedy Triangulation

A triangulation method suitable for use by hand as well as by computer is known as "greedy" triangulation (Watson and Philip 1984; Jones and Nelson 1992). The criterion is that the edge selected is the shortest line between vertices. No candidate edge is included if there is a shorter candidate edge that would intersect it (Jones and Nelson 1992). Figure 3.5a shows a network of candidate edges with the longer edges dashed. The TIN produced by this method is shown in Fig. 3.5b. This method is both logical and convenient for use by hand and is the approach used here as the first step in the interpretation.

A TIN network may contain extremely elongated triangles, implying a relationship between widely separated points. This is particularly likely to occur at the edges of the network, for example points a and b in Fig. 3.5b. Such widely separated points are not necessarily related and probably should not influence the shape of the surface between them. The TIN network should be edited to examine and possibly remove such long-distance connections. Some computer programs eliminate triangles for which one of the angles is smaller than some threshold value.

3.2.4.3
Biasing the Network

The choice of nearest neighbors is the major factor in determining the shape of the contoured surface. Changing the nearest neighbors is a simple way to change the shape of the surface. In Fig. 3.6a, points a and c are connected instead of points b and d (as in Fig. 3.5b). As will be seen below, this will change the shapes of the resulting contours. Another method for biasing the contouring is to introduce new points (Fig. 3.6b) that force the contours to conform to the desired shape.

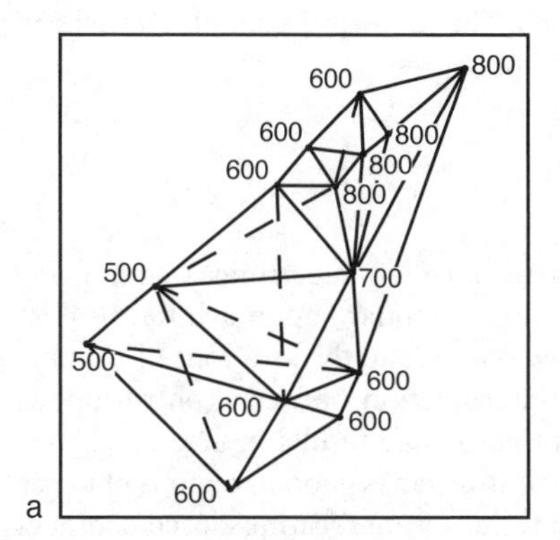

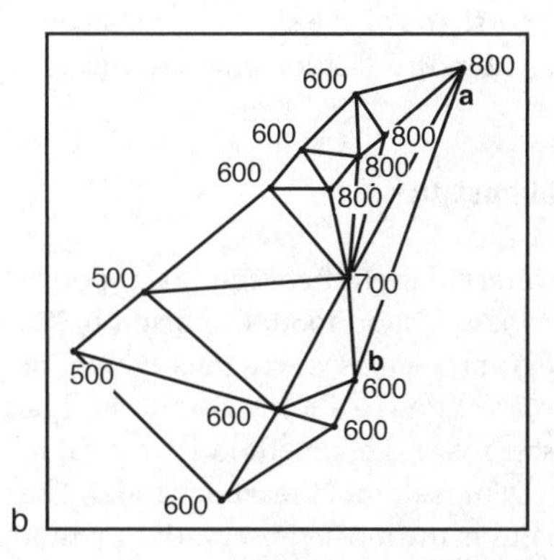

Fig. 3.5. Greedy triangulation. **a** Longer edges *dashed*. **b** Longer edges removed to define nearest neighbors

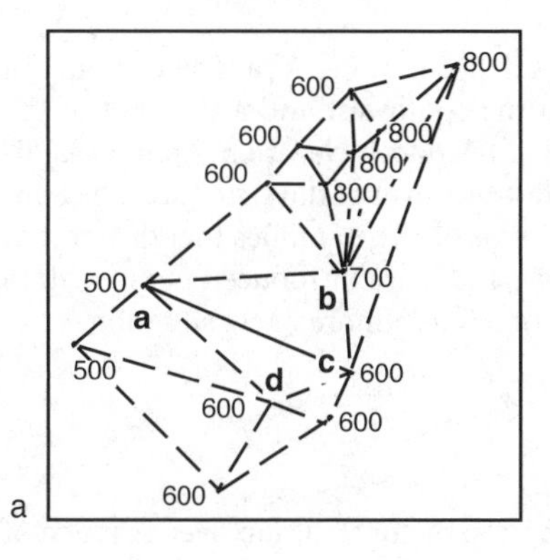

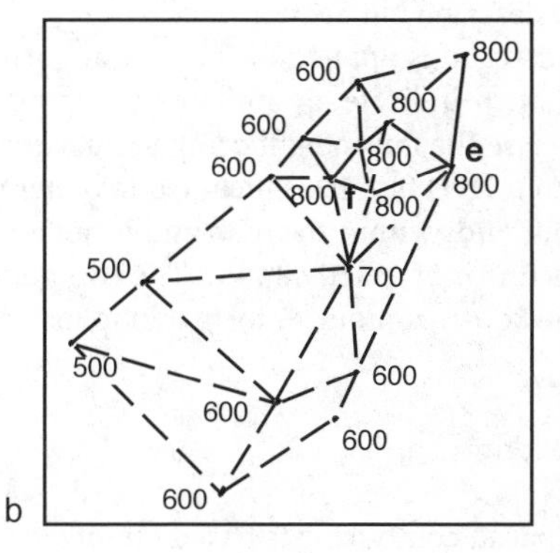

Fig. 3.6. Biased triangulation networks. **a** Different choice of nearest neighbors among points *a–d* from those chosen in Fig. 3.5b (*dashed lines*). **b** Points introduced at *e* and *f* and the resulting changes of neighbors

3.3 Contouring Styles

Contouring may be done using different styles, each of which produces its own characteristic pattern (Handley 1954). With a large amount of evenly spaced data, the difference between maps produced by different styles will usually be small. For precision structural interpretation it is recommended that contouring be done by hand, or very carefully reviewed, so that all the assumptions and uncertainties in the interpretation are recognized. Contouring by any method must be viewed as a preliminary interpretation because unknown structures can always occur between widely spaced control points. The characteristics of the common styles of contouring are summarized next. Contours are usually shown as smooth curves although this should depend on the structural style (Sect. 3.3.5). In constructing contours by computer drafting, I prefer to draw the preliminary contour lines as polygons, that is, as points joined by straight line segments. The polygons can be converted to smooth curves, if appropriate, once the general shape of the surface has been defined. The following examples have been constructed in this fashion and have been left as polygons.

3.3.1 Linear Interpolation

Linear interpolation between data points is also called mechanical contouring (Rettger 1929; Bishop 1960; Dennison 1968). This is a standard approach for producing topographic maps where the high points, low points, and the locations of changes in slope are known, allowing accurate linear interpolation between control points (Dennison 1968). This method may produce unreasonable results in areas of sparse control (Dennison 1968; Tearpock 1992). The resulting map is good in regions of dense control and is the most conservative method in terms of not creating closed contours that represent local culminations or troughs. The method tends to de-emphasize closed structures into noses, is good for gently dipping structures with no prominent fold axes, and is often used in litigation, arbitration and oil-field unitization (Tearpock 1992). Locate the positions of the contours between neighboring control points by the method described in Section 2.4.2.1.

The method is applied to the Blount Springs data set in Fig. 3.7a. The contours of the top Mtfp indicate an anticline plunging to the southwest and a flat spot in the south where the four neighboring points are all at the 600-ft elevation. An intermediate contour level (550 ft) is included to better show the form of the structure. The 800-ft contour ends within the map and is not a likely result. This implies that the crest of the anticline is at exactly 800 ft. The two points a and b are introduced in Fig. 3.7b to force the 800-ft contour to form a loop instead of a line, a more likely situation.

3.3.2 Equal Spacing

Equal-spaced contouring is based on the assumption of constant dip over as much of the map area as possible. This technique can be used in areas of sparse control. In the traditional approach, the dip selected is the greatest or the least that can be determined in an area of tight control. The same dip is then used over the entire map

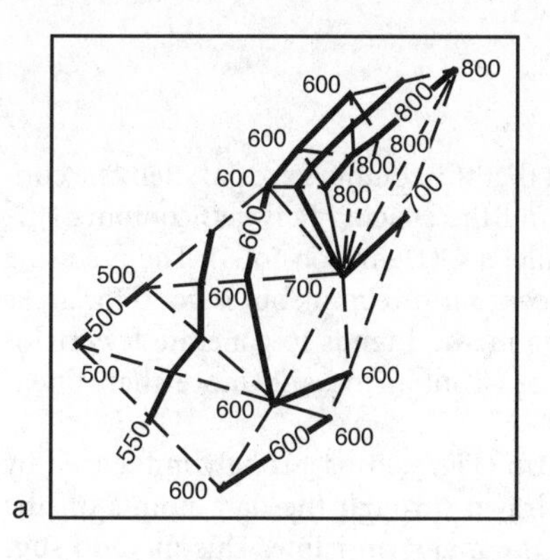

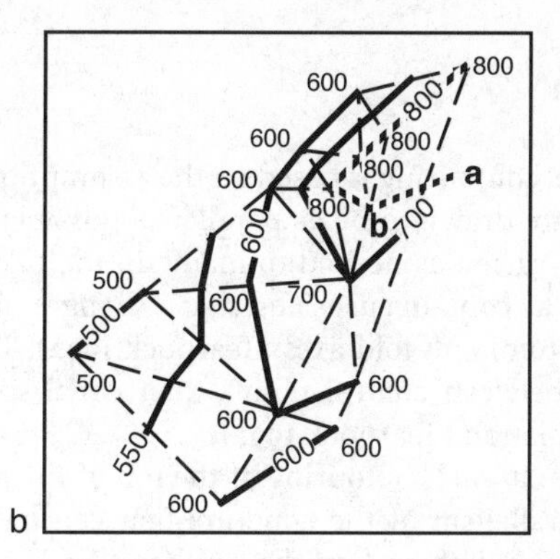

Fig. 3.7. Linear interpolation contouring of the top Mtfp in the Blount Springs area. *Heavy lines* are structure contours. **a** Original data only. **b** Two points, *a* and *b*, introduced to force the 800 elevation contour (dashed) into a more realistic shape

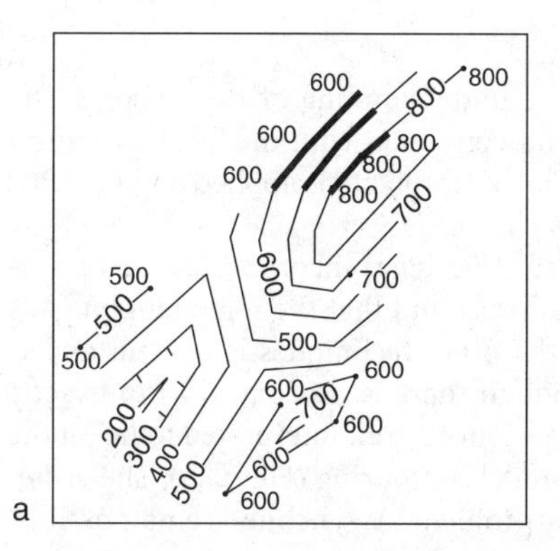

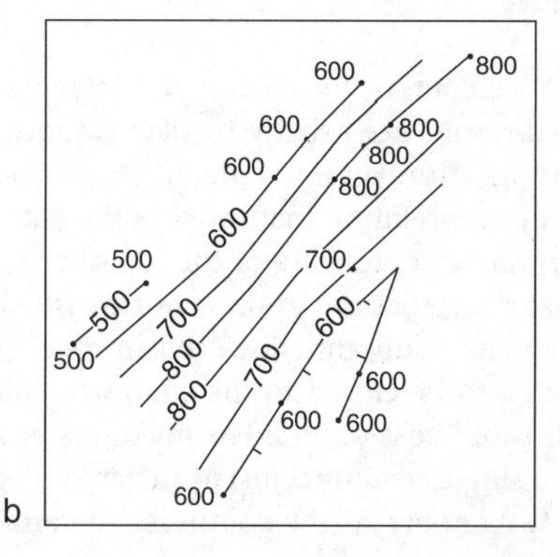

Fig. 3.8. Different contouring methods applied to the top of the Mtfp from the Blount Springs map area. **a** Equal-spaced contouring. The dip to be maintained (*heavy contours*) is chosen in the region of tightest control. **b** Parallel contouring

(Handley 1954; Dennison 1968). Find the dip from the control points with the three-point method (Sect. 2.4.1), or, if the dip is known, as from an outcrop measurement or a dipmeter, find the contour spacing from the dip (Sect. 2.4.2.2). This approach projects dips into areas of no control or areas of flat dip and so will create large numbers of structures that may be artifacts (Dennison 1968). Because of the great potential for producing nonexistent structural closures, this method is usually not preferred (Handley 1954).

Equal-spaced contouring of the Mtfp (Fig. 3.8a) produces multiple closures, two anticlines and a syncline that is much lower than any of the data points. The anticlinal nose defined by the 800-ft contour is reasonable but the closed 700-ft anticline to the south and the syncline bounded by the 200-ft contour on the southwest are forced into regions of low dip or low control.

3.3.3 Parallel

Parallel contouring is based on the assumption that the contours are parallel. The contours are drawn to be as parallel as possible and the spacing between contours (the dip) is varied as needed to maintain the parallelism (Dennison 1968). The resulting map may contain cusps and sharp changes in contour direction, but is good for areas with prominent fold axes (Tearpock 1992). The method tends to generate fewer closures between control points than equal-spaced contouring and more than linear interpolation (Tearpock 1992).

The parallel contouring of the top of the Mtfp (Fig. 3.8b) is strongly influenced by the parallelism of the contours that can be drawn through the data points on the northwest limb and on the southeast limb of the major anticline. This method suggests an elongate northeasterly trending anticline in the center of the map and a southwest-plunging syncline on the southeast. The syncline is in the same position the anticline predicted by the equal-spaced method (Fig. 3.8a).

3.3.4 Interpretive

Interpretive contouring reflects the interpreter's understanding of the geology. The preferred results are usually regular, smooth and consistent with the local structural style and structural grain. Generally, the principle of simplicity is applied and the least complex interpretation that satisfies the data is chosen.

Interpretive contouring of the top of the Mtfp (Fig. 3.9a) incorporates the knowledge that the structural grain is northeast–southwest and that the regional plunge is very low. The main anticline seen in each of the other techniques is present and is interpreted to be closed to the southwest, although there is no control as to exactly where the fold nose occurs. The northwestern contours are all interpreted to lie on the limb of a single structure, just as inferred by parallel contouring (Fig. 3.8b). The group of 600-ft contours in the southeast remains a problem. A syncline seems possible.

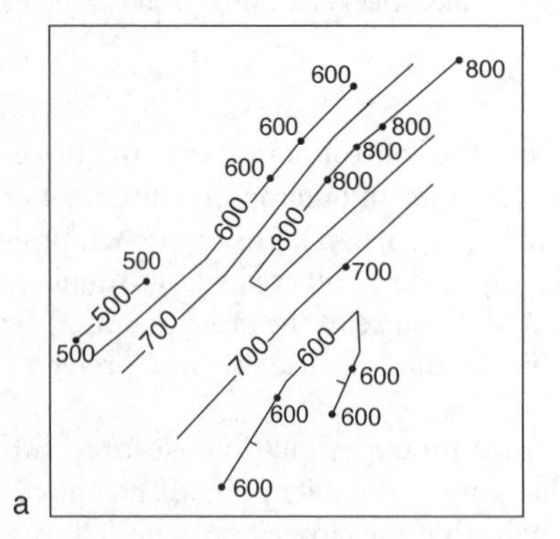

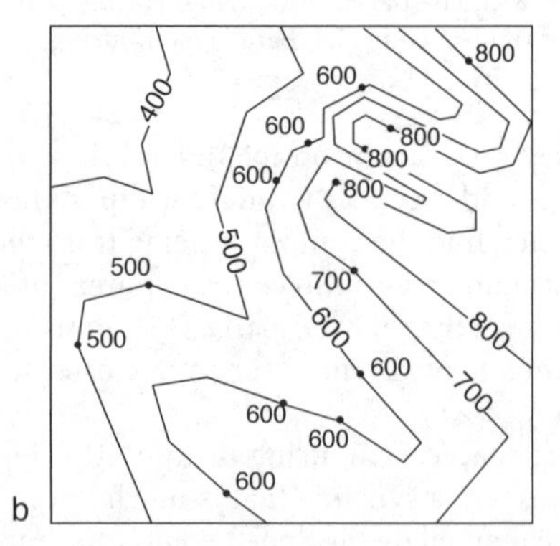

Fig. 3.9. Interpretive contouring. **a** Based on interpretation that regional trends are northeast–southwest. **b** Based on the interpretation that the regional trends are northwest–southeast

Although Fig. 3.9a may not be the best answer for the map (as will be seen later), it is better than the result of any of the previous methods alone. Clearly the interpreter will probably need to customize the map if it is obtained by any of the previous techniques.

It is possible to obtain dramatically different results by the interpretive contouring of sparse data. The two maps in Fig. 3.9 are completely different, although they are derived from exactly the same data. The difference between the two maps reflects the different assumed regional trends. There is no basis for choosing which map is better, given the available information. The simplest and most useful additional information for selecting the best interpretation is the bedding attitude, because the attitudes should indicate the trend direction (Sect. 3.4.1).

3.3.5 Smooth vs. Angular

Structure contours are usually drawn as smooth curves. This is appropriate for circular-arc and other smoothly curved fold styles. Many folds, however, are of the dip-domain

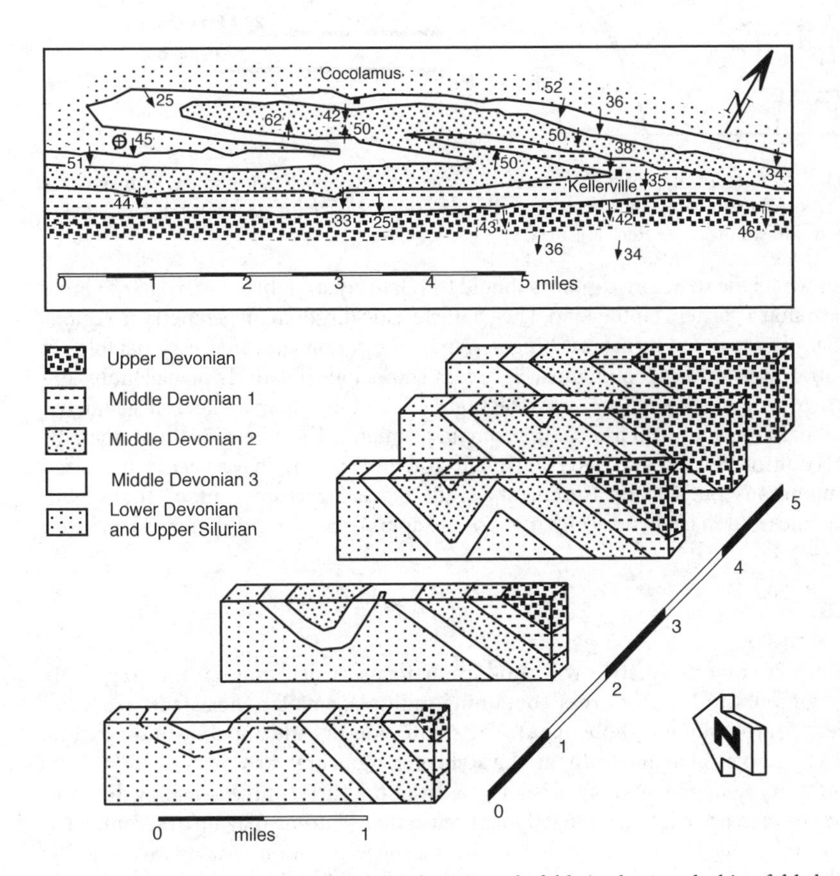

Fig. 3.10. Map and cross sections of dip-domain style folds in the Appalachian fold-thrust belt in Pennsylvania. (After Faill 1969)

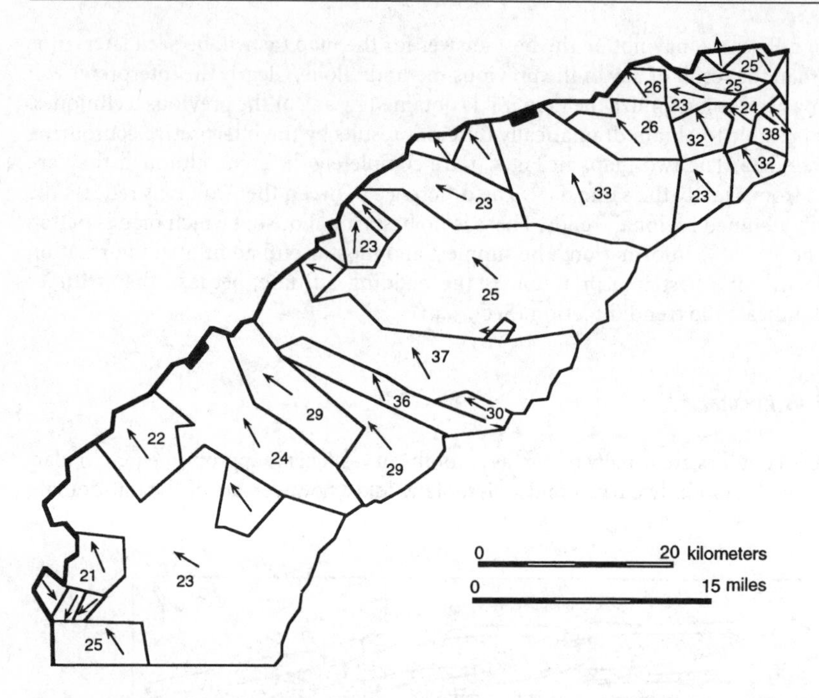

Fig. 3.11. Dip-domain map of a portion of the Triassic Gettysburg half graben, Pennsylvania. *Numbers* are domain dips and *arrows* are dip directions. *Heavy line* is a normal fault, downthrown to the southeast. (After Faill 1973)

style for which the structure contours should be relatively straight between sharp hinges and have sharp corners on the map. The characteristic dip-domain geometry is regions of planar dip separated by narrow hinges. A map of dip-domain compressional folds in the central Appalachian Mountains (Fig. 3.10a) shows long, relatively planar limbs and very narrow, tight hinges. The cross sections (Fig. 3.10b) show a chevron geometry. Extensional folds may also have a dip-domain geometry. The extensional dip domains in a portion of the Newark-Gettysburg half graben (Fig. 3.11) have been synthesized from numerous outcrop measurements (Faill 1973). Structure contour maps with straight lines and sharp bends (Figs. 3.7–3.9) are appropriate for dip-domain structures.

3.3.6 Artifacts

An incorrect contouring style will produce artifacts, which range from excessively wiggly contours to lines or areas containing multiple, small, nonexistent structural closures (Krajeweski and Gibbs 1994). The small, side-by-side anticline and syncline in Fig. 3.8 are typical artifacts due to the style of contouring.

Another type of artifact may arise if data from multiple sources, such as different seismic surveys, are contoured together, because their datums may be different. Each data set can be internally consistent, but one may have elevations systematically shifted with respect to the other. Cloverleaf patterns of highs and lows (Fig. 3.12) characterize this type of problem (Jones et al. 1986; Jones and Krum 1992). Two-dimension-

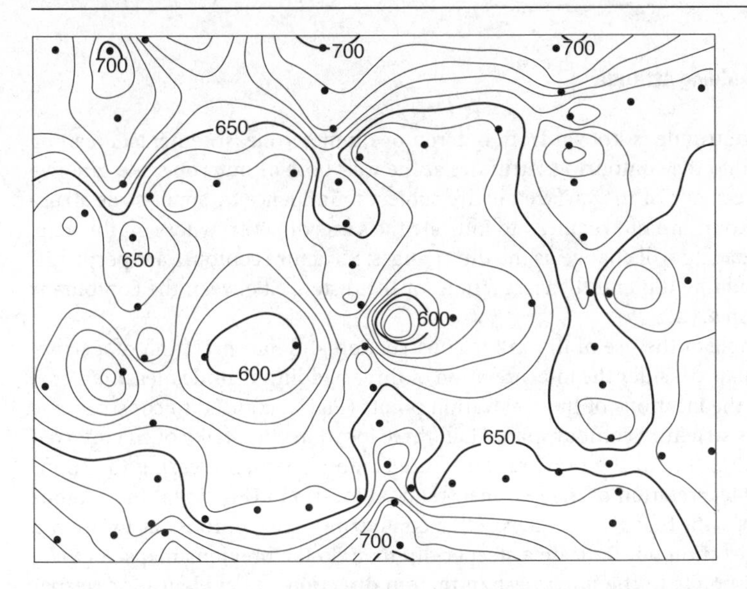

Fig. 3.12. Cloverleaf structure contour pattern produced by data along three north–south seismic lines mis-tied to three east–west lines. Control points are indicated by *dots*. Contours are concentrated along lines of data collection. (After Jones et al. 1986)

al seismic lines in areas of dipping units have the additional problem that the reflections on lines parallel to strike may be shifted laterally from the surface location of the line, leading to a similar datum shift (Fig. 1.48) between lines that are at right angles to one another (Oliveros 1989). Mis-ties between reflectors and incorrect stacking velocities can lead to similar cloverleaf patterns.

Before a map is completed it should always be examined for edge effects and data errors. A map may be unrealistic at the edges because of the inclusion of elongate triangles in the TIN network or because zeros have been included at the outer nodes of a grid network. Data errors are a likely possibility where single points fall far from the average surface. Problem points may be recognized by the presence of small closed highs or lows, usually defined by multiple contours, that surround an individual point. Such errors commonly arise from mistakes in the interpretation of the unit boundaries or as transcription errors in transferring data to the map. Gridding algorithms smooth the surface and usually reduce the effect of data errors. Problem points are recognized on a map produced from gridded data as original data points that lie far from the contoured surface.

3.4 Additional Sources of Information

Structure contour maps can be based on a significant amount of information in addition to the elevations on a single horizon. The shape of the contoured surface can be controlled using the attitudes of bedding, data from multiple stratigraphic horizons, and from pore-fluid behavior such as different groundwater levels or the presence or absence of hydrocarbon traps as indicated by shows of oil and gas in wells.

3.4.1 Including the Bedding Attitude

If the bedding attitude is known from outcrop or dipmeter measurements, it can be incorporated into the contouring. Attitudes at the well-bore or outcrop scale can give insight into the shape of the surface but are subject to influence by small-scale structures. The contours are not required to indicate the same dip everywhere on the map. The contour spacing will change as the dip changes. Structure contours are perpendicular to the bedding dip and the calculation for the distance between the contours is given in Section 2.4.2.2.

As an example of the use of dip in the construction and interpretation of a structure contour map, consider the interpretation of three bedding attitude measurements made close to the locations of three elevation points (Fig. 3.13a) used to construct the Blount Springs structure contour map. At all three locations the strike of the inferred contours matches that of the previous structure contour maps, increasing the confidence in the interpretation of the regional trend. At location 1 (Fig. 3.13a) the contour spacing agrees with that of the interpretive contouring. The contours at location 2 agree in dip direction but indicate a steeper dip than do the previous maps. As there are no control points to the northwest, in the dip direction, the evidence for steeper dip is important additional information. The inferred contours at point 3 slope much less than the contours inferred from the elevations, and in the opposite direction. The structure contours can be revised to be as compatible as possible with the elevation and dip data (Fig. 3.13b). The circles show the locations of the remaining incompatibilities. Both incompatibilities arise from the fact that the dip measurements represent the attitude at a point defined by the size of a field compass or a borehole diameter. Local dip may differ from regional dip. The attitude measured at location 3 is probably near the crest or trough of a minor fold. This inference is very likely because there

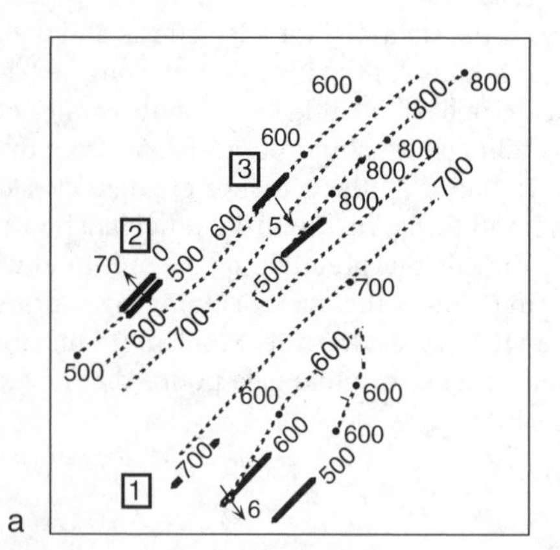

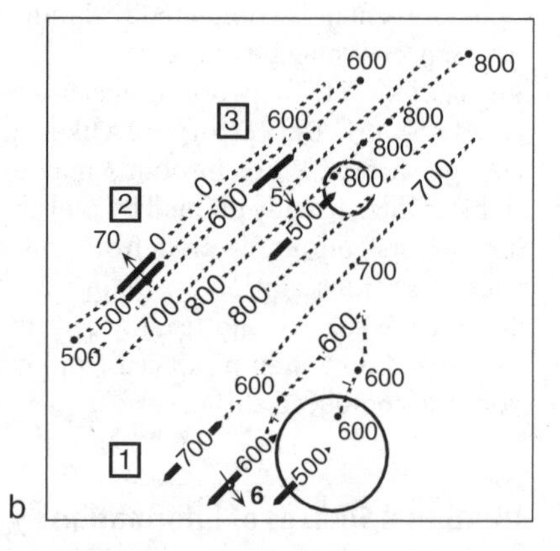

Fig. 3.13. Structure contour directions and spacings from attitude measurements, Blount Springs map area. *Heavy lines* are structure contours determined from bedding attitudes at points *1–3*. **a** *Dotted lines* are structure contours determined by interpretive contouring (from Fig. 3.9b). **b** *Dotted lines* are revised structure contours based on dip information. *Circles* show locations of incompatibility between contour elevations determined from the dips and contours determined from the elevations

are a number of minor folds incompletely exposed in the map area. In general, nearly horizontal dips should be located in the crest or trough regions of folds and need not apply over large areas. In the final interpretation, the attitude information must conform to the elevation control.

3.4.2 Projected and Composite Surfaces

A projected surface is a structure contour map derived entirely by projecting data from other stratigraphic levels. Usually a projected surface is below the lowest control points. A typical use is to project the subsurface location of an aquifer or an oil reservoir from outcrop or shallow subsurface information.

A composite surface is a structure contour map derived using data from multiple stratigraphic horizons, including the horizon being mapped. One horizon is selected as the reference surface and data from other stratigraphic horizons are projected upward or downward to this horizon, using the known stratigraphic thicknesses. The best choice of a reference horizon is one for which there is already a significant amount of control and that minimizes the projection distance. Usually a reference horizon that is stratigraphically in the middle of the best-controlled units should be selected. The data from multiple horizons provide increased control on the interpretation of the shape of the reference horizon. This type of map is particularly useful in the interpretation of outcrop data because all locations of all formation boundaries can be used to provide control points, greatly increasing the areal distribution of data.

An elevation on a marker surface is transformed into an elevation on a projected reference surface by adding or subtracting the vertical separation between the two (Fig. 3.14). The projection is made from a point where the elevation of the marker is known. This may be the location of a surface outcrop (as in Fig. 3.14) or the elevation of a contact in a well. The distance to the projected horizon is derived from the thickness of the unit by:

$$d = t / \cos \delta, \quad (3.1)$$

where d = vertical distance between the surfaces, t = true thickness, and δ = true dip (Badgley 1959). The projection can be either up or down from the known point, that is, from the marker horizon in Fig. 3.14 to either a or b. Be sure to use the same datum (i.e. sea level) for all measurements. Projections from a surface map need to use the topographic elevation to find the elevation with respect to sea level. Projections above the surface of the earth are as valid as projections below the surface; it is not necessary that the reference surface be confined to the subsurface.

If regional thickness variations are present, the thickness used for projection must be adjusted according to the location. An isopach map (Sect. 2.8) provides the information necessary to determine the thickness at specific points. In regions of low dip, the difference between the vertical distance and the true thickness is small. In this situation an approximate projected surface can be derived by simply adding or subtracting the thickness between the units to or from the elevation of the marker to obtain the projected surface (Handley 1954; Jones et al. 1986; Banks 1993).

Projected data can greatly augment the information on a single horizon and can lead to a significant improvement in the interpreted geometry of the structure. The increase in data available for contouring may significantly improve the map on the

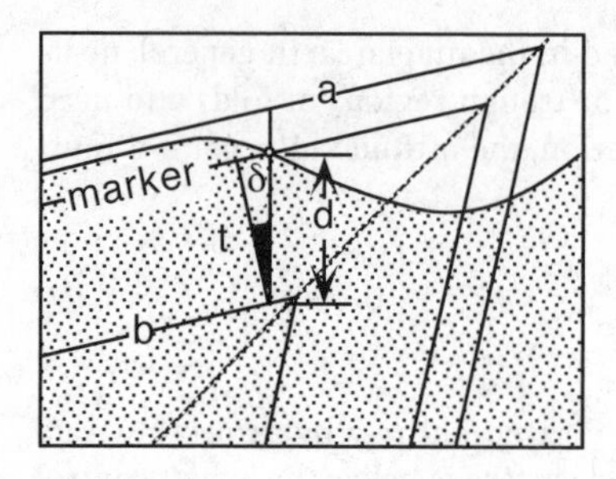

Fig. 3.14. Vertical cross section showing the projected distance from a point (*small circle*) on a marker horizon to reference surface. Projections may be done either upward (to *a*) or downward (to *b*). The region below ground level is *patterned*. *d* Vertical distance between surfaces; *t* true thickness; δ true dip

reference horizon. Inconsistent data on different horizons can be more easily recognized when all data are projected to the same surface. Accurate projection requires accurate knowledge of the unit thickness and the dip, both of which are likely to contain uncertainties and so a certain amount of "noise" is to be expected in the projected data set. The interpreted surface and the data will be iteratively improved as the inconsistencies are eliminated.

The creation of a composite surface map allows utilization of stratigraphic markers that are not formation boundaries. The location of any marker horizon separated from the reference surface by a known stratigraphic interval can be converted to an elevation on the reference horizon. Even if the marker is not usually mapped, it will provide important information.

The construction of a projected or composite-surface includes assumptions that must be considered in each application. Projection with Eq. (3.1) requires that the dip and thickness remain constant and the units unfaulted over the projection distance. The most definitive check on the validity of a projected or composite surface is to construct a cross section that shows all the horizons from which data have been obtained (Chapter 7), as in Figs. 3.15 and 3.16. Any projection problems should be reasonably obvious on the cross section.

Folding may produce thickness changes that are a function of position within the fold and the mechanical stratigraphy. Equation (3.1) is based on the assumption that bed thickness remains constant throughout the fold for all units being projected (Fig. 3.15a), in other words, that all the horizons are parallel. This is called parallel folding, and is only one of the possible fold styles. Dip-domain folds tend to maintain a more nearly parallel geometry than the circular-arc style (Sect. 1.5). Deformation can change the thicknesses, especially the thicknesses of thick soft units between stiffer units. The direction of constant thickness is an element of the fold style. A similar fold maintains constant thickness parallel to the axial surfaces. Constant vertical thickness projection (Handley 1954; Banks 1993) is strictly appropriate only for similar folds that have vertical axial surfaces (Fig. 3.15b). The resulting thinning on steep limbs is a common feature of compressional folds, even in those that maintain constant bed thickness elsewhere. If the axial surfaces of a similar fold are inclined, the direction of constant thickness is inclined to the vertical (Fig. 3.15c). Projection of surfaces is probably best restricted to situations in which bed thicknesses are approximately constant (Fig. 3.15a).

Projection of thickness is based on the further assumption that the stratigraphy between the projection point and the composite surface is an unbroken sequence of uniform dip. If the vertical line of projection crosses a fault (Fig. 3.16a) or an axial sur-

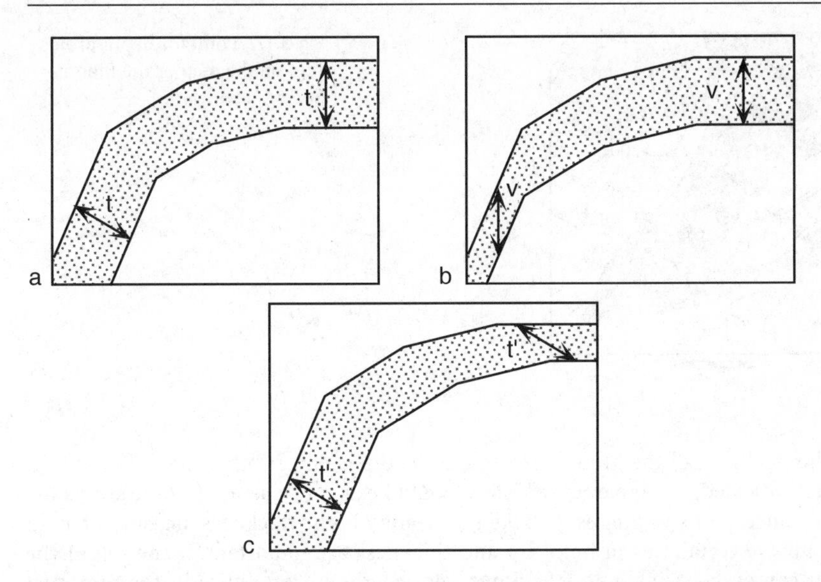

Fig. 3.15. Cross sections showing directions of constant thickness. The dips are identical in each cross section. **a** Constant bed thickness (t). **b** Constant vertical thickness (v) parallel to a vertical axial plane. **c** Constant apparent thickness (t') parallel to an inclined axial plane

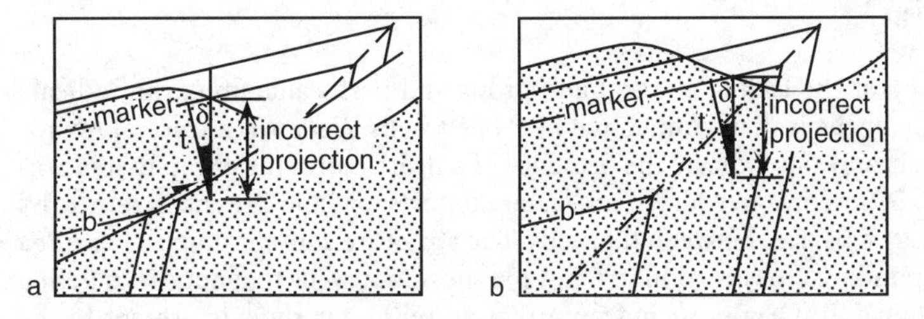

Fig. 3.16. Vertical cross sections showing incorrect projections across discontinuities. **a** Projection across a fault. **b** Projection across an axial surface. The region below ground level is patterned. δ Dip; t thickness of the interval being projected

face (Fig. 3.16b), then the projection will be incorrect. The shorter the projection distance, the less likely these problems are to occur.

As an example of the technique, find the elevation of the top of the Mtfp below the point shown in Fig. 3.17. For simplicity and accuracy, the point has been selected to be at a location where the Mpm–Mh contact crosses a topographic contour. Finding the depth to the top of the Mtfp requires the thickness of the Mpm and the local dip. The best values have been found from the local structure contours (Sect. 2.6.3) and are 97 ft and 04°. The direction of the dip is not required for the calculation. From Eq. (3.1), d = 97.2 ft. For such a low dip, the difference between the true thickness and the apparent thickness is clearly negligible. To find the elevation of the top of the Mtfp, the apparent thickness of the Mpm must be subtracted from the topographic elevation. The projected elevation of the top of the Mtfp is thus 600 ft minus 97 ft = 503 ft.

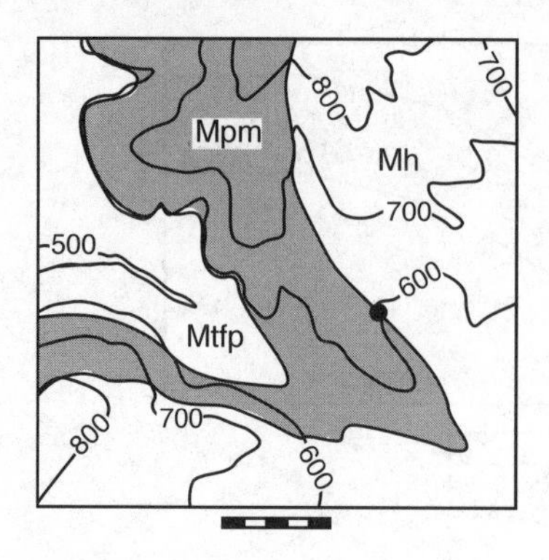

Fig. 3.17. Point for depth projection on the map of the Blount Springs area

If the thickness of the Mpm is 108 ft and the dip is 08°, as other measurements in the area indicated, the apparent thickness could be as much as 109 ft. At low dips the primary uncertainty in the result is the uncertainty in the thickness measurement. At steep dips, uncertainties in both dip and thickness are important. If the top of the Mtfp is contoured using a 20-ft contour interval, the uncertainties in the projected thicknesses are not very important.

3.4.3 Fluid-Flow Barriers

Fluid movement, or the lack of it, through porous and permeable units can indicate the connectivity or the lack of connectivity between wells. A *show* is a trace of hydrocarbons in a well, and can indicate the presence of a nearby structural trap that is otherwise unseen. Different water levels, oil–water contacts, or fluid pressures in nearby wells can indicate a barrier between the wells. The structure contour map in Fig. 3.18a and the corresponding cross section in Fig. 3.19a show four wells that appear to define a region of uniform dip. Suppose, however, that an oil or gas show is present in the downdip well but not in the updip wells. (We assume that the show is natural, not the result of a spill.) The show suggests proximity to a hydrocarbon trap, yet the map does not indicate a trap. The map must be revised to include some form of barrier because of this additional information (Sebring 1958). Possible alternatives that could produce an oil or gas show in the downdip well include a hydrocarbon-filled structural closure farther up the dip (figs. 3.18b; 3.19b), different types of faults between the downdip well and the updip wells (Figs. 3.18c,d,e; 3.19c,d,e) or a stratigraphic permeability barrier (Figs. 3.18f; 3.19f). A stratigraphic barrier does not necessarily require the structure to be changed from the original interpretation. The structural configuration is the same in Figs. 3.18f and 3.19f as in Figs. 3.18a and 3.19a.

3.5 Multiple-Unit Map Validation

Two powerful tools for map validation are composite-surface maps and contour compatibility testing. Both tools provide methods for testing the map for internal consistency. Use of these criteria is discussed in the following two sections.

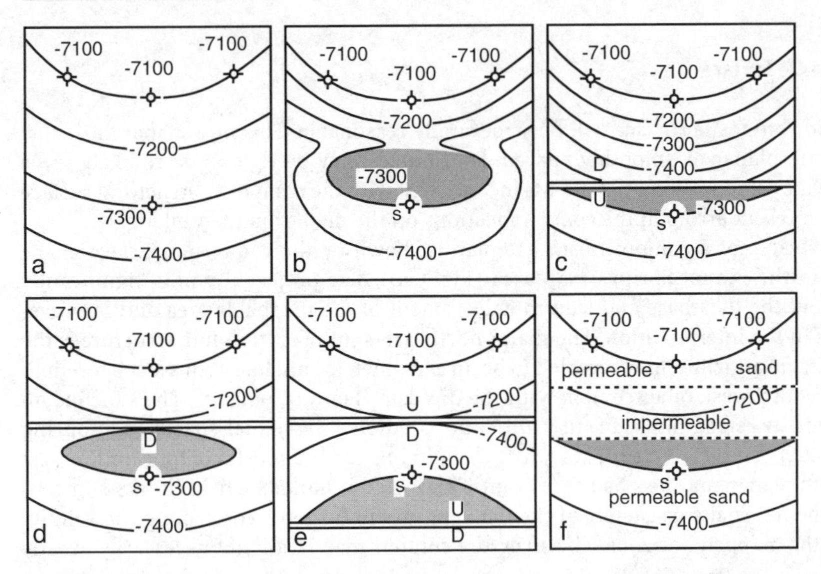

Fig. 3.18. Alternative maps honoring the same data points. **a** Map based on the four wells being dry holes. **b–f** Maps based on presence of an oil show in the well labeled *s*, implying presence of a barrier between this well and the three updip dry holes. Hydrocarbon accumulations are *shaded*. Fig. 3.19 shows the corresponding cross sections. (After Sebring 1958)

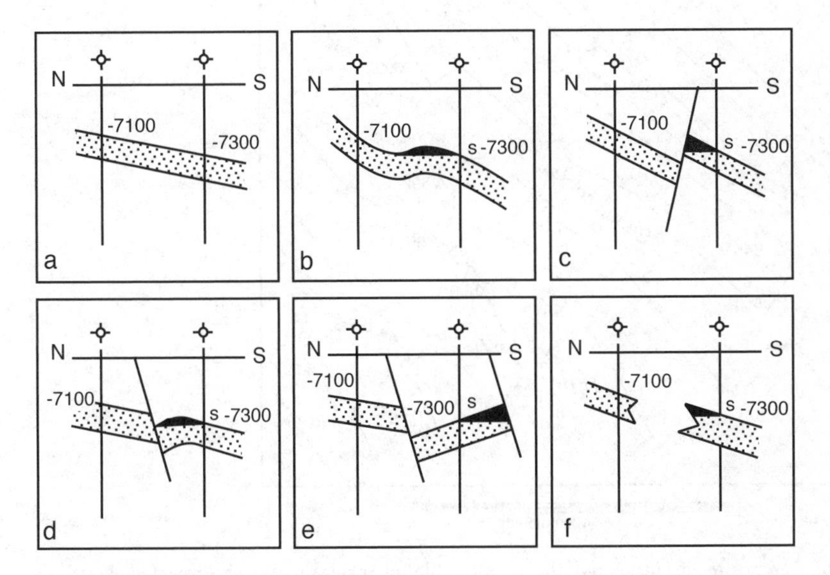

Fig. 3.19. Alternative cross sections in the dip direction honoring the same data points. **a** Section based on the four wells being dry holes. **b–f** Sections based on presence of an oil show (*s*) in the downdip well, implying a barrier between this well and the updip wells. Hydrocarbon accumulations are *black*. Fig. 3.18 shows the corresponding structure contour maps. (After Sebring 1958)

3.5.1 Composite Surface

A valid composite surface will be structurally reasonable. This means that most surfaces are planar or smoothly curved. Fold hinges may be tight and are likely to be straight or smoothly curved. Points inconsistent with the composite structural surface may represent errors in the contact locations on the map or in the well log.

The value of a composite-surface map is shown by the composite surface of the Mtfp in the Blount Springs map area (Fig. 3.20). The projected points significantly augment the data base and lead to enlargement of the mappable area and improvements in the interpretation. The major northeast–southwest trending structure is the Sequatchie anticline, now revealed to be an asymmetric anticline with a steep forelimb on the northwest, in agreement with the dip data (Fig. 3.13, point 2). The syncline on the northwestern limb of the anticline predicted by equal-spaced contouring (Fig. 3.8a) can be rejected. The -341-ft elevation inside the 300-ft contour on the west side of the map may be a bad data point. This location should be field checked. In general, the internal consistency of the data appears to be good, confirming the validity of all the mapped horizons. The structure contour map is still not necessarily exactly

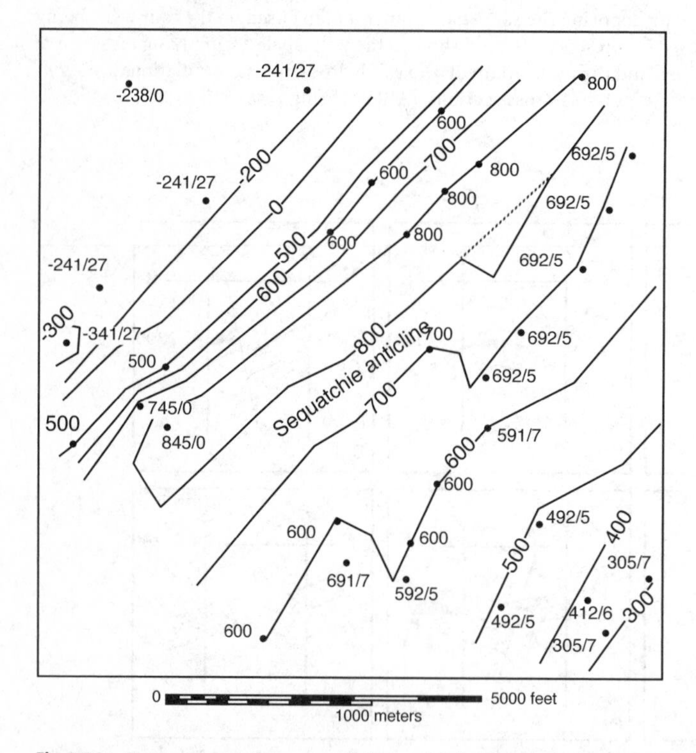

Fig. 3.20. Composite-surface map of the top Mtfp, Blount Springs map area. Projected elevations of the top Mtfp are given as elevation/dip; elevations without dip are from the outcrop. Thicknesses used for projection are: Mpm = 108 ft and Mh = 105 ft, calculated from the geologic map (Fig. 3.1); Mtfp = 245 ft from a well in the area; Mb = 625 ft from the outcrop just southeast of the map area. The *dotted* 800-ft contour is an alternative interpretation

correct, however. Small systematic errors will occur in contour locations if the dips or thicknesses used for projection are in error.

The crest of the Sequatchie anticline clearly extends to the southwest beyond the locations suggested by Fig. 3.9. Whether the 800-ft contour of the anticline should be closed around the 845-ft data point as shown on Fig. 3.20, continue unclosed off the southwestern end of the map, or be a separate closure, remains a matter of conjecture. If a stratigraphic marker within one of the formations was found that could give a few elevations near the southwestern part of the 800-ft contour, the confidence in the interpretation would be significantly increased. Fold plunge analysis (Chapter 4) will provide critical information for improving this aspect of the interpretation.

The southeastern area of the map, which was previously interpreted as being either flat (linear-interpolation contouring; Fig. 3.7), a separate anticline (equal-spaced contouring; Fig. 3.8a) or a separate closed syncline (parallel contouring and interpretive contouring, Figs. 3.8b, 3.9a) appears to be a gentle anticline–syncline fold pair that trends more northerly than the Sequatchie anticline itself (Fig. 3.20). Second-order structures like these superimposed on the regional dip could form structural traps for hydrocarbons or lead to fracturing that increases the flow of groundwater, hence the evidence for and against the minor folding should be critically examined. The inference that second-order folds are present could be due to moderate mis-locations of contacts in the field; perhaps the mapped trace should be straighter. (Further analysis of the map area given in Sect. 4.4 will provide support for second-order folds oblique to the main fold axis.) The nature of the northeastern termination of the fold pair is also open to interpretation. The structure is continued farther to the northeast by the solid 800 ft contour line than by the dashed contour. Either interpretation is possible. The presence of the second-order folds as well as their extent should be field checked. If a stratigraphic marker were available within one of the mapped formations that could provide a few elevations near the northeastern part of the minor folds, confidence in the interpretation could be greatly improved.

The interpretation of Fig. 3.20 should be viewed as a working hypothesis. The composite-surface map indicates that most of the map is internally consistent. Field checking a few specific aspects of the interpretation is suggested. Without this insight, validation of a map is either impossible or requires complete re-mapping.

3.5.2 Contour Compatibility

The structure of nearby stratigraphic horizons is usually fairly similar. If the thicknesses are approximately constant, the shapes of nearby horizons will be nearly the same. The compatibility of the maps on closely spaced horizons is indicated by structure contours that are nearly parallel (Fontaine 1985) and separated by approximately constant distances in gently dipping beds. A significant difference in trends on adjacent surfaces suggests a possible misinterpretation of one or both of the surfaces. The structure contours on different horizons cannot intersect without implying a structural or stratigraphic discontinuity. Only faults and unconformity surfaces can cut across stratigraphic boundaries.

Independently mapping different horizons can easily lead to incompatible surfaces and so the compatibility must be checked. Figure 3.21 is a map of the top of the Pride Mountain Formation (Mpm) derived directly from elevations on the geologic map in

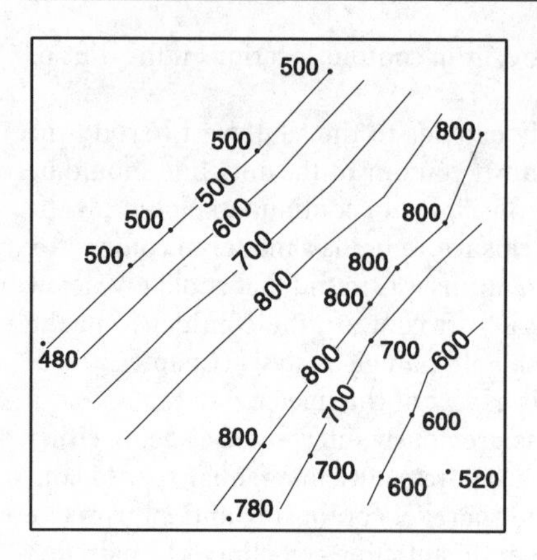

Fig. 3.21. Interpreted structure contours on the top of the Mpm, Blount Springs map area. Control points are labeled. Contour interval is 100 ft

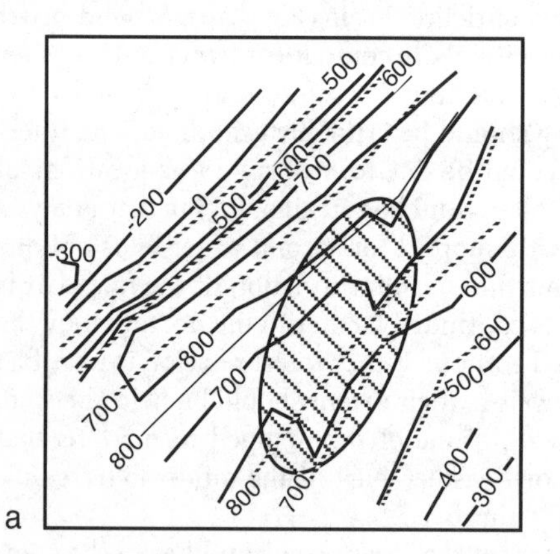

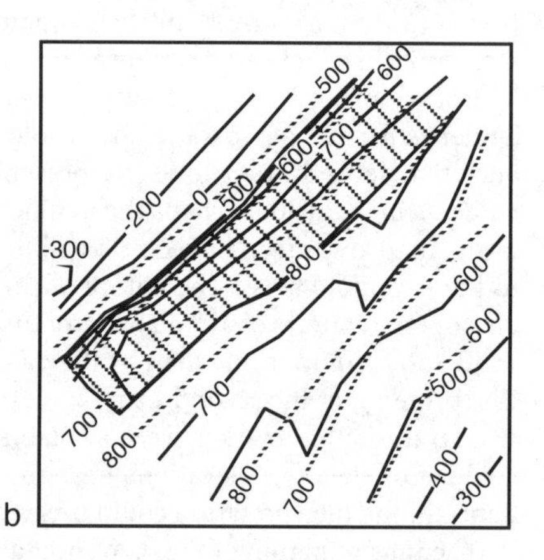

Fig. 3.22. Testing for compatible surfaces by superimposing structure contour maps. Structure contour map of the top Mpm (*dashed contours*) overlain on the composite map of the top Mtfp (*solid contours*). **a** *Diagonal-line pattern* shows an area of contour shape incompatibility. **b** *Diagonal-line pattern* shows approximate area where the stratigraphically higher surface (Mpm, *dashed lines*) is implied to lie below the stratigraphically lower surface (Mtfp, *solid lines*)

Fig. 3.1. It is 108 ft above the top of Fort Payne Formation shown in previous structure contour maps (i.e. Fig. 3.20). Given the control on the top of the Pride Mountain alone, the map is reasonable. By superimposing this map on a composite map of the top of the Fort Payne from Fig. 3.20, the two contour surfaces can be tested for compatibility (Fig. 3.22). The contours are approximately parallel (Fig. 3.22a) and so are compatible by this criterion, except in the region of second-order folding on the southeast limb of the Sequatchie anticline. A check of the elevations, however, reveals that the upper stratigraphic surface lies below the lower stratigraphic surface over a substantial area (Fig. 3.22b). This indicates a glaring lack of compatibility between the two surfaces. A cross section would also reveal this problem and make it easy to visualize, one of the reasons that cross sections should always be included as part of the interpretive process. A better map of the top Mpm would be obtained by creating a composite-surface map as described previously.

3.5.3 Trend Compatibility

Consistency between the map geometry and the structural trend derived from independent data, such as the attitude of bedding, provides strong confirmation of the interpretation expressed in a structure contour map. Conversely, a lack of consistency implies that the interpretation may require revision. Figure 3.23a is an accurate outcrop map of a sandstone bed. Elevations of the upper and lower contacts of the sandstone are shown in Fig. 3.23b at points where the contacts cross contours; these points could represent formation boundaries determined from drilling. Preliminary structure contours, based on the concept that the bed is planar or gently folded, suggest a northwest-southeast trend to the strike. There is nothing about the data in Fig. 3.23b to definitively argue against this interpretation, although a greater degree of parallelism of the contours might be expected. Problems are revealed when these same structure contours are drawn on the geologic map (Fig. 3.24a). The question marks show locations where one of the unit boundaries should occur in outcrop, but does not. The actual strike trend in the area is north-south, which clearly invalidates the interpretation of Figs. 3.23b and 3.24a. The correct structure of the map area is a non-plunging anticline with an axis that trends north–south (Fig. 3.24b).

It may be necessary to know the structural trend before a reasonable map of the surface can be constructed. The data of Fig. 3.23b are never likely to be contoured as in Fig. 3.24b without independent evidence for the trend of the fold axis. The map of Fig. 3.23a must be considered to be seriously incomplete without at least a few bedding attitudes on it. The next chapter describes methods for finding the fold trend from the attitudes of bedding and applying this information to map construction.

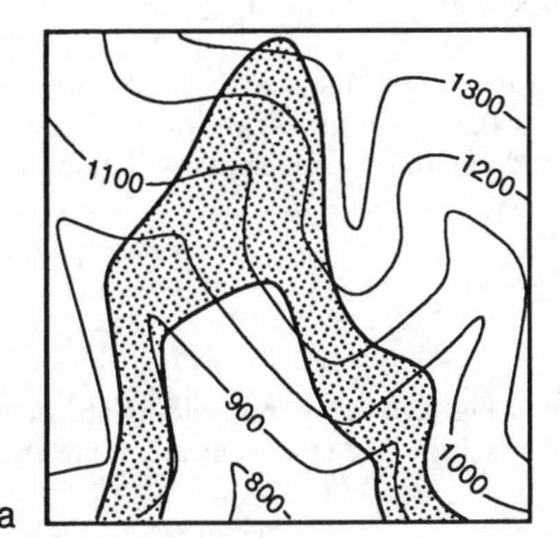

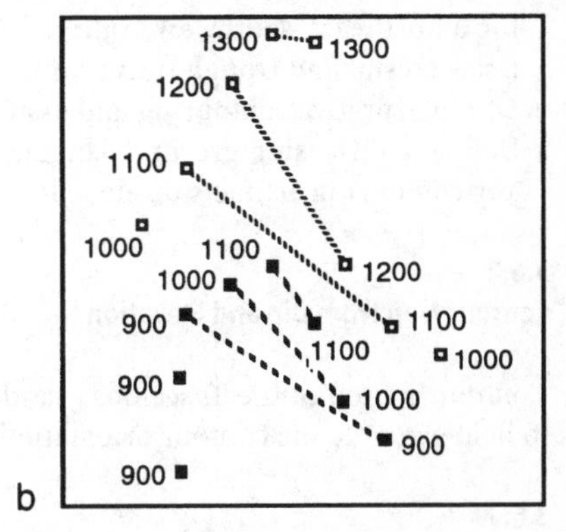

Fig. 3.23. Elevation data for constructing a structure contour map. **a** Outcrop trace of a sandstone bed (*shaded*) on a topographic base (after Weijermars 1997). **b** Point elevations on upper surface (*open squares*) and lower surface (*filled squares*) of the map unit in **a**. *Dotted contours* are on the upper surface and *bold dashed contours* are on the lower surface

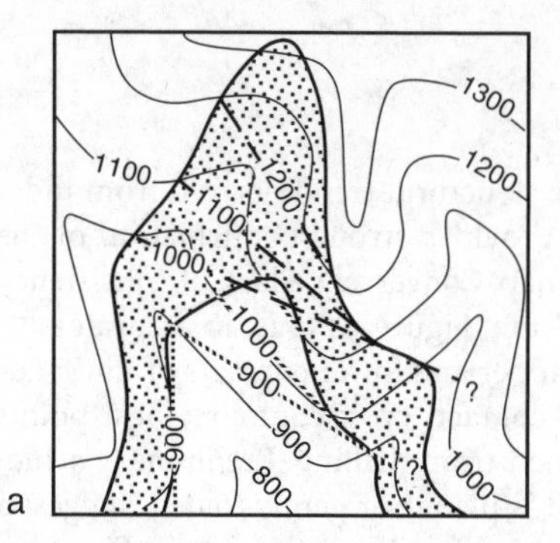

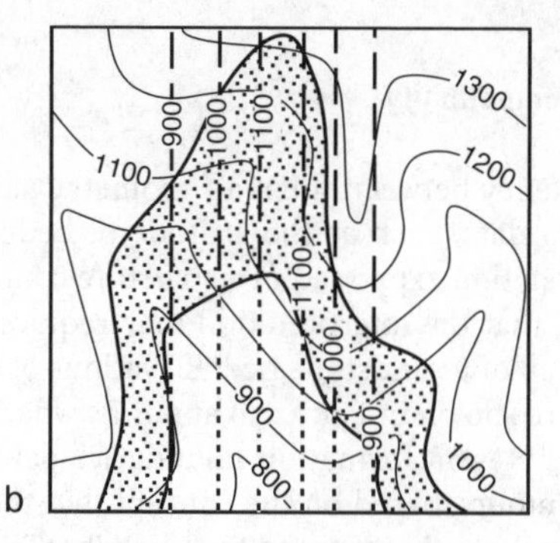

Fig. 3.24. Structure contours interpreted from the geologic map of Fig. 3.23. **a** Structure contours drawn between closest corresponding elevations across the V formed by the outcrop trace. *Long dash contours* are on top of the bed, *dotted contours* are on the base of the bed. *Question marks* show where structure contours indicate that the bed surface should intersect the topographic surface but no intersection is observed. **b** Structure contours on the base of the bed, based on interpretation that the structure is a north–south trending, non-plunging fold (after Weijermars 1997). *Long dashes* indicate contours below ground, *short dashes* indicate contours projected above ground

3.6 Problems

3.6.1 Contouring Styles

Use the data from the Weasel Roost Formation (Fig. 3.25) to try out different contouring techniques and to see the effect of trend biasing. The elevations are in feet.

1. Use interpretive contouring and assume a surface with no grain.
2. Contour by parallel contouring: (a) assuming a northwest–southeast grain; (b) assuming a northeast–southwest grain.
3. Draw crestal and trough traces on the structure contour maps just completed.
4. Use interpretive contouring and assume a northeast–southwest grain.
5. Define a TIN using greedy triangulation and contour by linear interpolation. Are any contouring artifacts produced?

3.6.2 Contour Map from Dip and Elevation

Contour the top of the Tuscaloosa sandstone in Fig. 3.26 using the bedding attitudes to help generate the contour orientations and spacings. The elevations are in meters.

3.6.3 Depth to Contact

Find the elevation of the top of the Mtfp below the dot in Fig. 3.27. The thickness of the Mpm is 97 ft and the dip is 04°.

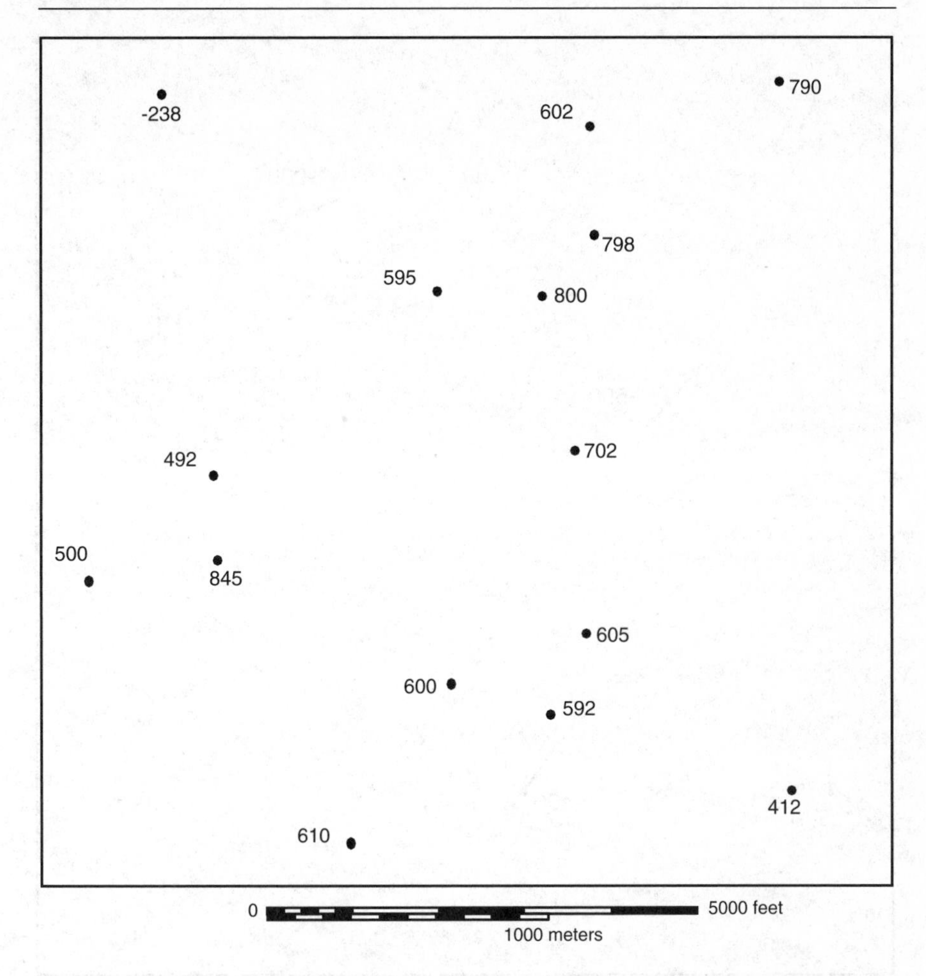

Fig. 3.25. Map of elevations (feet or meters) of the top of the Weasel Roost Formation

3.6.4 Composite-Surface Map

Create a composite-surface map for the top of the Mtfp in the Blount Springs Quadrangle (Fig. 3.1). Use the stratigraphic thicknesses given in the caption to Fig. 3.20. For other thicknesses refer to values found in the exercises of Chapter 2. Are there any inconsistencies in the map?

3.6.5 Projected-Surface Map and Map Validation

Use the geologic map of Fig. 3.28 to construct a projected structure contour map of the top of the Fairholme, a potential hydrocarbon reservoir. Use every point where a

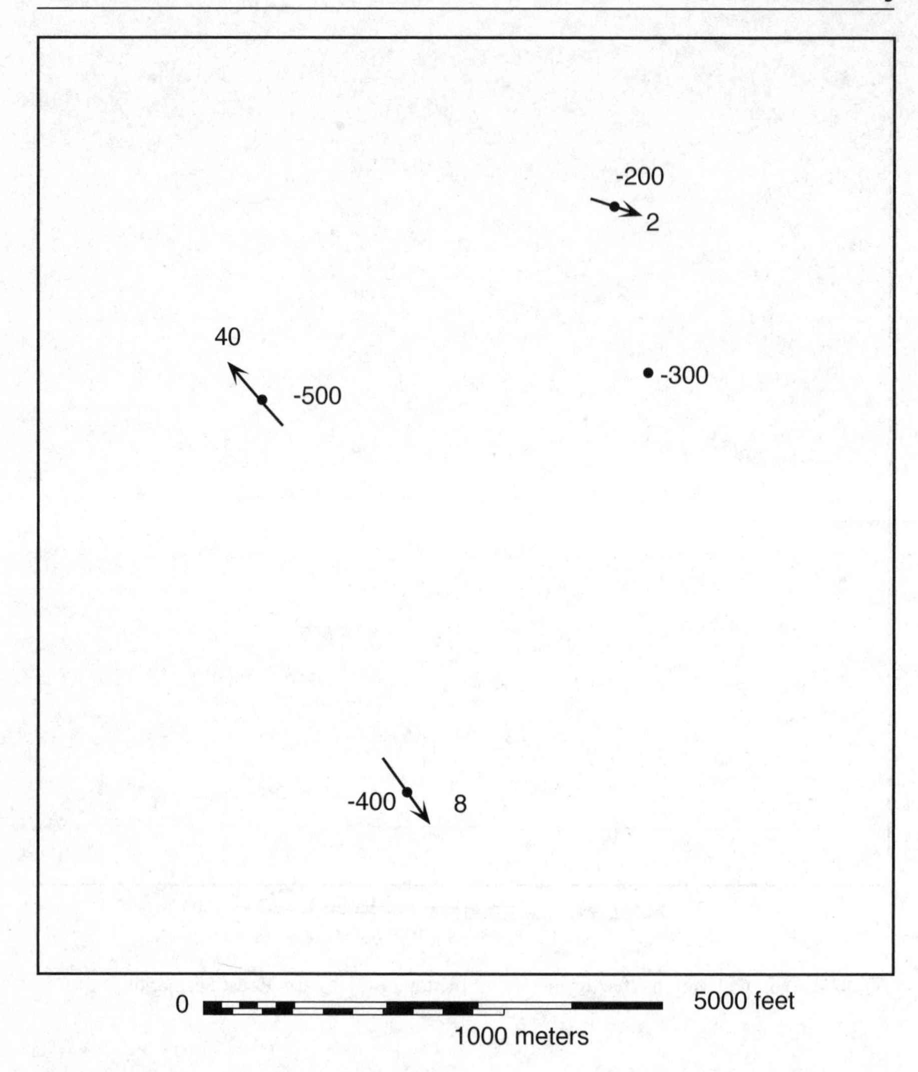

Fig. 3.26. Map of the top of the porous Tuscaloosa sandstone. Negative elevations are below sea level; azimuth of bedding dip is indicated by *arrows*

formation boundary crosses a topographic contour. Post all the elevations on your map before contouring.

1. What is the best method for contouring this map? Explain your reasons.
2. Is the geological map correct? Why or why not?
3. Does the projected structure-contour map agree with the drilled depths to the top of the Fairholme?
4. The wells to the Fairholme were drilled to find a hydrocarbon trap but were not successful. What is a structural reason for drilling the wells and what is a structural reason why they were unsuccessful?

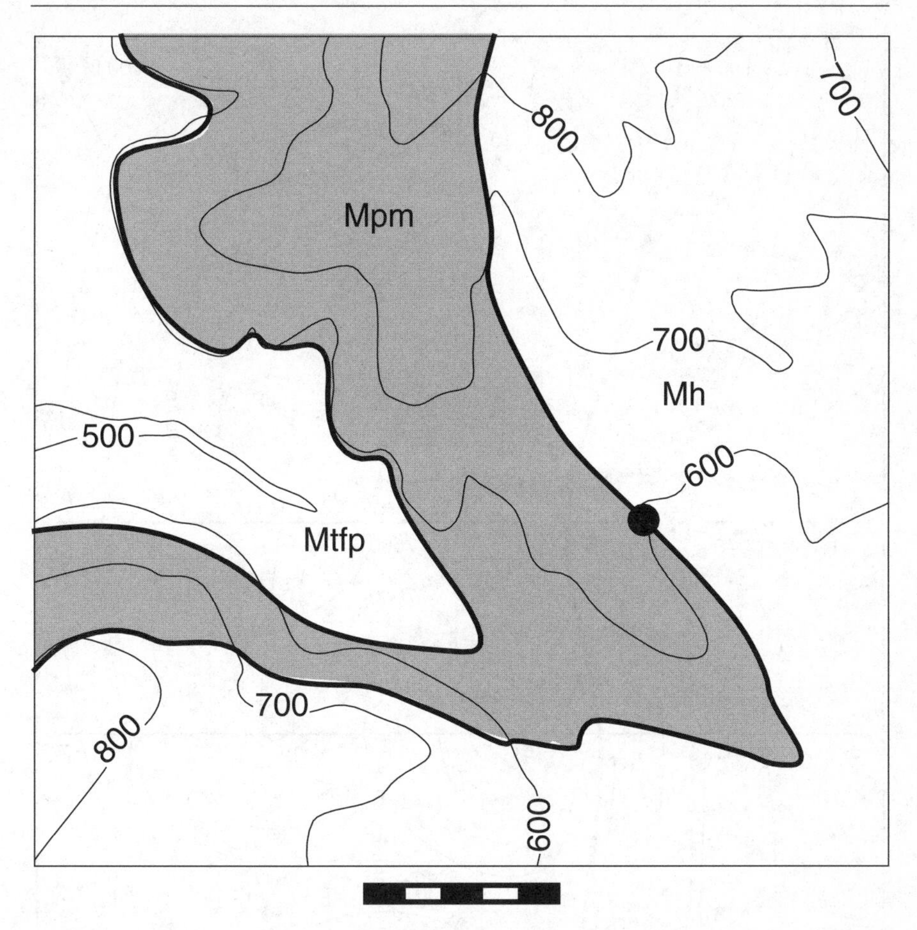

Fig. 3.27. Geologic map of the Mill Creek area. Topographic elevations are in feet and the scale bar is 1000 ft

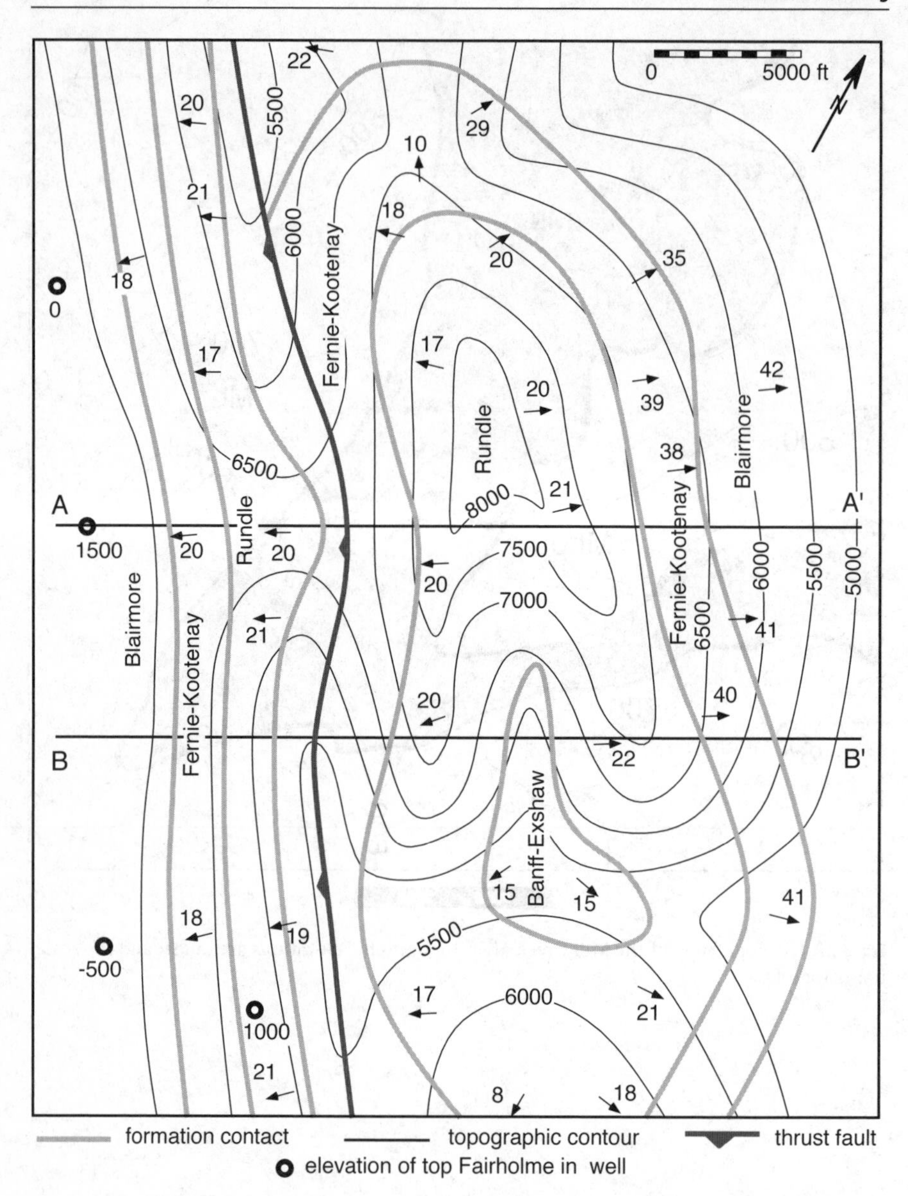

Fig. 3.28. Geologic map from the Canadian Rocky Mountains. Dimensions are in feet. The stratigraphic column (with thickness in feet) from top to base is: Blairmore (2400), Fernie-Kootenay (700), Rundle (900), Banff-Exshaw (900), Palliser (800), Fairholme(1200). (After Badgley 1959)

References

Badgley PC (1959) Structural methods for the exploration geologist. Harper & Brothers, New York, 280 pp
Banks R (1991) Contouring algorithms. Geobyte 6: 15–23
Banks R (1993) Computer stacking of multiple geologic surfaces. Petro Systems World. Winter: 14–16
Bishop MS (1960) Subsurface mapping. John Wiley, New York, 198 pp
Cherry BA (1990) Internal deformation and fold kinematics of part of the Sequatchie anticline, southern Appalachian fold and thrust belt, Blount County, Alabama. MS Thesis, Univ. Alabama, Tuscaloosa, 78 pp
Dennison JM (1968) Analysis of geologic structures. WW Norton, New York, 209 pp
Faill RT (1969) Kink band structures in the Valley and Ridge Province, central Pennsylvania. Geol. Soc. Am. Bull. 80: 2539–2550
Faill RT (1973) Tectonic development of the Triassic Newark-Gettysburg Basin in Pennsylvania. Geol. Soc. Am. Bull. 84: 725–740
Fontaine DA (1985) Mapping techniques that pay: Oil Gas J. 83, 12: 146–147
Hamilton DE, Jones TA (1992) Algorithm comparison with cross sections. In,: Hamilton DE, Jones TA (eds) Computer modeling of geologic surfaces and volumes. American Association of Petroleum Geologists, Tulsa, Oklahoma, pp 273–278
Handley EJ (1954) Contouring is important. World Oil 183, 4: 106–107
Jones NL, Nelson J (1992) Geoscientific modeling with TINs. Geobyte 7: 44–49
Jones TA, Hamilton DE (1992) A philosophy of contour mapping with a computer. In: Hamilton DE, Jones TA (eds) Computer modeling of geologic surfaces and volumes. American Association of Petroleum Geologists, Tulsa, Oklahoma, pp 1–7
Jones TA, Krum GL (1992) Pitfalls in computer contouring, Part II. Geobyte 7: 31–37
Jones TA, Hamilton DE, Johnson CR (1986) Contouring geologic surfaces with the computer. Van Nostrand Reinhold, New York, 314 pp
Krajewski SA, Gibbs BL (1994) Computer contouring generates artifacts. Geotimes 39: 15–19
Oliveros RB (1989) Correcting 2-D seismic mis-ties. Geobyte 4: 43–47
Rettger RE (1929) On specifying the type of structural contouring. Am. Assoc. Pet. Geol. Bull. 13: 1559–1561
Sebring L Jr (1958) Chief tool of the petroleum exploration geologist: The subsurface structural map. Am. Assoc. Pet. Geol. Bull. 42: 561–587
Tearpock DJ (1992) Contouring: Art or science?. Geobyte 7: 40–43
Thomas WA (1986) Sequatchie anticline, the northwesternmost structure of the Appalachian fold-thrust belt in Alabama: Geological Society of America centennial field guide – southeastern section. Geological Society of America, Boulder, Colorado, pp 177–180
Walters RF (1969) Contouring by machine: a user's guide. Am. Assoc. Pet. Geol. Bull. 53: 2324–2340
Watson DF, Philip GM (1984) Triangle-based interpolation. Math. Geol. 16: 779–795
Weijermars R (1997) Structural geology and map interpretation. Alboran Science Publishing, Amsterdam, 378 pp

Chapter 4
Fold Geometry

4.1
Introduction

This chapter describes methods for defining the geometry of folded surfaces in three dimensions, primarily based on the attitudes of bedding. The fold shapes so defined are applied to the problems of fitting the correct fold surface to the data and in projecting the fold into areas of little or no data. Folds are divided into domains where the shapes are cylindrical or conical, smoothly curved or planar. The geometries within domains are efficiently described in terms of the orientations and properties of fold axes, plunge lines, crest lines and trough lines. The relationship of these elements to bed attitudes has implications for bed thickness changes and the persistence of the folds along their trend. Dip-sequence analysis is introduced as a tool for interpreting three-dimensional fold geometry from outcrop traverses and dipmeter data. Minor fold geometry is described as an aid to the interpretation of map-scale structure and for its potential to create problems in determining the shapes of the major folds.

4.2
Form and Trend from Bed Attitudes

Folds are characterized by their trend, plunge, and form within domains of relatively homogeneous structure. The fold trend is a key element in making and confirming the grain in a map. The change in shape along plunge is given by the fold form. A fold in a cylindrical domain continues unchanged along plunge, whereas a fold in a conical domain will die out along plunge. The trend, plunge, and style of a fold are determined from the bedding attitudes as plotted on stereograms and tangent diagrams. The bedding attitude data are collected from outcrop measurements or from dipmeters. Dip-domain style folds may combine both cylindrical and conical elements. If the data show too much scatter for the form to be clear, the size of the domain under consideration can usually be reduced until the domain is homogeneous and has a cylindrical or conical geometry.

4.2.1
Cylindrical Fold

A cylindrical fold is defined by the property that the poles to bedding all lie parallel to the same plane regardless of the specific cross-sectional shape of the fold (Fig. 4.1a).This property is the basis for finding the fold axis. On a stereogram the poles to bedding fall on a great circle (Fig. 4.1b). The pole to this great circle is the fold axis, known as the π axis. The trend of a cylindrical fold is parallel to its axis. A cylindrical fold maintains constant geometry along its axis as long as the trend and plunge remain constant. The stereogram is the best method for fold axis determination for folds with very steep dips. Overturned folds always contain vertical dips between the upright and overturned limbs (Rowland and Duebendorfer 1994) and are more accurately treated on a stereogram.

An alternative method for finding the axis is to plot the bedding attitudes on a tangent diagram. The method is based on the principle that intersecting planes have the same apparent dip in a vertical plane containing their line of intersection (Bengtson 1980). Let $\boldsymbol{\delta}$ represent the dip vector of a plane. In Fig. 4.2, planes $\boldsymbol{\delta}_1$ and $\boldsymbol{\delta}_2$ are plotted and connected by a straight line. The perpendicular to this line through the origin, $\boldsymbol{\delta}_3$, gives the bearing and plunge of the line of intersection. In a cylindrical fold, all bedding planes intersect in the straight line ($\boldsymbol{\delta}_3$) which is the fold axis.

Each bedding attitude is plotted on the tangent diagram as a point at the appropriate azimuth and dip. The straight line through the bed attitude points goes through

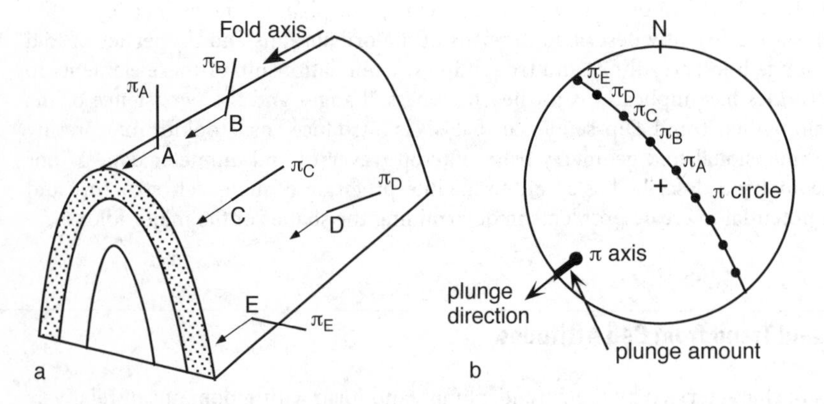

Fig. 4.1. Axis of a cylindrical fold. **a** Fold geometry. *A–E* are measurement points, π_A–π_E are poles to bedding. **b** Axis (π axis) determined from a stereogram, lower-hemisphere projection. (After Ramsay 1967)

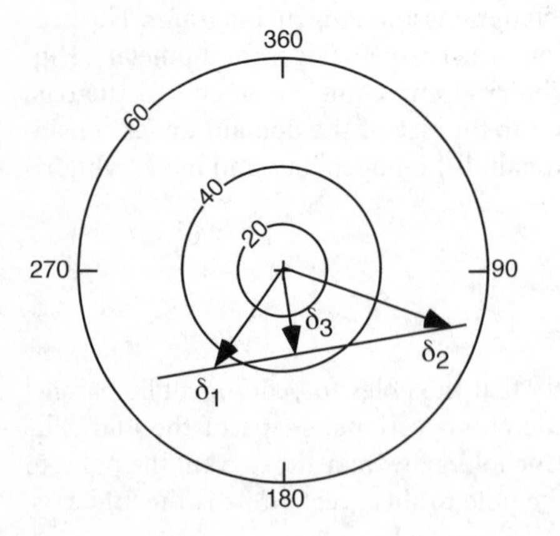

Fig. 4.2. Fold axis ($\boldsymbol{\delta}_3$) found as the intersection line between two bedding planes ($\boldsymbol{\delta}_1$ and $\boldsymbol{\delta}_2$) on a tangent diagram. (After Bengtson 1980)

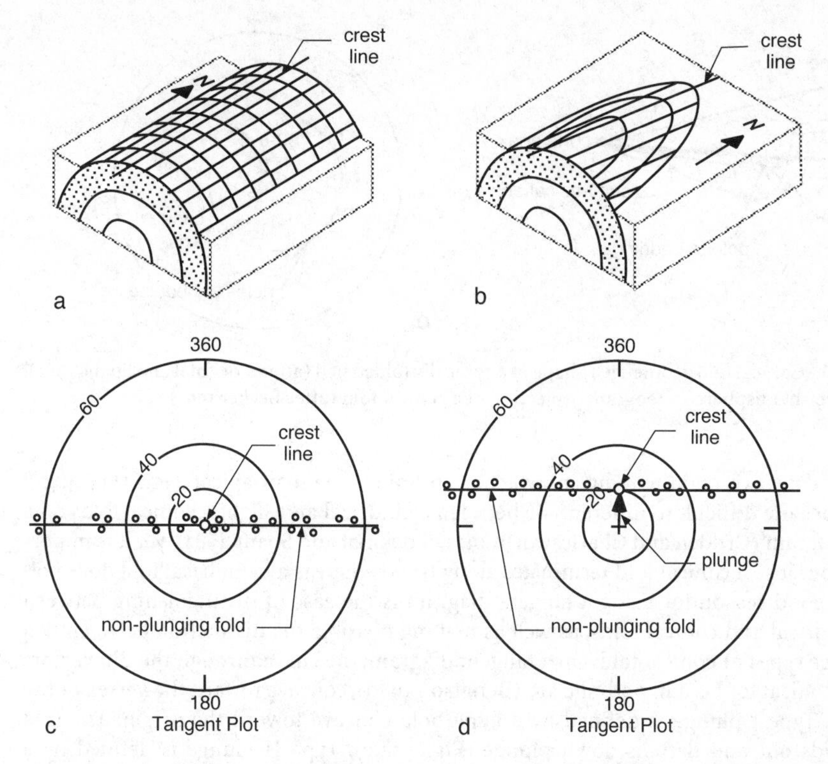

Fig. 4.3. Cylindrical folds showing trend and plunge of the crest line. **a** Non-plunging. **b** Plunging. **c** Tangent diagram of bed attitudes in a non-plunging fold. **d** Tangent diagram of bed attitudes in a plunging fold. (After Bengtson 1980)

the origin for a non-plunging fold (Fig. 4.3a,c) and is a straight line offset from the origin for a plunging fold (Fig. 4.3b, d).

A drawback to the tangent diagram is that dips over 80° require an unreasonably large piece of paper, the tangent of 90° being infinity. The non-linearity of the scale also exaggerates the dispersion of steep dips. A practical solution for folds defined mainly by dips under 80° is to plot dips over 80° on the 80° ring (Bengtson 1981b). This will have no effect on the determination of the axis and plunge of the folds (Bengtson 1981b) and will have the desirable effect of reducing the dispersion of the steep dips. It is primarily the positions of the more gently dipping points that control the location of the axis and its plunge.

4.2.2 Conical Fold

A conical fold is defined by the movement of a generatrix that is fixed at the apex of a cone (Fig. 4.4a); the fold shape is a portion of a cone. A conical fold terminates along its trend. On a stereogram the bedding poles fall on a small circle, the center of which

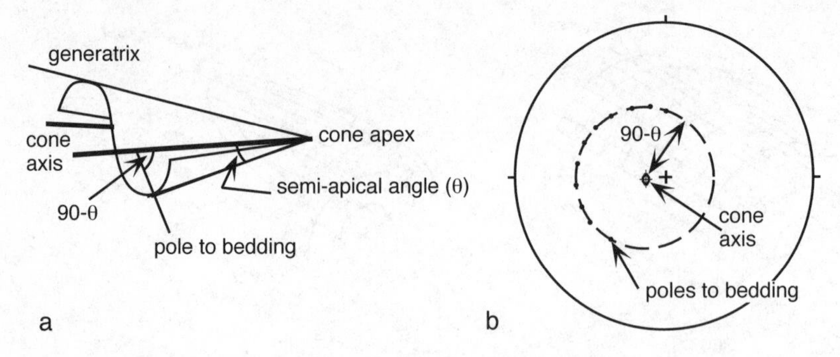

Fig. 4.4. Conical fold geometry. **a** Shape of a conically folded bed (after Stockmal and Spang 1982). **b** Lower-hemisphere stereogram projection of a conical fold (after Becker 1995)

is the cone axis and the radius of which is 90° minus the semi-apical angle (Fig. 4.4b). It is usually difficult to differentiate between cylindrical and slightly conical folds on a stereogram (Cruden and Charlesworth 1972; Stockmal and Spang 1982), yet it is important because a conical fold terminates along trend whereas a cylindrical fold does not.

A good reason for using a tangent diagram is the ease of distinguishing between cylindrical and conical folds, as well as making possible the distinction between two plunge types of conical fold. On a tangent diagram, the curve through the dip vectors of a conical fold is a hyperbolic arc (Bengtson 1980), concave toward the vertex of the cone. Type I plunge is defined by a hyperbola concave toward the origin. The fold spreads out and flattens down plunge (Fig. 4.5a,c). Type II plunge is defined by a hyperbola convex toward the origin on the tangent diagram (Fig. 4.5b,d) and the fold comes to a point down plunge. In the strict sense, a bed in a conical fold does not have an axis and therefore does not have a plunge like a cylindrical fold. The orientation of the crest line in an anticline (Fig. 4.5) or trough line in a syncline provides the direction that describes the orientation of a conically folded bed, in a manner analogous to the axis of a cylindrical fold. The plunge of a conical fold as defined on a tangent diagram is the plunge of the crest or trough line, not the plunge of the cone axis. If the crest line is horizontal, the fold is non-plunging but nevertheless terminates along the crest or trough in the direction of the vertex of the cone. In general, each horizon in a fold will have its vertex in the same direction but at a different elevation from the vertices of the other horizons in the same fold.

The plunge in a conical fold is the magnitude and direction of the vector from the origin in the direction normal to the tangent of the bedding hyperbola at the point to be projected (figs. 4.6, 4.7) in the direction toward the tangent line (figs. 4.6, 4.7). For points on the fold that are not on the crest or trough, draw a line tangent to the data hyperbola at the dip representing the point to be projected. A line drawn from the origin to the tangent line and that is perpendicular to the tangent line gives the plunge amount and direction (Bengtson 1980). In a type I fold (Fig. 4.6) the minimum plunge angle is that of the crest line and all other plunge lines have greater plunge angles. In a type II fold (Fig. 4.7) the plunge line at some limb dip has a plunge of zero. The trend of this plunge line is normal to the tangent line (Fig. 4.7), the same as for all the other plunge lines. At greater limb dips in a type II fold, the down-plunge direction can be away from the vertex of the cone (Fig. 4.7).

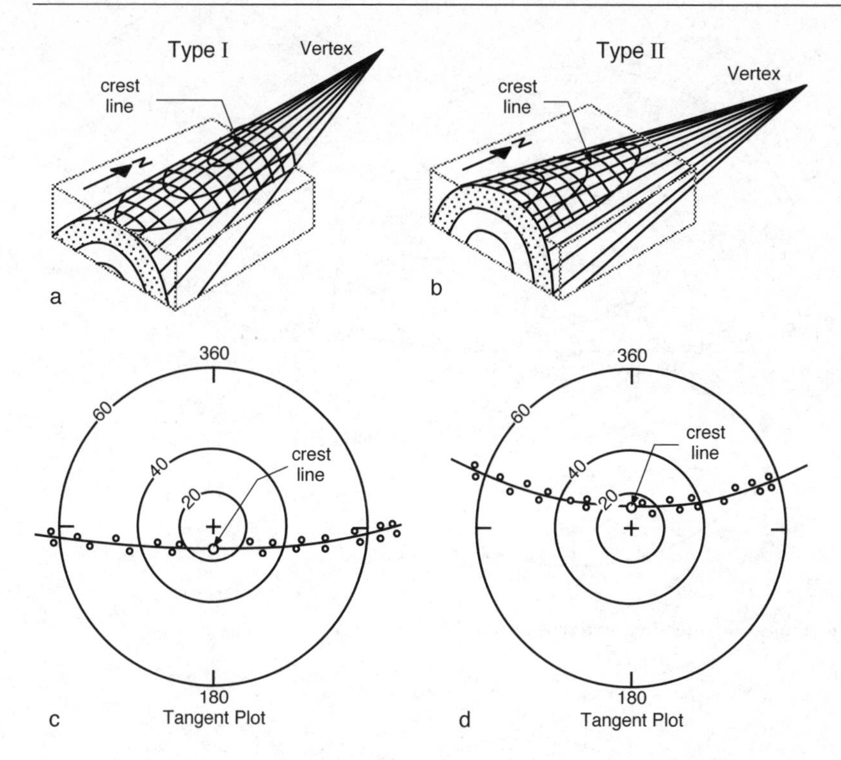

Fig. 4.5. Conical folds showing trend and plunge of crest line. **a** Type I plunge. **b** Type II plunge. **c** Tangent diagram of bed attitudes in a type I plunging fold. **d** Tangent diagram of bed attitudes in a type II plunging fold. (After Bengtson 1980)

4.2.3 Planar Dip Domains and Hinge Lines

A folded surface may be divided into domains of relatively uniform dip, separated by hinge lines (Fig. 4.8). A hinge line is a line of locally sharp curvature. On a structure contour map, hinge lines are lines of rapid changes in the strike of the contours. In a fixed-hinge model of fold development, the hinges represent the axes about which the folded layers have rotated.

The orientations of the hinge lines can be found on a stereogram by finding the intersection between the great circles that represent the orientations of the two adjacent dip domains (Fig. 4.9). If the intersection line is horizontal, as for hinge lines 1 and 2 in Fig. 4.8, the hinge line is not plunging. Hinge line 3 between the two dipping domains plunges 11° to the south (Fig. 4.9). The fold geometry between any two adjacent dip domains is cylindrical with the trend and plunge being equal to that of the hinge line. If the bed attitudes from multiple domains intersect at the same point on the stereogram, the fold is cylindrical, and the line is the fold axis, called the β-pole when determined this way (Ramsay 1967).

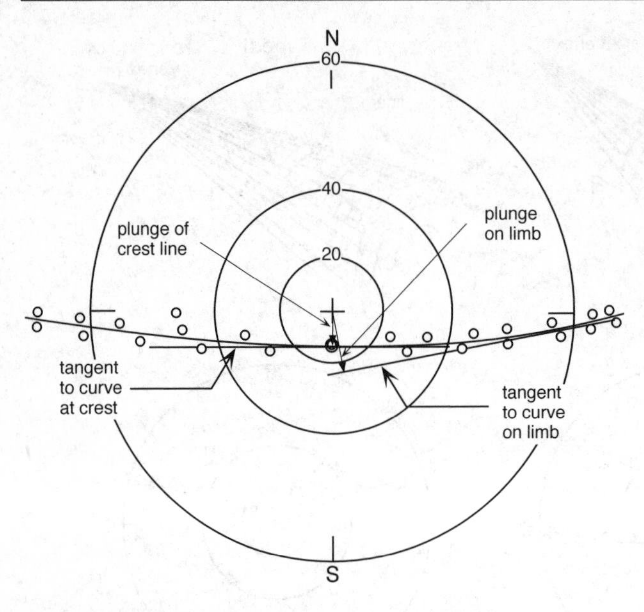

Fig. 4.6. Projection directions in a type I conical fold

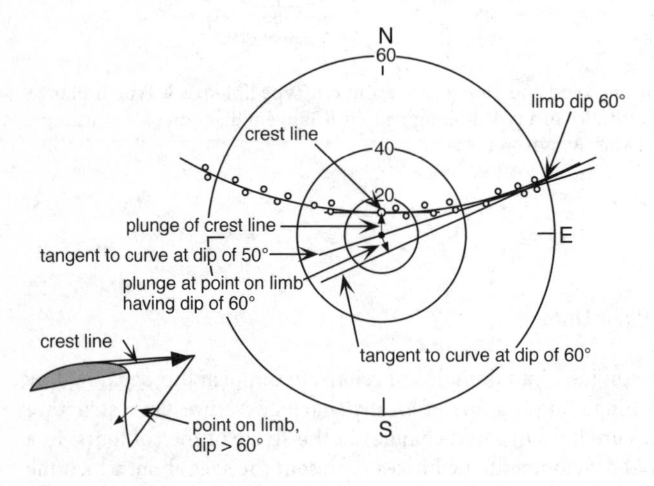

Fig. 4.7. Projection directions in a type II conical fold

The tangent diagram (Fig. 4.10) shows the dip-domain fold from Fig. 4.8 to be conical. The crest line is horizontal and the concave-to-the-south curvature of the line through the dip vectors indicates that the vertex of the cone is due south. This means that the crest of the structure is horizontal but that the fold terminates to the south, in accord with the geometry shown in Fig. 4.8. The hinge line between the two fold limbs is perpendicular to the straight line joining the two dip vectors (Fig. 4.10) and plunges 11° to the south.

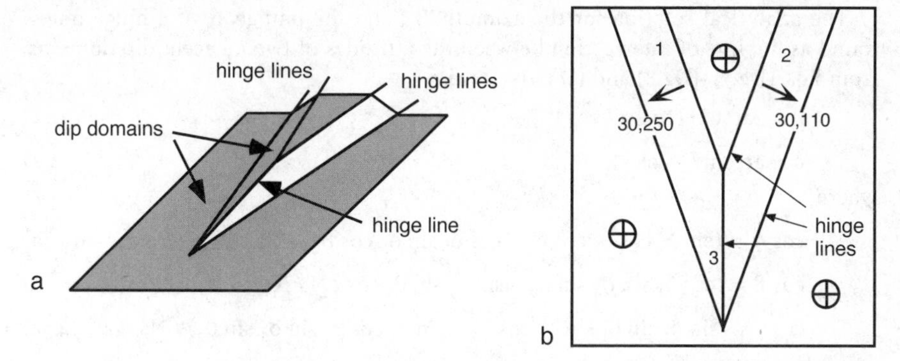

Fig. 4.8. Dip-domain fold. Hinge lines separate domains of constant dip. **a** Perspective diagram. **b** Map of bed attitudes and hinge lines

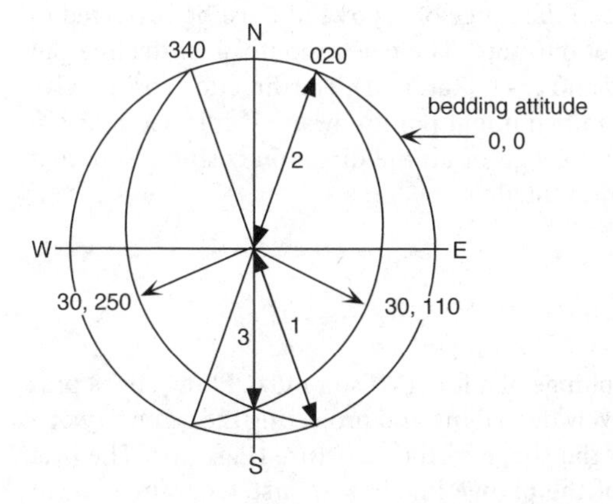

Fig. 4.9. Hinge line orientations for Fig. 4.8b, found from the bed intersections on a lower-hemisphere equal-area stereogram. The primitive *circle* that outlines the stereogram is the trace of a horizontal bed. One of the bedding dip domains has an attitude of 30, 250 and the other of 30, 110. *Numbered arrows* give orientations of hinge lines. Hinge line *1* has the direction 0, 340; hinge line *2* is 0, 020; hinge line *3* is 11, 180

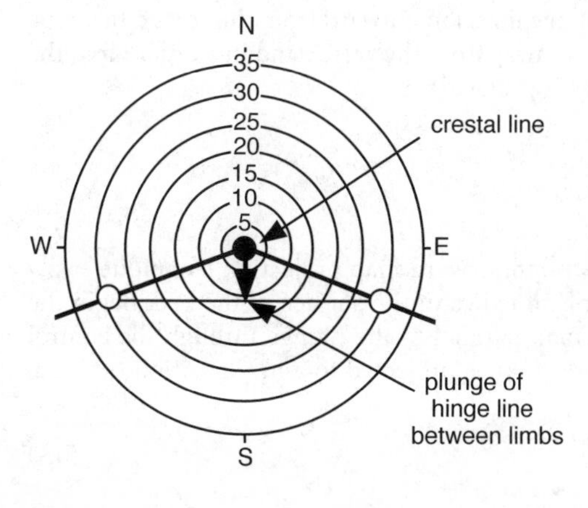

Fig. 4.10. Tangent diagram of the conical dip-domain fold in Fig. 4.8. *Circles* give attitudes of the three dip domains present in the fold

The analytical solution for the azimuth, θ', and the plunge, δ, of a hinge line is found as the line of intersection between the attitudes of two adjacent dip domains, from Eqs. (D2.7)–(D2.8) and (D4.9)–(D4.11):

$$\theta' = \arctan(\cos\alpha / \cos\beta), \tag{4.1}$$

$$\delta = \arcsin(-\cos\gamma), \tag{4.2}$$

where

$$\cos\alpha = (\sin\delta_1 \cos\theta_1 \cos\delta_2 - \cos\delta_1 \sin\delta_2 \cos\theta_2) / N; \tag{4.3a}$$

$$\cos\beta = (\cos\delta_1 \sin\delta_2 \sin\theta_2 - \sin\delta_1 \sin\theta_1 \cos\delta_2) / N; \tag{4.3b}$$

$$\cos\gamma = (\sin\delta_1 \sin\theta_1 \sin\delta_2 \cos\theta_2 - \sin\delta_1 \cos\theta_1 \sin\delta_2 \sin\theta_2) / N; \tag{4.3c}$$

$$N = [(\sin\delta_1 \cos\theta_1 \cos\delta_2 - \cos\delta_1 \sin\delta_2 \cos\theta_2)^2 + (\cos\delta_1 \sin\delta_2 \sin\theta_2 - \sin\delta_1 \sin\theta_1 \cos\delta_2)^2 + (\sin\delta_1 \sin\theta_1 \sin\delta_2 \cos\theta_2 - \sin\delta_1 \cos\theta_1 \sin\delta_2 \sin\theta_2)^2]^{1/2}, \tag{4.4}$$

and the azimuth and plunge of the first plane is θ_1, δ_1 and of the second plane is θ_2, δ_2. The value θ' given by Eq. (4.1) will be in the range of ± 90° and must be corrected to give the true azimuth over the range of 0 to 360°. The true azimuth, θ, of the line can be determined from the signs of cos α and cos β (Table 2.1). The direction cosines give a directed vector. The vector so determined might point upward. If it is necessary to reverse its sense of direction, reverse the sign of all the direction cosines. Note that division by zero in Eq. (4.1) must be prevented.

4.3 Plunge Lines

A plunge line is a line parallel to the plunge of a fold (Wilson 1967). Plunge lines provide an effective means for quantitatively describing and projecting the geometry of a fold. A series of plunge lines defines the shape of the structure (Fig. 4.11). The best method for finding the orientation of the plunge line is with a stereogram, tangent diagram or equations described in Section 4.2.

The plunge line for a cylindrical fold is parallel to the fold axis and is the same for every point within the fold (Fig. 4.11). Each plunge line in a conical fold has its own bearing and plunge (Fig. 4.12). The plunge lines fan outward from the vertex. In a type I conical fold (Fig. 4.12a,b) the plunge is away from the vertex and in a type II fold the plunge is generally toward the vertex (Fig. 4.12c,d).

4.3.1 Projection Along Plunge Lines

Plunge lines provide the basis for a straightforward map projection technique easily done by hand (De Paor 1988). A plunge line lies in the surface of the bed. Begin the projection by drawing a line on the map parallel to the plunge through the control point to be projected. A point (1 in Fig. 4.13), is projected to a new position (point 2, Fig. 4.13) with the equation:

$$v = h \tan\phi, \tag{4.5}$$

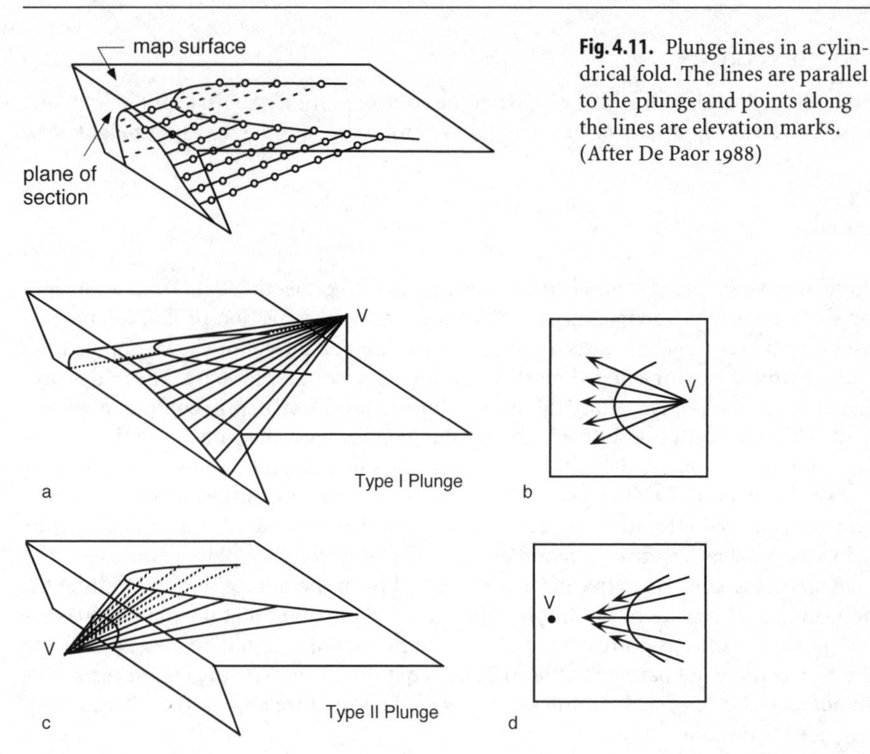

Fig. 4.11. Plunge lines in a cylindrical fold. The lines are parallel to the plunge and points along the lines are elevation marks. (After De Paor 1988)

Fig. 4.12. Plunge lines in conical folds. *V* Fold vertex. **a** Perspective view of type I fold. **b** Map view of plunge lines in type I fold. *Arrows* point down the plunge direction. **c** Perspective view of type II fold. **d** Map view of plunge lines in type II fold. *Arrows* point down the plunge direction

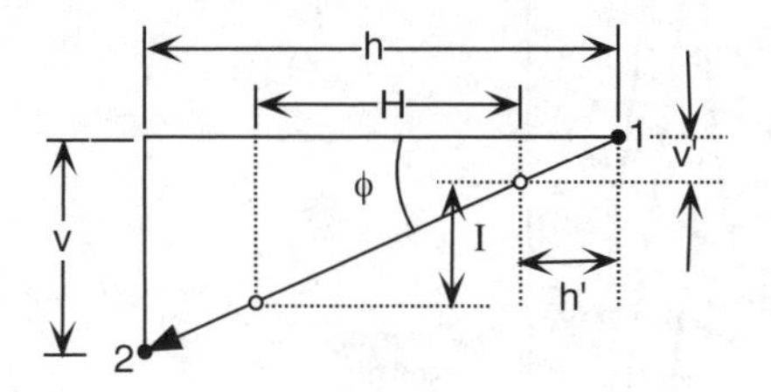

Fig. 4.13. Projection along plunge in a vertical cross section. The projection is parallel to plunge along the plunge line from point *1* to point *2*. *Open circles* are spot heights along the plunge line. For explanation of symbols, see text (Sect. 4.3.1)

where h = the horizontal distance on the map in the plunge direction, v = the vertical distance from the elevation of point 1 to its projection at point 2, and ϕ = the plunge angle. It is assumed that the plunge is constant between points 1 and 2. Starting from the known elevation of the control point, mark spot heights spaced according to

$$H = I / \tan \phi, \tag{4.6}$$

where H = horizontal spacing of points, I = contour interval, and ϕ = plunge. If the control point is not at a spot height, the distance from the control point to the first spot height is

$$h' = H\,v' / I, \tag{4.7}$$

where h' = the horizontal distance from the control point to the first spot height and, v' = the elevation difference between the control point and the first spot height.

4.3.2 Example

Projection along plunge lines is particularly suited to projecting data from an irregular surface, such as a map, onto a surface, such as a cross section or fault plane, that itself can be represented as a structure-contour map. Figure 4.14 shows plunge lines derived from a map of a folded marker horizon on a topographic base. The fold is projected south, up plunge, along the plunge lines onto the structure contour map of a fault. The intersection points where the plunge lines have the same elevation as the fault contours are marked and then connected by a line that represents the trace of the marker horizon of the fault plane (Fig. 4.14). A curved fault surface would be represented by curved structure contours. This method is modified for a conical fold by using the appropriate direction and amount of plunge for each plunge line.

A structure contour map can be constructed from the plunge lines by joining the points of equal elevation (Fig. 4.15). Figure 4.15 demonstrates that the plunge lines are not parallel to the structure contours and that projections should be made parallel to the plunge lines, not parallel to the structure contours. The structure contours provide an additional cross check on the geometry of the structure and on the internal consistency of the data.

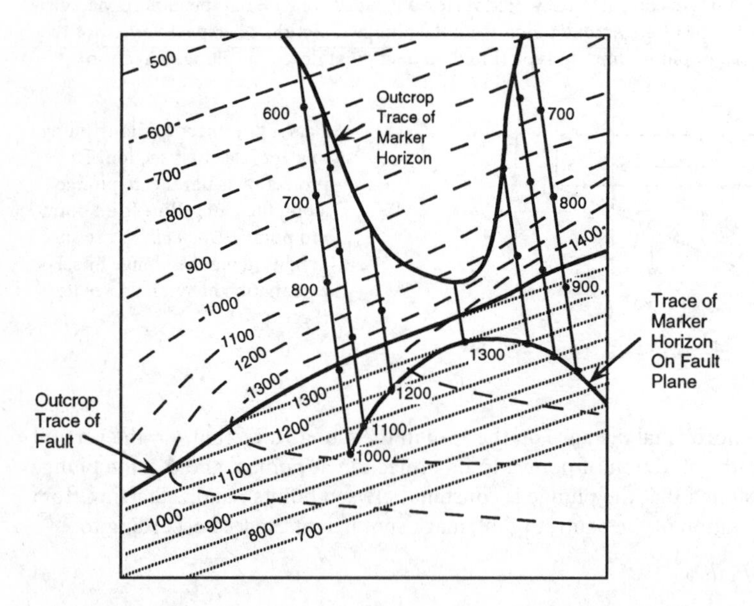

Fig. 4.14. Projection of a marker horizon to a fault plane along plunge lines. *Dashed lines* are topographic contours above sea level. *Dotted lines* are subsurface structure contours on the fault. Plunge lines are *solid* and marked by spot elevations. (After De Paor 1988)

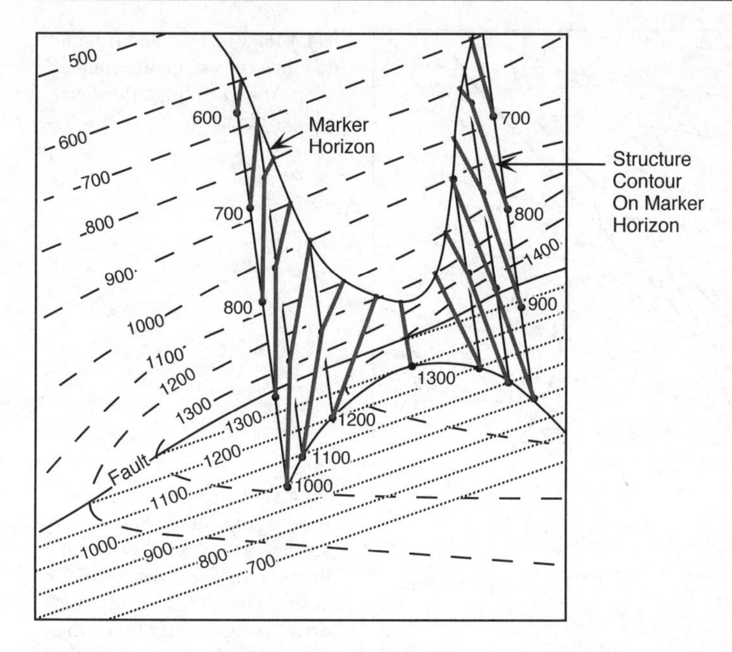

Fig. 4.15. Structure contours on a marker horizon (*wide gray lines*) added to Fig. 4.14

4.4 Crest and Trough

A consistent definition of the fold trend, applicable to cylindrical or conical folds, is the orientation of the crest or trough line (Fig. 4.16). In both cylindrical and conical folds (figs. 4.3, 4.5) the crest line is a line on a folded surface along the structurally highest points on successive cross sections (Dennis 1967). The trough line is the trace of the structurally lowest line on successive cross sections. In cross section, the crest and trough traces are the loci of points where the apparent dip changes direction. In cylindrical folds the crest and trough lines are parallel to the fold axis and to each other, but in conical folds the crest and trough lines are not parallel. Crest and trough surfaces connect the crest and trough lines on successive horizons. The trace of a crest or trough surface is the line of intersection of the surface with some other surface such as the ground surface or the plane of a cross section. The crests and troughs of folds are of great practical importance because they are the positions of structural traps. Light fluids will migrate toward the crests and heavy fluids will migrate toward the troughs.

The U-shaped trace of bedding made by the intersection of a plunging fold with a gently dipping surface, such as the surface of the earth, is called a fold nose. Originally a nose referred only to a plunging anticline (Dennis 1967; Bates and Jackson 1987) and a chute is the corresponding feature of a plunging syncline (Dennis 1967). Today, common usage refers to both synclinal and anticlinal fold noses. The term nose is also applied to the anticlinal or synclinal bend of structure contours on a single horizon (Fig. 4.16). The dip of bedding at the crest or trough line is the plunge of the line.

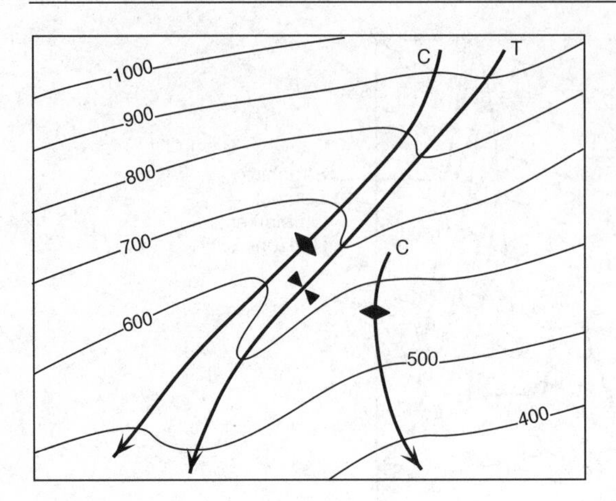

Fig. 4.16. Crest (*C*) and trough (*T*) lines on a structure contour map. Arrows point in the down-plunge direction

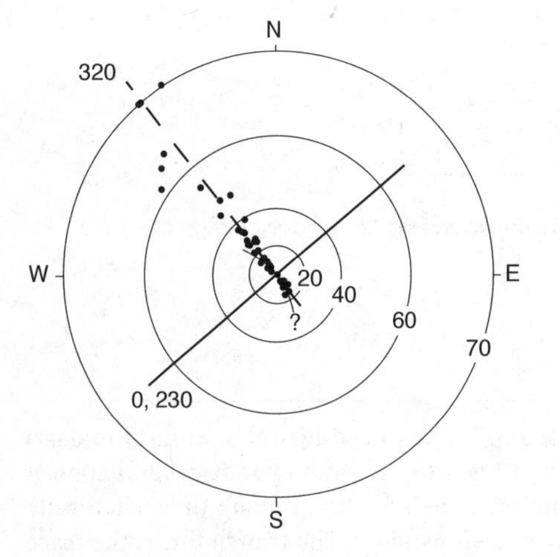

Fig. 4.17. Tangent diagram of bedding dips in the Blount Springs area (from Fig. 3.1 and the adjacent area). The *heavy dashed line* is the best-fit cylindrical fold geometry for all the dips; the fold axis plunges 0, 230. The *thin curved line* with a *question mark* represents possible conical folding at low dips

If the orientation of the crest or trough lines can be determined from bedding attitudes, then the orientation can be used to add trend control to a structure contour map. The spacing of contours parallel to the trend of the crest or trough is given by Eq. (4.6). From a structure contour map, the plunge of the crest or trough line is

$$\phi = \arctan (I / H), \quad (4.8)$$

where ϕ = plunge, I = contour interval, and H = map spacing between the structure contours along the plunge.

The Sequatchie anticline (Fig. 3.1) provides an example of fold plunge interpreted using a tangent diagram of the bedding attitudes (Fig. 4.17). The best fit cylindrical fold geometry for the whole map is indicated by the thick lines and shows the anticline to have a horizontal fold axis that trends southwest (230°). The regional geology clearly indicates that the main anticline maintains a relatively constant geometry for a long distance along its axis, as expected for a cylindrical fold, and has a plunge to the southwest of less than 1°. As is typical for real structures, there is a fair amount of scatter of the data points. The scatter can be caused by measurement error (usually small), by sedimentologic scatter (can be large in crossbedded units), or by minor folds with trends oblique to the major fold. Both the Pottsville sandstone and the Bangor limestone contain large-scale crossbedding, making sedimentologic scatter a strong possibility in this area. In small outcrops of the coarse-grained lithologies it can be difficult to determine whether a bed surface is that of a crossbed or of an originally horizontal bed surface. This is also a common problem with dipmeter data.

At low dips the data suggest the possibility of type 2 conical folding (thin line, Fig. 4.17). The crest/trough line should be in about the same direction as the axis of the major structure and the plunge should be near zero. The composite structure contour map (Fig. 3.20) shows second-order folding on the gently dipping backlimb. The dips obtained from outcrops could be sampling these folds.

Revisions are made to the structure contour map of the Blount Springs area (Fig. 4.18) to incorporate the information provided from the interpretation of the tangent diagram. The average plunge of the Sequatchie anticline (Fig. 4.17) is zero and so the structure contours should not close to form a nose along the axis. The revised map (Fig. 4.18) removes the connection between the 800-ft contours on the fold limbs. Because the second-order folds are suggested by both the previous structure contour map (Fig. 3.20) and the tangent diagram, they are accepted as being real and their crest and trough traces are shown. Most of the low dips that define these folds come from the central portion of the map and so in that area the axial traces are drawn to trend as close to that of the Sequatchie anticline as possible. The difference in crest/trough trace directions indicated by the map and the tangent diagram may be significant and should be confirmed or revised by more detailed mapping. Finally, the structure contours are smoothed to present a more traditional appearance.

4.5 Axial Surfaces

The axial surface geometry is important in defining the complete three-dimensional geometry of a fold (Section 1.5) and in the construction of cross sections (Chap. 7).

4.5.1 Characteristics

The axial surface of a fold is defined as the surface that contains the hinge lines of all horizons in the fold (Fig. 4.19; Dennis 1967). The axial trace is the trace of the axial surface on another surface such as on the earth's surface or on the plane of a cross section. The orientation of an axial surface within a fold hinge is related to the layer thicknesses on the limbs. If, for example, the layers maintain constant thickness, the axial surface bisects the hinge. An axial surface may have one or both of the additional attributes: (1) it may divide the fold into two limbs (Fig. 4.19), and (2) it may be the surface at which the sense of layer-parallel shear reverses direction (Fig. 4.19). In a fold with only one axial surface, the axial surface necessarily divides the fold into two limbs; however, many folds have multiple axial surfaces, none of which bisect the whole fold (Fig. 4.20). The relationship between a fold hinge and the sense of shear on either side of it depends on the movement history of the fold. The movement history of a structure is called its kinematic evolution and a model for the evolution is called a kinematic model. According to a fixed-hinge kinematic model, the hinge lines are

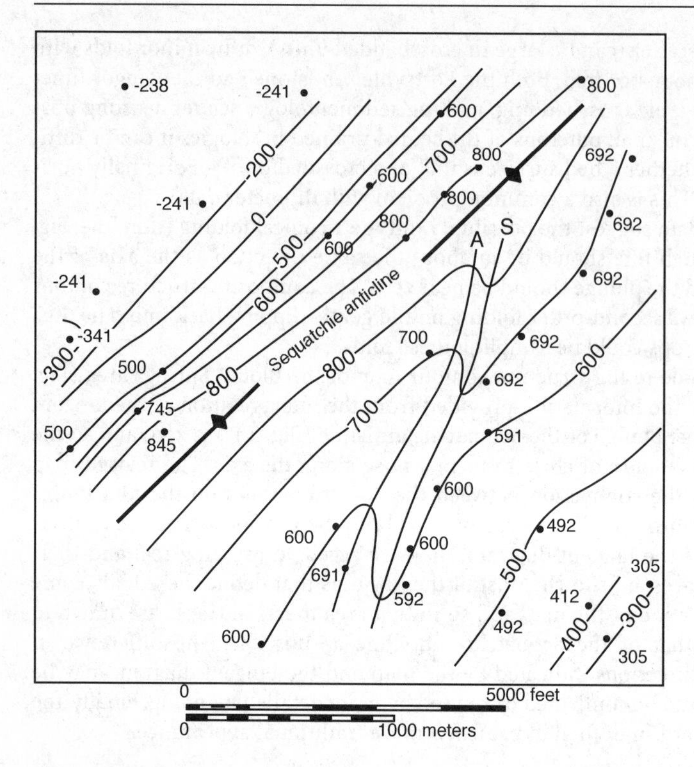

Fig. 4.18. Revised composite structure contour map of the Sequatchie anticline in the Blount Springs area. Contours on the top of the Mtfp. Contours and spot elevations are in feet. The *wide line* represents the crest of Sequatchie anticline; *A* crest of second-order anticline; *S* trough of second-order syncline

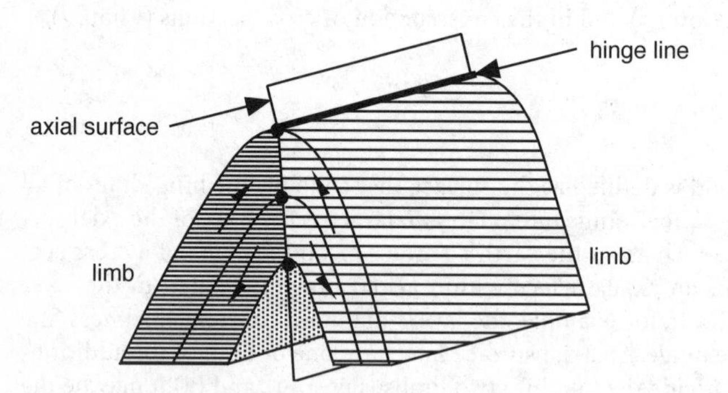

Fig. 4.19. Characteristics that may be associated with an axial surface. The axial surface contains the hinge lines of successive layers (the defining property). The surface may be the boundary between fold limbs (different *patterns*) and may be the plane across which the sense of layer-parallel shear (*double arrows*) reverses direction

fixed to material points within the layer and form the rotation axes where the sense of shear reverses, as shown in Fig. 4.19. Other kinematic models do not require the sense of shear to change at the hinge (Fig. 4.20b).

Dip-domain folds commonly have multiple axial surfaces (Fig. 4.20) which form the boundaries between adjacent dip domains. These axial surfaces are not likely to have the additional attributes described in the previous paragraph. The folds are not split into two limbs by a single axial surface (except in the central part of Fig. 4.20a). In the fixed hinge kinematic model, the sense of shear does not necessarily change across an axial surface although the amount of shear will change (i.e., axial surface 4, Fig. 4.20b). Horizontal domains may be unslipped and so an axial surface may separate a slipped from an unslipped domain (i.e., axial surfaces 2 and 3, Fig. 4.20b), not a change in the shear direction.

In the strict sense, a round-hinge fold does not have an axial surface because it does not have a precisely located hinge line. In a round-hinge fold it must be decided what aspect of the geometry is most important before the axial surface can be defined (Stockwell 1950; Stauffer 1973). For the purpose of determining the three-dimensional fold geometry, the virtual axial surface in a rounded fold is defined here as the surface which contains the virtual hinge lines. In relatively open folds, virtual hinge lines can be constructed as the intersection lines of planes extrapolated from the adjacent fold limbs (Fig. 4.21a). In a tight fold the extrapolated hinges are too far from the layers to be of practical use. In a tight fold (Fig. 4.21b) the virtual hinges can be defined as the centers of the circles (or ellipses) that provide the best fit to the hinge shapes, after the method of Stauffer (1973). Defined in this fashion, the axial surface divides the hinge into two limbs, but does not necessarily divide the fold symmetrically nor is it necessarily the surface where the sense of shear reverses. The axial surface, whether defined by actual or virtual fold hinges, does not necessarily coincide with the crest surface (Fig. 4.22) or the trough surface.

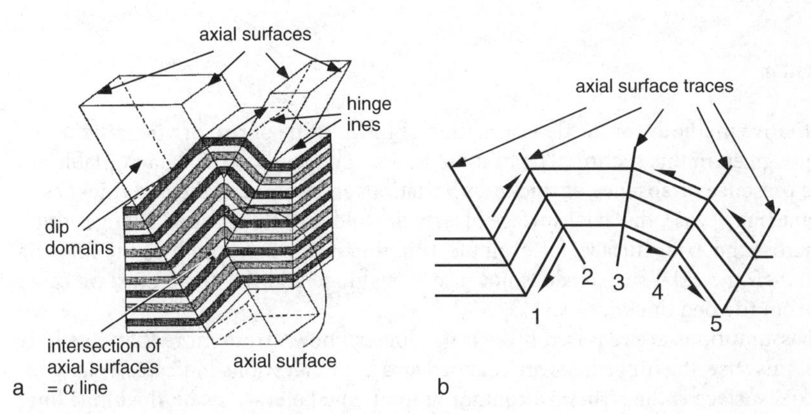

Fig. 4.20. Dip-domain folds. **a** Non-plunging fold showing hinge lines, axial surfaces, and axial surface intersection lines (modified from Faill 1969). Axial surface intersection lines are horizontal and parallel to the fold hinge lines. **b** *Half arrows* indicate the sense of shear in dipping domains. The horizontal domains did not slip. Axial surfaces are *numbered*

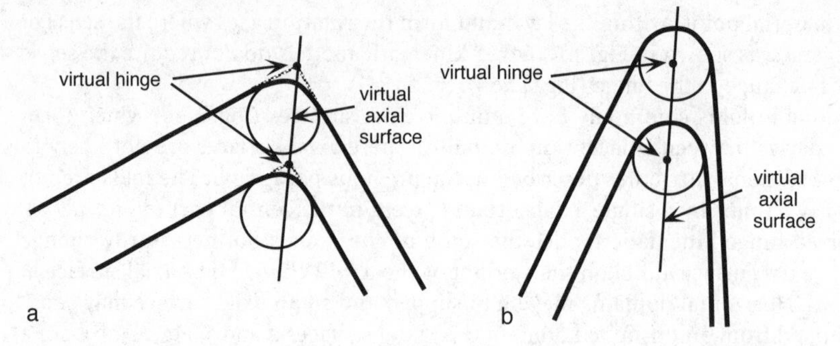

Fig. 4.21. Virtual hinges and virtual axial surfaces in cross sections of round-hinge folds. **a** Virtual hinges in an open fold found by extrapolating fold limbs to their intersection points. **b** Virtual hinges in a tight fold found as centers of circles tangent to the surfaces in the hinge. (After Stauffer 1973)

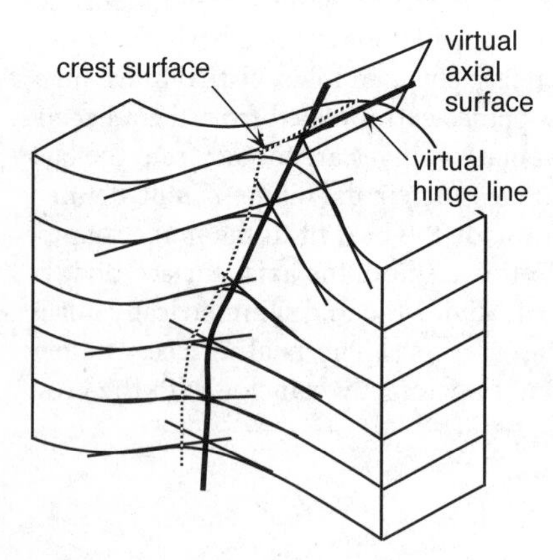

Fig. 4.22. A fold in which the crest surface does not coincide with the axial surface

4.5.2 Orientation

Quantitative methods for the determination of the attitude of real or virtual axial surfaces are given in this section. The best method will depend on the data available and on the presence or absence of thickness variations in the fold. If the bed thickness is constant (Fig. 4.23a), the axial surface bisects the fold hinge. If the bed changes thickness across the axial surface, for example, thinning in the steep limb of the fold (fig, 4.23b), then the axial surface does not bisect the hinge and the angle must be calculated from the bed thicknesses.

No assumptions are required if the hinge line is known from more than one horizon. In this case, the hinge lines are mapped and their elevations indicated (Fig. 4.24). The axial surface is the structure contour map of equal elevations on the hinge lines. The attitude of the axial surface can be found from the structure contours using Eq. (2.5).

The relationship between axial surface geometry and bed thickness is illustrated in the cross section perpendicular to the hinge line. If a bed maintains constant thick-

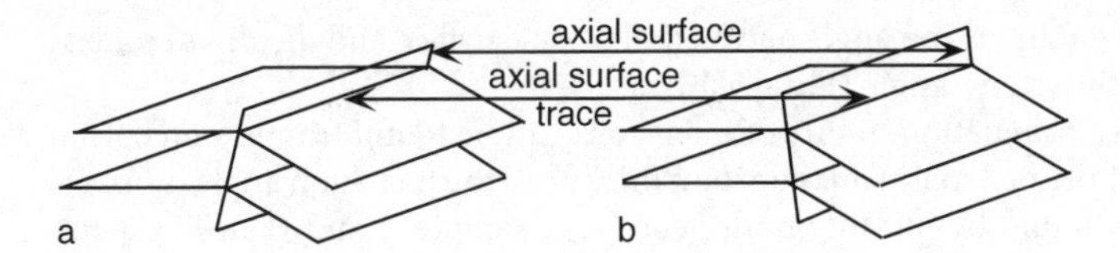

Fig. 4.23. Axial surface between two dip domains. In both **a** and **b**, orientation of the axial surface trace on bedding is the same, but the axial surface orientations are different in cross section. **a** Constant bed thickness. **b** Bed that is thinner in the steep limb

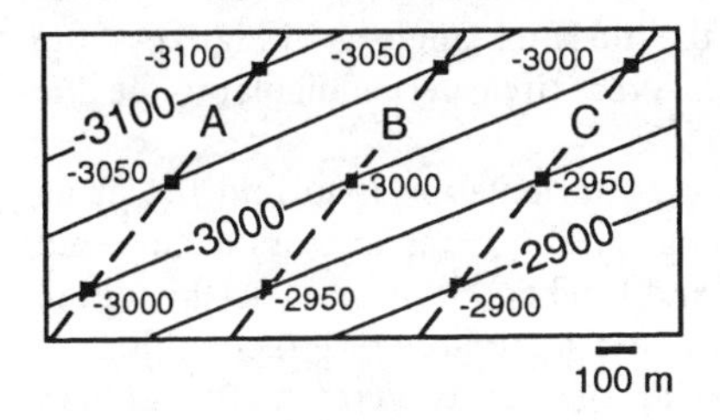

Fig. 4.24. Structure contour map of an axial surface. *Dashed lines* are hinge lines on three different horizons, *A*, *B*, and *C*. *Solid lines* are contours on the axial surface. Contours are in meters

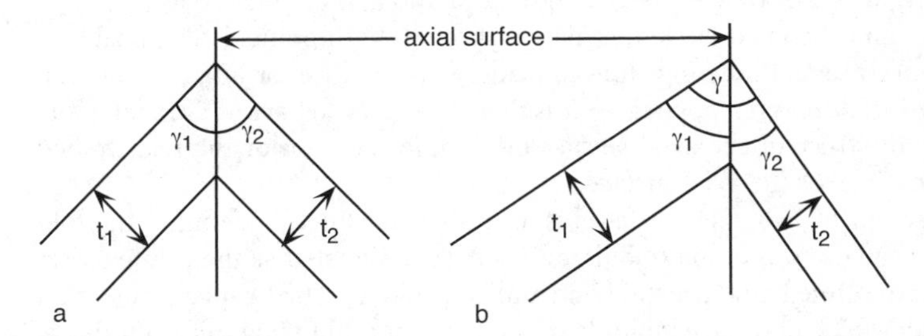

Fig. 4.25. Bed thickness in dip-domain fold hinges. The cross sections are perpendicular to the hinge line. **a** Constant bed thickness. **b** Thinning in one limb

ness across the hinge, the axial surface bisects the hinge. This is demonstrated by the following relationship, based on the geometry of Fig. 4.25a (from Eq. D4.22):

$$t_1 / t_2 = \sin \gamma_1 / \sin \gamma_2, \tag{4.9}$$

where t_1 and t_2 = the bed thicknesses in each limb and γ_1 and γ_2 = the corresponding partial interlimb angles. If bedding maintains constant thickness, $t_1 = t_2$ and so $\gamma_1 = \gamma_2$, and the axial surface bisects the hinge.

Some folds change thickness across the axial surface, typically becoming thinner in the steep limb (Fig. 4.25b; Gill 1953). If bed thickness is not constant across the axial surface, the method for determination of the orientation of the axial surface must be modified because it cannot bisect the hinge. Given the thicknesses and the whole interlimb angle, from Eq. (D4.24):

$$\gamma_2 = \arctan [t_2 \sin \gamma / (t_1 + t_2 \cos \gamma)], \tag{4.10}$$

where γ_2 = the partial interlimb angle between the axial surface and the dip of limb 2, and γ = the whole interlimb angle (Fig. 4.25b).

The axial surface orientation in three dimensions can be found from a stereogram plot of the limbs. If the bed maintains constant thickness, the interlimb angle is bisected. The bisector is found by plotting both beds (for example, 2 and H in Fig. 4.26), finding their line of intersection (I), and the great circle perpendicular to the intersection (dotted line). The obtuse angle between the beds is bisected by the double arrows. The great circle plane of the axial surface (shaded) goes through the bisection point and the line of intersection of the two beds. If bed thickness is not constant across the hinge, the partial interlimb angle is found from Eq. (4.10). To find the axial surface on a stereogram, instead of bisecting the interlimb angle as in Fig. 4.25a, the appropriate partial angle is marked off along the great circle perpendicular to the line of intersection of the bedding dips.

Another method to determine the axial surface orientation is to find the plane through two non-parallel lines: the trace of a hinge line on a map or cross section, and either the fold axis or the crest line (Rowland and Duebendorfer 1994). In the fold in Fig. 4.8b, for example, hinge 3 plunges south and the surface trace of the hinge is north-south and horizontal and so the axial surface is vertical and strikes north–south. In the general case, the axial trace and fold axis are plotted as points on the stereogram and the great circle that passes through both points is the axial surface. If the axial trace and the hinge line coincide, as is the case for hinges 1 and 2 in Fig. 4.8b, this procedure is not applicable. It is then necessary to have additional information, either the trace of the axial surface on another horizon or the relative bed thickness change across the axial surface.

The intersection of two axial surfaces is an important line for defining the fold geometry, especially in dip-domain folds, and is here designated as the α line (α for axial surface). An α line has a direction and, unlike a fold axis, has a specific position in space. Angular folds may have multiple α lines. In a non-plunging fold an α line is coincident with a hinge line (Fig. 4.20a), but in a plunging fold (Fig. 4.27) the hinge lines and the α lines are neither parallel nor coincident.

The orientation of the α line is found by first finding the axial surfaces, and then their line of intersection. The stereogram solution is to draw the great circles for both axial surfaces and locate the point where they cross (Fig. 4.28), which is the orientation of the

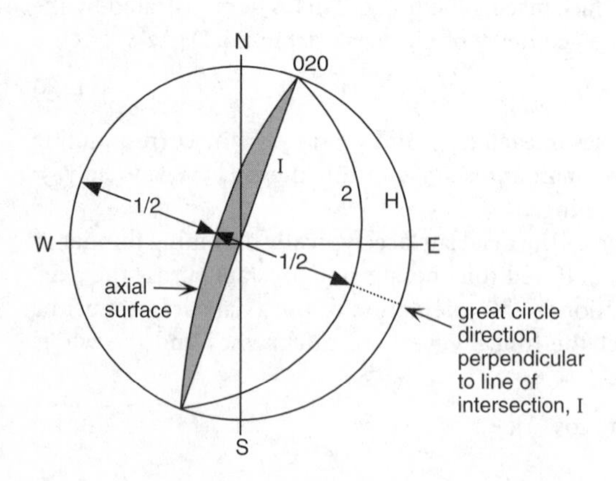

Fig. 4.26. Determination of the axial surface orientation using a stereogram. Angle between two constant thickness domains, one horizontal (bed *H*) and one 30, 110 (bed 2). The axial surface (*shaded great circle*) bisects the angle (*double arrows*) between bed 2 and the horizontal bed. The axial surface attitude is 75, 290. Lower-hemisphere, equal-area projection

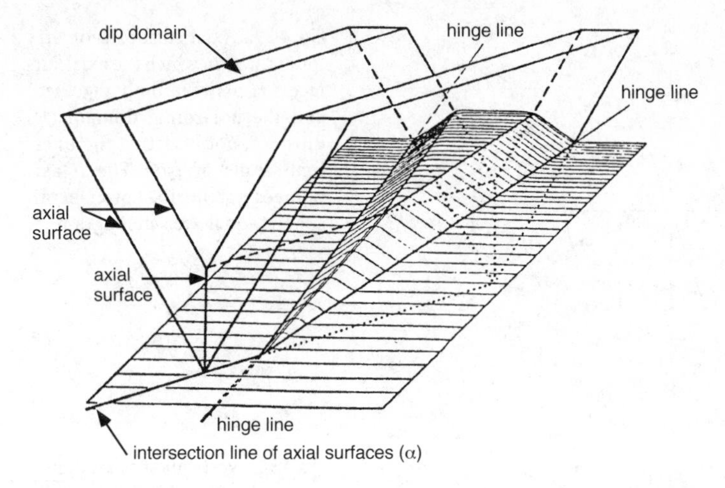

Fig. 4.27. Plunging dip-domain fold. The axial surface intersection line (*a*) plunges in a direction opposite to that of the hinge lines. (After Faill 1973)

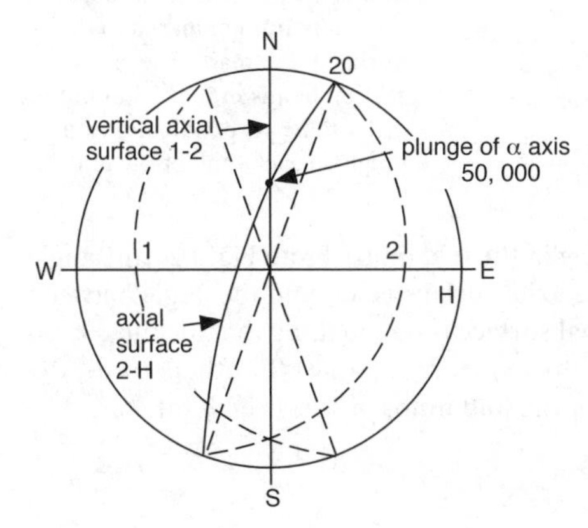

Fig. 4.28. The intersection line between axial surfaces. Axial surface *1-2* bisects bed *1* (30, 250) and bed *2* (30, 110). Axial surface *2-H* bisects bed *2* and bed *H* (0, 000). The line of intersection of axial surfaces (50, 000) is the α line. Lower-hemisphere, equal-area stereogram

line of intersection. The analytical solution for the α line is that for the line of intersection between two planes (Eq. 4.1–4.4), in which the two planes are axial surfaces.

4.5.3
Predicting Thickness Changes

The position of the axial surfaces in the hinge of a fold can be used to predict bed thickness changes. The axial surface orientations in Fig. 4.28 were derived by bisecting the limb attitudes, that is, with the assumption of constant bed thickness. The α axis for this geometry plunges 50° due north (Fig. 4.28). A geometry based on the same limb attitudes, but demanding bed thickness changes, is illustrated by the dip domains in Fig. 4.27. Here the axial surfaces that separate the flat top of the fold from the limbs dip approximately 60° and the corresponding α axis plunges 30° north, as shown in Fig. 4.29. The 60° dipping axial surfaces do not bisect the interlimb angle of

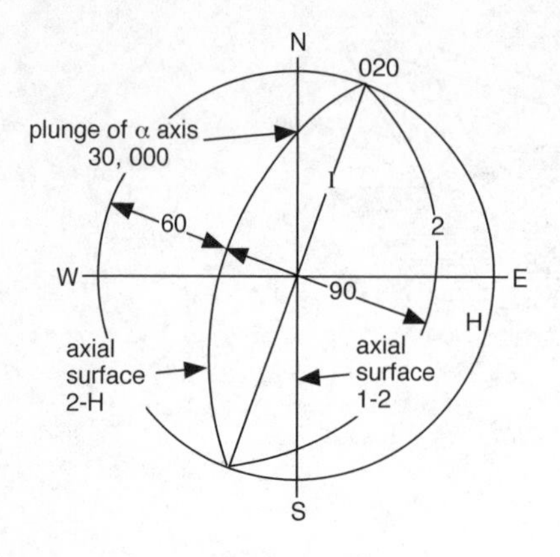

Fig. 4.29. Axial surface geometry for Fig. 4.27 in which the axial surface *2-H* between limb 2 (30, 110) and the horizontal domain (*H*) dips 60°, not bisecting the interlimb angle of 150°. The α axis plunges 30° north. Lower-hemisphere, equal-area stereogram

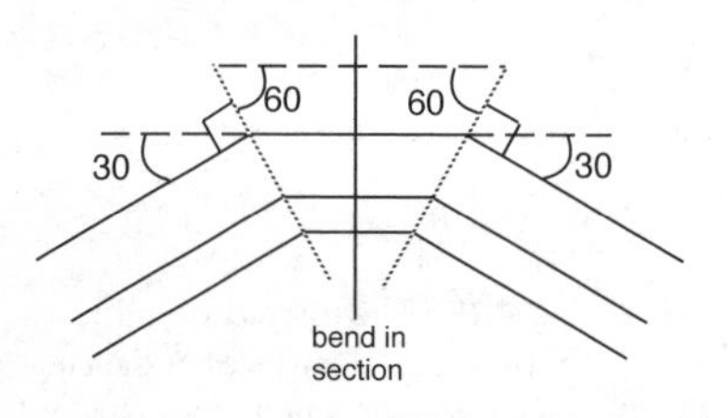

Fig. 4.30. Vertical cross section perpendicular to the horizontal hinge lines of Fig. 4.27. Axial surfaces (*dotted*) dip 60°. Bed thicknesses are 15% greater in the dipping domains compared to the horizontal domain. The cross section bends in the center in order to be perpendicular to the hinge on each limb of the fold

150°, implying thickness changes of bedding in the fold limbs. From Fig. 4.29, the angle between bedding on the fold limb and the axial surface is 90° and the angle between the horizontal crest of the fold and the axial surface is 60°, making the thickness ratio between folded and unfolded beds equal to 1.15 from Eq. (4.9). In other words, the geometry of Fig. 4.27 implies thickening of the fold limbs of 15% (Fig. 4.30).

4.6 Dip Sequence Analysis

The three-dimensional geometry of a structure can be determined from the bedding attitudes measured in a single well bore or on a traverse through a structure. The method of dip sequence analysis presented here was developed for the structural analysis of dipmeter logs by Bengtson (1981a) but is equally informative whether the traverse is down a well or along a stream. A major problem with dip data is the high noise content and the potential complexity of the structures to be interpreted. The dip sequence analysis techniques, called statistical curvature analysis techniques (SCAT or SCAT analysis) by Bengtson (1981a), are particularly good for extracting the structural signal from the noise. Using SCAT it is possible to determine the plunge of folds, the locations of fold axial surfaces, crests, and troughs, to infer the strike and dip directions of faults, and to separate regional fold trends from local fault trends. Techniques applicable to folds are given in this chapter; those for faults are in Chapter 5. The power of the technique derives from (1) the noise-reduction strategy of examining the data as dip components in both the strike and dip directions of folding and

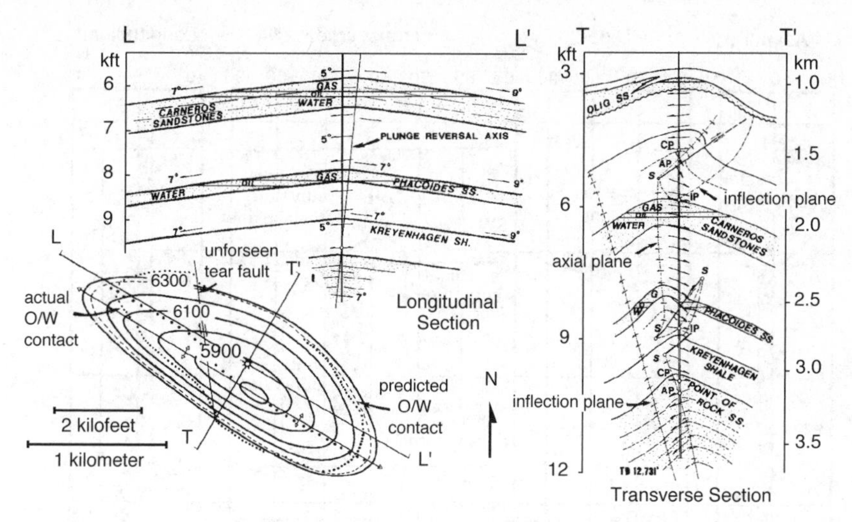

Fig. 4.31. Railroad Gap Field, California, predicted longitudinal and transverse cross sections and structure contour map on the top Carneros sandstone, based on the SCAT analysis of a single well at the crest of the anticline. (After Bengtson 1981a). *O/W* oil–water contact

(2) providing models for the SCAT responses of the geometry to be interpreted. Computer programs for the preparation of SCAT diagrams have been published by Elphick (1988).

The potential of the method is indicated by the interpretation of the Railroad Gap oil field on the basis of the SCAT analysis of a single, favorably located well (Fig. 4.31). SCAT analysis (Bengtson 1981a) was used to predict the structure on perpendicular cross sections from which the map was generated. The map view shows the close correspondence between the predicted and observed oil–water contact.

Multiple folds and faults can be present in a single well, as in the Railroad Gap field example (Fig. 4.32). The dipmeter interpretations are transformed into cross sections in the L and T directions from which the map (Fig. 4.31) was made. Such a complete interpretation is possible because the crest well penetrated both plunge directions of the doubly plunging fold. Without SCAT analysis such a dipmeter log might appear to be hopelessly complex.

4.6.1 Curvature Models

Figure 4.33 illustrates the basic curvature geometries. The first step in the analysis is to differentiate a monoclinal dip sequence from a fold. This is accomplished with an azimuth histogram and/or with a tangent diagram. An azimuth histogram is a plot of the azimuth of the dip versus the amount of the dip (Fig. 4.34). The natural variation of dips around a monoclinal dip gives a horizontal distribution of noise. Thus a zero true dip gives a small false positive average on the azimuth histogram because all dips are recorded as positive (Fig. 4.34a). As the dip increases the dips form point concentrations that become better defined as the true dip increases (Fig. 4.34b,c). Non-plunging and uniformly plunging folds give vertical concentrations of points corresponding

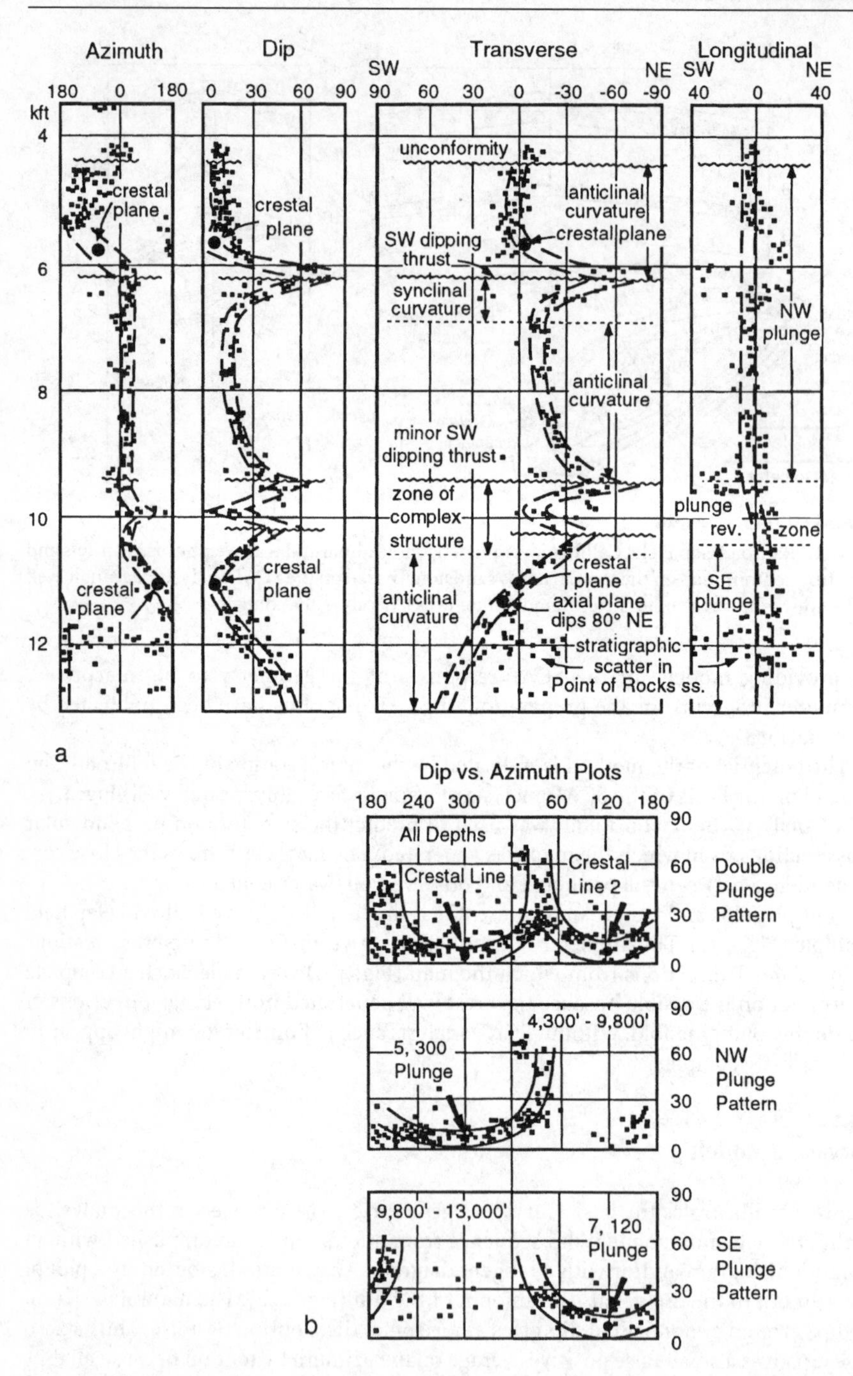

Fig. 4.32. SCAT plots for the discovery well of Railroad Gap Field, California. For the map and cross sections see Fig. 4.31. **a** Dip component vs. depth plots. **b** Dip vs. azimuth plots. (After Bengtson 1981a)

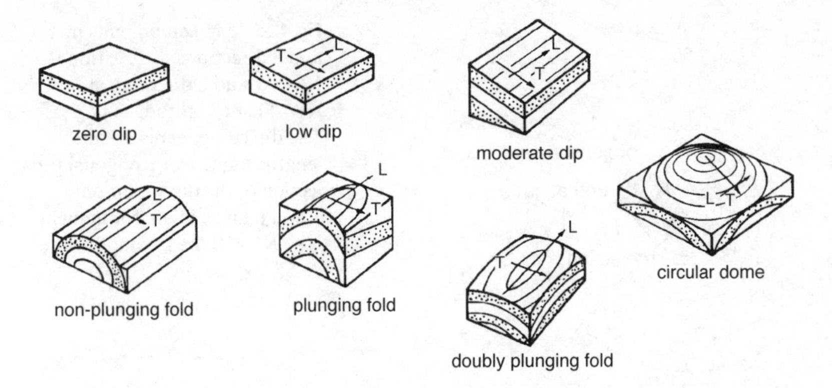

Fig. 4.33. Models of structural curvature geometries. (Bengtson 1981a). *L* Longitudinal; *T* transverse directions

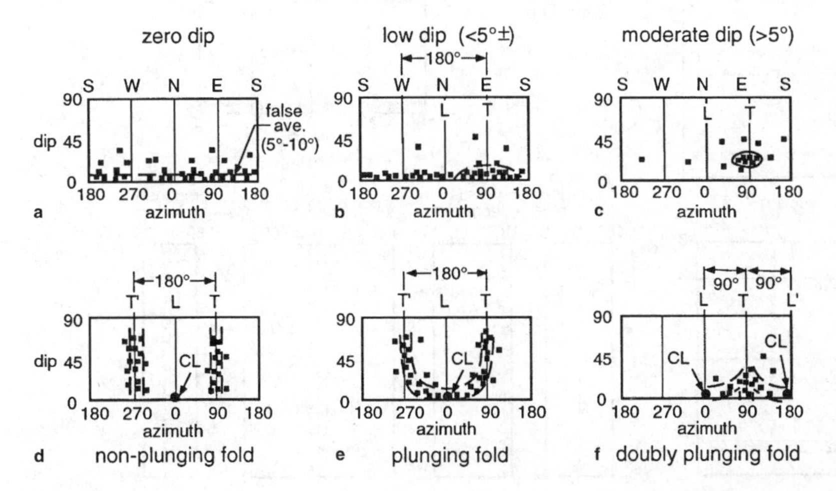

Fig. 4.34. Dip vs. azimuth patterns corresponding to the models of Fig. 4.33. *CL* crestal line. (After Bengtson 1981a)

to the limbs (Fig. 4.34d,e) and a doubly plunging fold produces an arrow-head-shaped distribution of points (Fig. 4.34f). On a tangent diagram a monocline plots as a diffuse point concentration of dips, and a fold (figs. 4.4 ,4.5) will produce a linear or curvilinear concentration of points.

4.6.2 Dip Components

The next step of the analysis is to determine the dip component in the transverse direction (T = regional dip) and the longitudinal direction (L = regional strike) which is at right angles to it. The T and L directions are found from a dip vs. azimuth histogram (Fig. 4.34) or from the plot of bedding on a tangent diagram. For monoclinal dip, the center of the point concentration on the dip-azimuth histogram is the T direc-

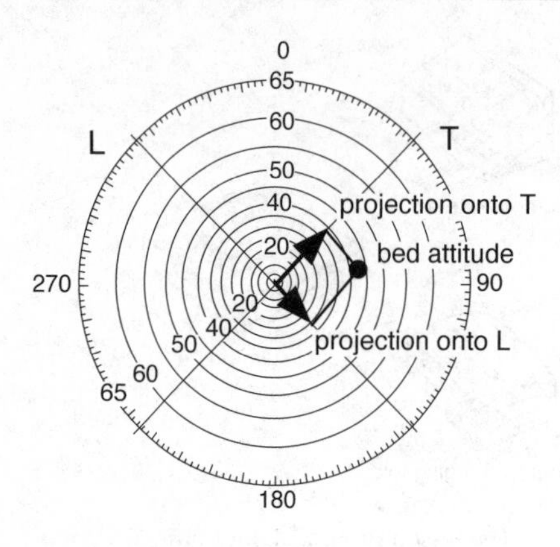

Fig. 4.35. Dip components in T and L directions. T direction is NE–SW and L direction is NW–SE. Bed attitude is 55, 082. The dip components are the lengths found by orthogonal projection of the dip vector onto the T and L lines. The T component is 50 NE and the L component is 40 SE

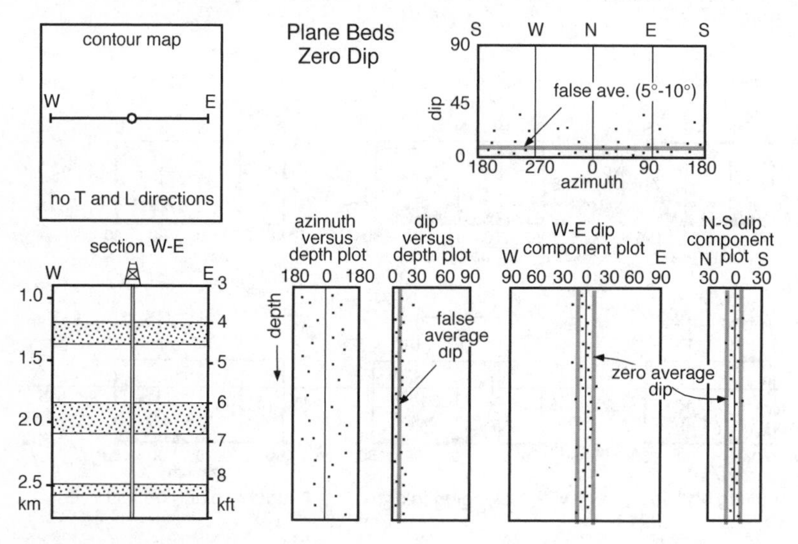

Fig. 4.36. Model map, cross section and SCAT plots for zero dip. (After Bengtson 1981a)

tion and the L direction is 90° away from it (Fig. 4.34b,c). For a fold, the center of the limb concentrations on an azimuth histogram is the T direction and the midpoint between the concentrations is the L direction (Fig. 4.34d–f). On a tangent diagram (Figs. 4.3, 4.5), the orientation of the crest (or trough) line is the L direction and the T direction is at right angles to the crest (or trough) line. The dip components are the apparent dips in the T and L directions. They are found as the projections of the dip onto the T and L axis lines (Fig. 4.35). Histograms of the various dip components versus distance (or depth) are then prepared.

The dip component plots are the primary noise reduction strategy. For zero dip the azimuth of the dip is random (actually stratigraphic scatter) and the dip amount shows a false positive average (Fig. 4.36). As the amount of homoclinal dip increases (Figs. 4.37, 4.38) the concentration of points become sharper. The component plots for zero dip show the correct zero average (Fig. 4.36). Low and moderate planar dips (Figs.

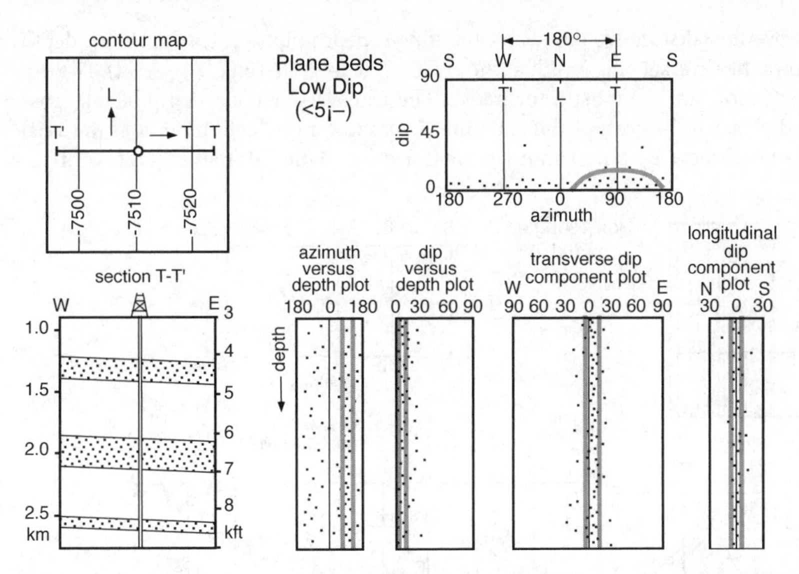

Fig. 4.37. Model map, cross section and SCAT plots for low monoclinal dip. (After Bengtson 1981a)

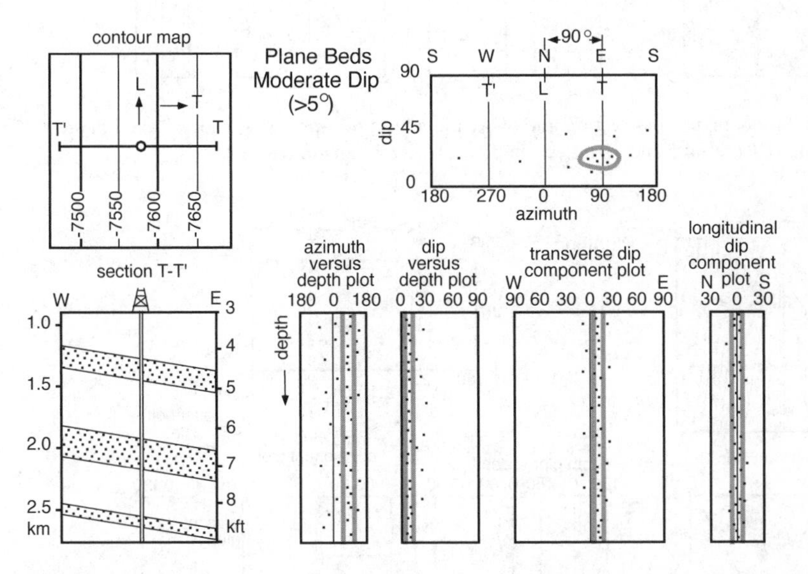

Fig. 4.38. Model map, cross section and SCAT plots for moderate to steep monoclinal dip. (After Bengtson 1981a)

4.37, 4.38) show their true dip values on the transverse component plots because these are in the dip direction. The longitudinal dip components average zero because they are in the strike direction. The zero L component average (Figs. 4.34, 4.35) confirms the choice of the L and T directions.

Folds produce distinctive curves on the dip vs. depth plots. The azimuth vs. depth plots shows the reversal of azimuth at the crest of the fold, CP (Figs. 4.39–4.41). The dip component plots are the most informative. The transverse component plots all cross the zero dip line at the crest of the anticline (CP), show an inflection point at the axial plane (AP) and show a dip maximum at the inflection plane (IP) that separates anticli-

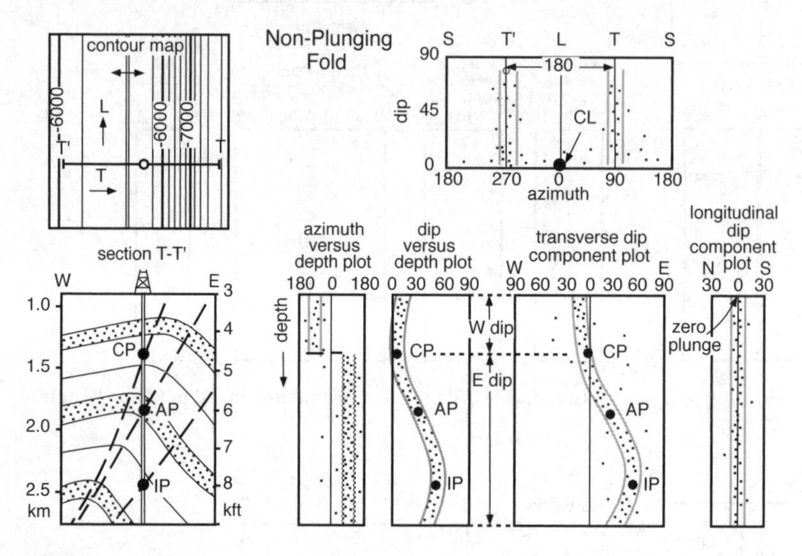

Fig. 4.39. Model map, cross section and SCAT plots for a non-plunging fold. *AP* axial plane; *CL* crestal line; *CP* crestal plane; *IP* inflection plane. (After Bengtson 1981a)

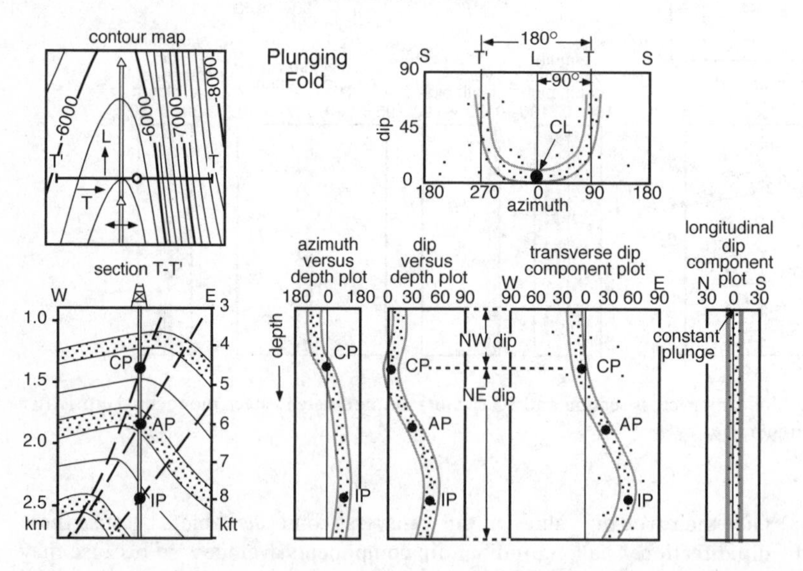

Fig. 4.40. Model map, cross section and SCAT plots for a plunging fold. *AP* axial plane; *CL* crestal line; *CP* crestal plane; *IP* inflection plane. (After Bengtson 1981a)

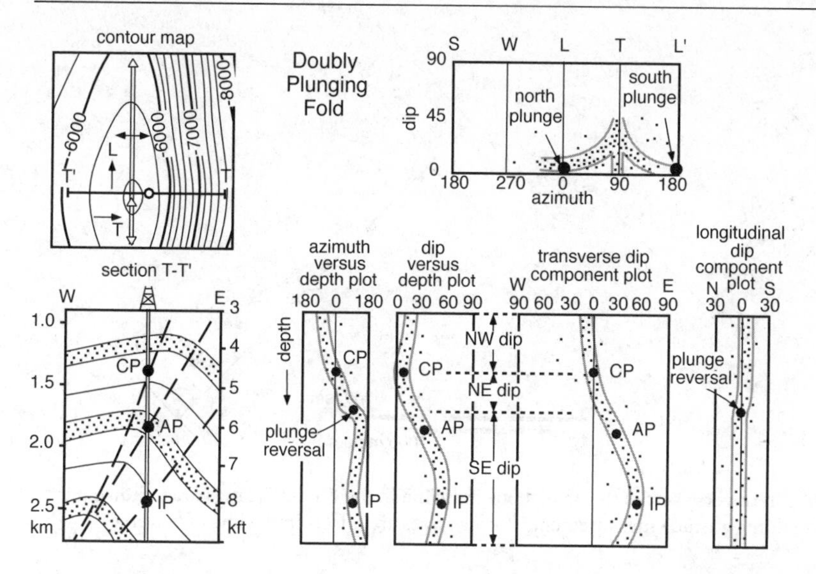

Fig. 4.41. Model map, cross section and SCAT plots for a doubly plunging fold. *AP* axial plane; *CP* crestal plane; *IP* inflection plane. (After Bengtson 1981a)

nal curvature from synclinal curvature. Any variations in plunge are apparent on the longitudinal component plot. The non-plunging fold (Fig. 4.39) is defined by a straight line on the plot of L dip with depth that gives the average plunge of zero. A singly plunging fold (Fig. 4.40) plots as a line of constant plunge with depth. A doubly plunging fold (Fig. 4.41) shows the plunge reversal with depth on the L component plot.

4.6.3 Example

Dip-sequence analysis can be performed on a traverse in any direction through a structure. As an example, the method is applied to a horizontal traverse across the map of the Sequatchie anticline originally presented in Fig. 3.1. The traverse runs from northwest to southeast at right angles to the fold axis along a stream valley that provides the best exposure and therefore the most data (Fig. 4.42). The traverse is broken into three straight-line segments at the dashed lines in order to follow the valley. The attitudes of bedding are located on the SCAT diagrams according to their distance from the northwest end of the traverse.

The dip-azimuth diagram (Fig. 4.43) shows two vertical lines of points, indicating, in comparison to Fig. 4.34, a non-plunging fold with a crest line that trends 50, 230. The trend of the crest line and the lack of significant plunge is in agreement with the tangent diagram for the entire map area (Fig. 4.17). The synoptic tangent diagram shows additional complexity in part because it samples additional portions of the structure and in part because it is a more sensitive indicator of the plunge style and direction. The azimuth-dip diagram should always be checked against the tangent diagram before finally deciding on the plunge direction and amount. The direction of the crest line, here equal to the fold axis, is the L direction to be used in the next stage of the analysis. The T direction is at 90° to L, parallel to the azimuth of the limb dip.

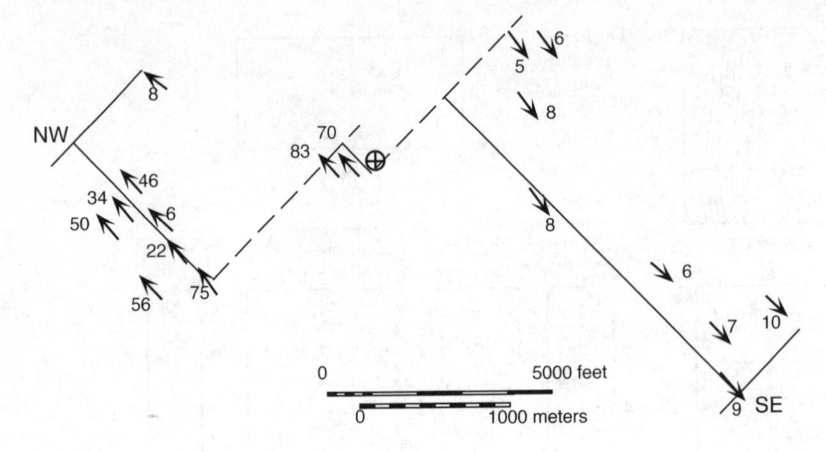

Fig. 4.42. Dip traverse across the Sequatchie anticline in the Blount Springs area, showing locations of bedding attitude measurements. *Dashed lines* are offsets in the line of traverse

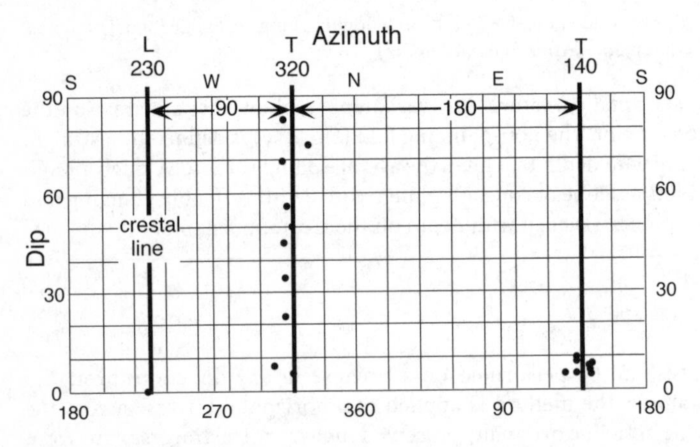

Fig. 4.43. Azimuth-dip diagram for the traverse across the Sequatchie anticline. *T* transverse dip direction; *L* longitudinal dip direction; crest line is at 0, 230

The dip-sequence diagrams reveal the details of the structure. The bedding azimuths and dip components from the traverse are given in Table 4.1 and plotted in Fig. 4.44. The azimuth and dip diagrams (Fig. 4.44a,b) show the locations of the crest plane, axial plane, and inflection plane (compare Fig. 4.44 with Fig. 4.39). The locations of the crest plane and inflection plane are well defined in Fig. 4.44b,c, and the axial plane falls between the two. Note that in the dip-distance plot (Fig. 4.44b) all dips are plotted to the right, whereas in the T-component plot the dips are plotted by their quadrant direction. The dip data for the northwest limb is noisy, even on the T component plot, suggesting some of the complexity seen previously on the tangent diagram (Fig. 4.17). Most of the dips on the L component diagram (Fig. 4.44d) are zero or close

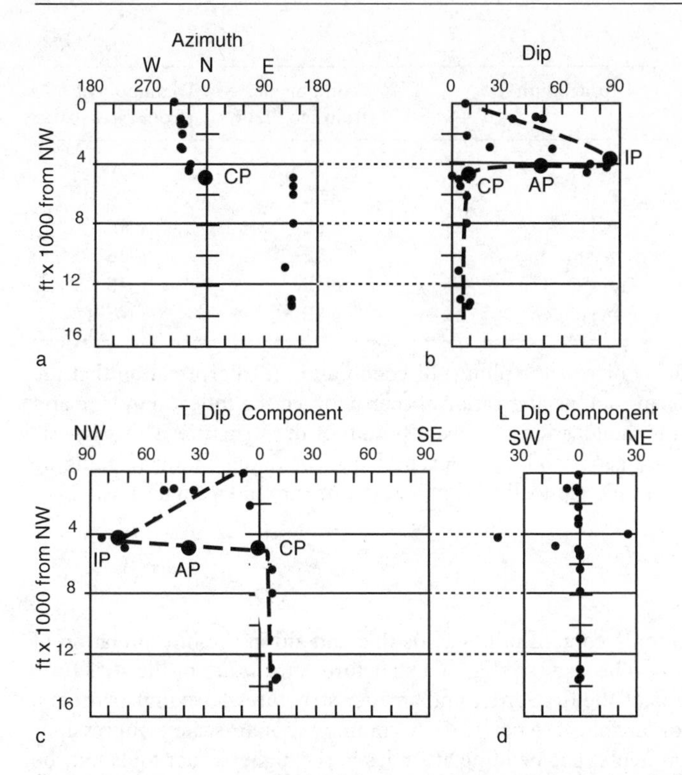

Fig. 4.44. Dip-sequence analysis of the Sequatchie anticline. **a** Azimuth-distance diagram. **b** Dip-distance diagram. **c** T component dip-distance diagram. **d** L component dip-distance diagram. *AP* axial plane; *CP* crestal plane; *IP* inflection plane

Table 4.1. Dip traverse across Sequatchie anticline

Distance from NW end of traverse (ft)	Dip, azimuth	T component (from 320 / 140)	L component (from 230 / 50)
256	8, 308	8 NW	1 SW
1384	46, 315	45 NW	7 SW
1640	34, 316	34 NW	2 SW
1660	50, 320	50 NW	0
2328	6, 320	6 NW	0
3143	22, 316	22 NW	1 SW
3261	56, 318	56 NW	1 SW
4096	75, 330	75 NW	25 SW
4253	Break		
4253	83, 315	83 NW	~45 SW
4528	70, 315	70 NW	13 SW
4891	0, 0	0	0
5147	Break		
5323	5, 145	5 SE	1 SW
5815	6, 144	6 SE	1 SW

Table 4.1. (continued)

Distance from NW end of traverse (ft)	Dip, azimuth	T component (from 320 / 140)	L component (from 230 / 50)
6404	8, 145	8 SE	1 SW
8005	8, 144	8 SE	1 SW
10942	6, 127	6 SE	1 NE
12789	7, 136	7 SE	1 NE
13466	10, 136	10 SE	1 NE
13692	9, 136	9 SE	1 NE

to zero, confirming the choice of the plunge direction and the interpretation that the plunge is zero. Significant plunge aberrations occur between the inflection plane and the crest plane which is the location of the steep limb of the structure. This suggests that the structure of the steep limb is complex, perhaps containing obliquely plunging minor folds, not just a simple monoclinal dip or curvature around a single axis.

4.7 Minor Folds

Map-scale folds commonly contain minor folds that can aid in the interpretation of the map-scale structure. The size ranking of a structure is the *order* of the structure. The largest structure is of the first-order and smaller structures have higher orders. Second- and higher-order folds are particularly common in map-scale compressional and wrench environments. The bedding attitudes in the higher-order folds may be highly discordant to the attitudes in the lower-order folds (Fig. 4.45). Minor folds provide valuable information for interpreting the geometry of the first-order structure but, if unrecognized, may complicate or obscure the interpretation. A pitfall to avoid is interpreting the geometry of the first-order fold to follow the bedding attitudes of the minor folds.

If produced in the same deformation event, the lower-order folds are usually coaxial or nearly coaxial to the first-order structure and are termed parasitic folds (Bates and Jackson 1987). Plots of the bedding attitudes of the minor folds on a stereogram or tangent diagram should be the same as the plots for the first-order structures and can be used to infer the axis direction of the larger folds. Seemingly discordant bedding attitudes seen on a map can be inferred to belong to minor folds if they plot on the same trend as the attitudes for the larger structure.

Asymmetric minor folds are commonly termed drag folds (Bates and Jackson 1987). A drag fold is a higher-order fold, usually one of a series, formed in a unit located between stiffer beds; the asymmetry is inferred to have been produced by bedding-parallel slip of the stiffer units. The sense of shear is indicated by the arrows in Fig. 4.45 and produces the asymmetry shown. The asymmetry of the drag folds can be used to infer the sense of shear in the larger-scale structure. In buckle folds, the sense of shear is away from the core of the fold and reverses on the opposite limbs of a fold (Fig. 4.45), thereby indicating the relative positions of anticlinal and synclinal hinges. The inclination of cleavage planes in the softer units (Fig. 4.45), in the orientation axial planar to the drag folds, gives the same sense of shear information. Cleavage in the stiffer units is usually at a high angle to bedding, resulting in cleavage dips that fan across the fold.

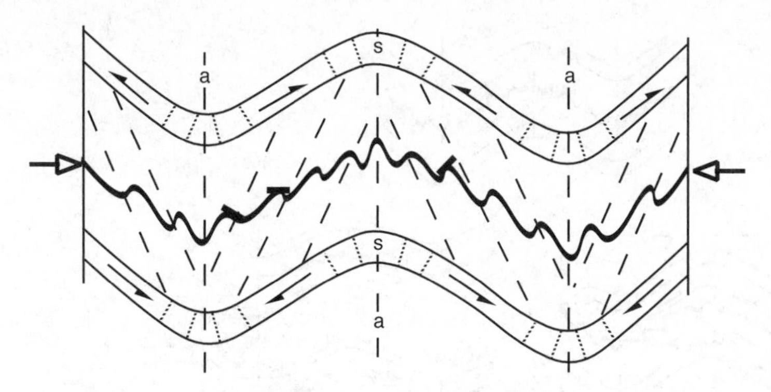

Fig. 4.45. Cross section of second-order parasitic folds in a thin bed showing normal (buckle-fold) sense of shear (*half arrows*) on the limbs of first-order folds. *Short thick lines* are local bedding attitudes. *Dashed lines* labeled *a* are axial-surface traces of first-order folds. *Unlabeled dashed lines* are cleavage, fanning in the stiff units (*s*) and antifanning in the soft units. *Large arrows* show directions of boundary displacements

Individual asymmetric higher-order folds are not always drag folds. The asymmetry may be due to some local heterogeneity in the material properties or the stress field, and the implied sense of shear could be of no significance to the larger-scale structure. Three or more folds with the same sense of overturning in the same larger fold limb are more likely to be drag folds than is a single asymmetric fold among a group of symmetric folds.

Folds in which the sense of shear of the drag folds remains constant from one fold limb to the next (Fig. 4.46a) require a different interpretation. One possibility is the presence of a bedding-plane fault with transport from left to right (Fig. 4.46b); the folds would be fault-related drag folds. Alternatively the beds may be recumbently folded and then refolded with a vertical axial surface. In this situation (Fig. 4.46c), the sense of shear would be interpreted as in Fig. 4.45, and Fig. 4.46a then represents the upright limb of a recumbent anticline, the hinge of which must be to the right of the area of Fig. 4.46a, as shown in Fig. 4.46c.

Fold origins other than by buckling may yield other relationships between the sense of shear given by drag folds and position within the structure. Structures caused by differential vertical displacements, for example salt domes or gneiss domes, could result in exactly the opposite sense of shear on the fold limbs from that in buckle folds (Fig. 4.47). The pattern in Fig. 4.47 has been called Christmas-tree drag.

Potential interpretation problems associated with minor folds are illustrated by the map in Fig. 4.48a. The observed contact locations and bedding attitudes could be explained by the maps in either Fig. 4.48b or 4.48c. The shape of the first-order fold honors the 34SW dip in figure 4.48b and ignores it in Fig. 4.48c. The dip oblique to the contact could be justified as being either a cross bed or belonging to a minor fold.

Plotting the data on a tangent diagram (Fig. 4.49a) shows that all the points, including the questionable point (34, 212), fall on the same line, indicating a cylindrical fold plunging 30° due south. This result leads to rejection of the cross bed hypothesis and indicates that the oblique bedding attitude is coaxial with the map-scale syncline. Both the interpretations of Fig. 4.48b and 4.48c are possible. If the map of Fig. 4.48c is supported by the contact locations, then the structure has the form given by Fig. 4.49b.

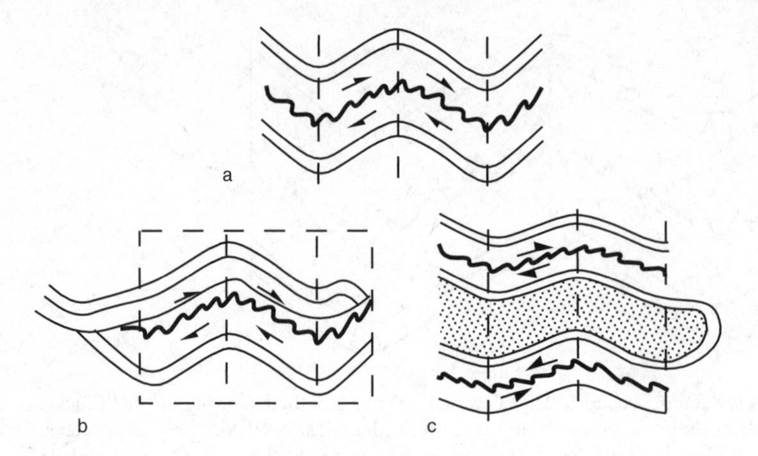

Fig. 4.46. Interpretations of drag folds having the same sense of shear on both limbs of a larger fold. **a** Observed fold with vertical axial traces. **b** Sense of shear interpreted as caused by a folded bedding-parallel thrust fault. **c** Sense of shear interpreted as being the upper limb of a refolded recumbent isoclinal fold

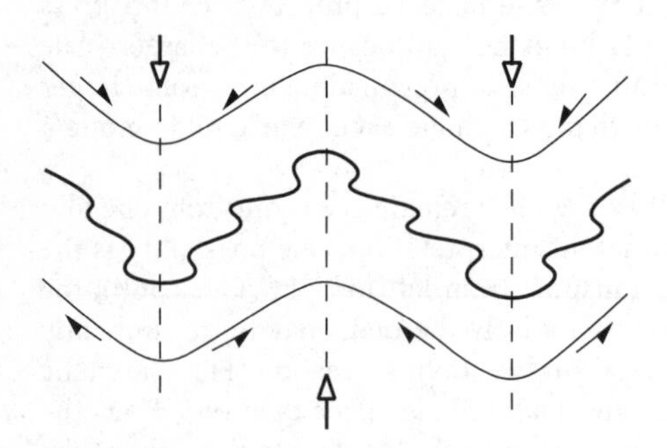

Fig. 4.47. Cross section of parasitic drag folds in the center bed as related to first-order differential vertical folding. *Large arrows* show directions of the boundary displacements

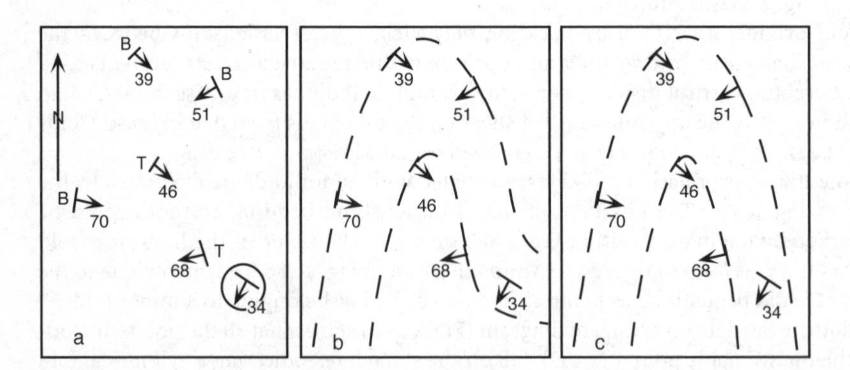

Fig. 4.48. Geological data and different possible geometries based on alternative interpretations of the significance of the *circled* attitude measurement. **a** Data. Bedding attitudes with B = observed base of formation; T = observed top of formation. **b** Interpretation honoring all contacts and attitudes. **c** Interpretation honoring all contacts but not all attitudes

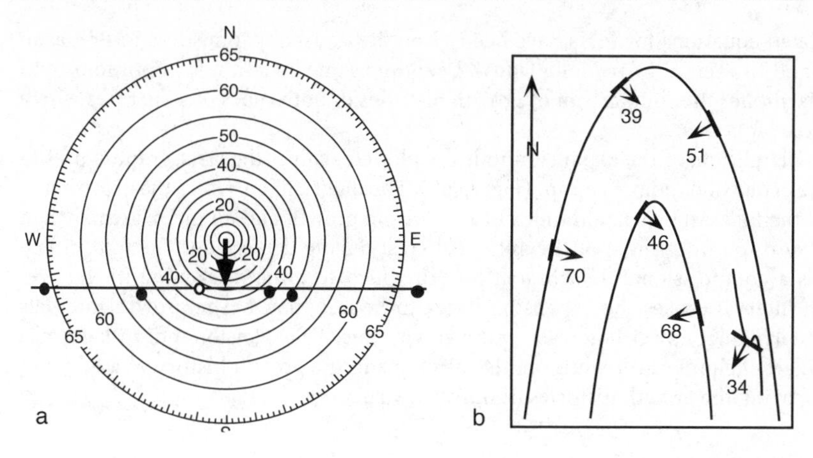

Fig. 4.49. Alternative interpretation of geological data in Fig. 4.48a. **a** Tangent diagram of bedding attitudes. *Open circle* is the 34, 212 point. Plunge of the fold is 30S. **b** Geological map of syncline with a coaxial minor fold on the limb. (After Stockwell 1950)

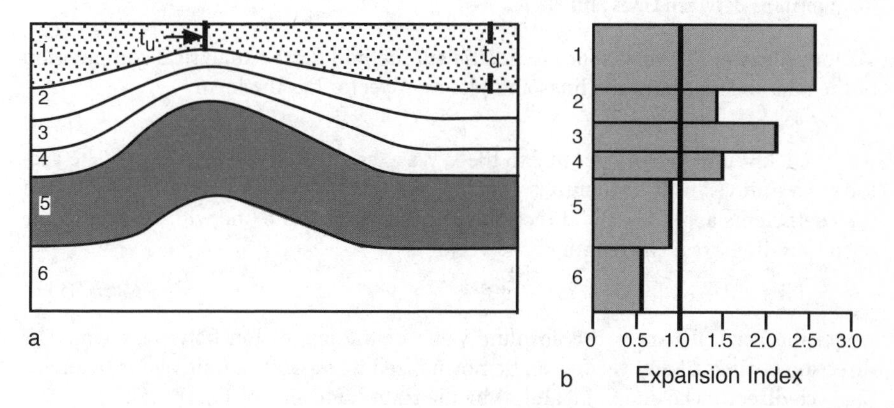

Fig. 4.50. Expansion index for a fold. Units *1–4* are growth units and units *5–6* are pre-growth units. **a** Cross section in the dip direction. **b** Expansion index diagram

4.8 Growth Folds

A growth fold develops during the deposition of sediments (Fig. 4.50a). The growth history can be quantified using an expansion index diagram, where the expansion index, E, is

$$E = t_d / t_u, \tag{4.11}$$

with t_d = the downthrown thickness (off structure) and t_u = the upthrown thickness (on the fold crest). The thicknesses should be measured perpendicular to bedding so as not to confuse dip changes with thickness changes (Fig. 4.50a). The expansion index given here is the same as for growth faults (Thorsen 1963; Sect. 5.7.2). Different but related equations for folds have been given previously by Johnson and Bredeson (1971) and Brewer and Groshong (1993). Using the same equation for both folds and faults facilitates the comparison of growth histories of both types of structures where they occur together.

The magnitude of the expansion index is plotted against the stratigraphic unit to give the expansion index diagram (Fig. 4.50b). The diagram illustrates the growth history of the fold. An expansion index of 1 means no growth and an index greater than 1 indicates upward growth of the anticlinal crest during deposition. The fold in Fig. 4.50a is a compressional detachment fold in which tectonic thickening in the pre-growth interval causes the expansion index to be less than 1. The growth intervals show an irregular upward increase in the growth rate. An expansion index diagram is particularly helpful in revealing subtle variations in the growth history of a fold and in comparing the growth histories of different structures.

4.9 Derivations

4.9.1 Relationships Between Lines and Planes

Vector geometry (Thomas 1960) is an efficient method for the analytical computation of the relationships between lines and planes. A vector has the form

$$\mathbf{v} = a\mathbf{i} + b\mathbf{j} + c\mathbf{k}, \tag{D4.1}$$

where **i, j, k** = unit vectors parallel to the x, y, z axes, respectively (Fig. D4.1). The vector **v** is a unit vector if its length is equal to one, and gives the orientation of a line, if the coefficients a, b, c are the direction cosines l, m, n of the line with respect to the corresponding axes. The length of the vector is

$$| l\mathbf{i} + m\mathbf{j} + n\mathbf{k} | = (l^2 + m^2 + n^2)^{1/2}, \tag{D4.2}$$

where the vertical bars = the absolute value of the expression between them. The direction cosines of any vector can be normalized to generate a unit vector by dividing each direction cosine (l, m, and n) by the right-hand side of Eq. (D4.2).

4.9.2 Angle Between Two Lines or Planes

The angle, Θ, between two lines, is given by the scalar or dot product of the two unit vectors with the same orientations as the lines. If the two vectors represent the poles

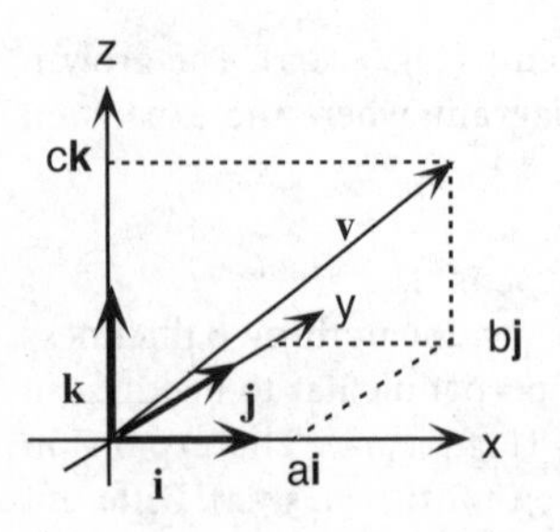

Fig. D4.1. Vector components in the xyz coordinate system. Vector components **i, j, k** each have lengths equal to one

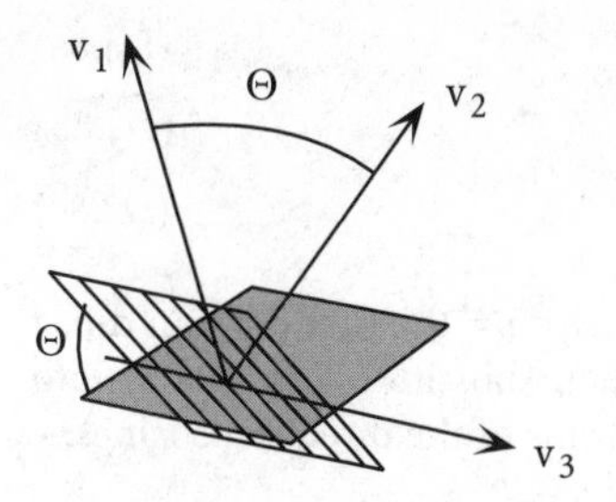

Fig. D4.2. Geometry of the dot product and cross product. Θ is the angle between $\mathbf{v}_1$ and $\mathbf{v}_2$; $\mathbf{v}_3$ is perpendicular to the plane of $\mathbf{v}_1$ and $\mathbf{v}_2$. The *shaded* plane is perpendicular to $\mathbf{v}_1$ and the *ruled* plane is perpendicular to $\mathbf{v}_2$

to planes or the dip vectors of planes, then the dot product is the angle between the planes, normal to their line of intersection (Fig. D4.2). The *dot product* is defined as:

$$\mathbf{v}_1 \cdot \mathbf{v}_2 = |\mathbf{v}_1|\,|\mathbf{v}_2| \cos \Theta, \tag{D4.3}$$

where $|\mathbf{v}| = 1$ for a unit vector and $\mathbf{v}_1 \cdot \mathbf{v}_2$ is evaluated as,

$$\mathbf{v}_1 \cdot \mathbf{v}_2 = l_1 l_2 + m_1 m_2 + n_1 n_2, \tag{D4.4}$$

where the subscript 1 = the direction cosines of the first vector and subscript 2 = the direction cosines of the second vector. Equating (D4.3) and (D4.4) gives the angle between two unit vectors:

$$\cos \Theta = l_1 l_2 + m_1 m_2 + n_1 n_2. \tag{D4.5}$$

4.9.3 Line of Intersection Between Two Planes

The orientation of a line perpendicular to two other vectors is given by the vector product, also called the cross product. Two planes can be defined by their poles and the cross product of the unit vectors parallel to these poles will have the orientation of the hinge line between them and the magnitude defined by the following equation (Fig. D4.2). The *cross product* of two vectors is defined as:

$$\mathbf{v}_1 \times \mathbf{v}_2 = \mathbf{n}\,|\mathbf{v}_1|\,|\mathbf{v}_2| \sin \Theta, \tag{D4.6}$$

where $\mathbf{n}$ = a unit vector perpendicular to the plane of $\mathbf{v}_1$ and $\mathbf{v}_2$. When the vectors are unit vectors given in terms of the direction cosines, the cross product is:

$$\mathbf{v}_3 = \begin{vmatrix} \mathbf{i} & \mathbf{j} & \mathbf{k} \\ l_1 & m_1 & n_1 \\ l_2 & m_2 & n_2 \end{vmatrix} = (m_1 n_2 - n_1 m_2)\mathbf{i} + (n_1 l_2 - l_1 n_2)\mathbf{j} + (l_1 m_2 - m_1 l_2)\mathbf{k}. \tag{D4.7}$$

In Eq. (D4.7), $\mathbf{v}_3$ is the vector perpendicular to both $\mathbf{v}_1$ and $\mathbf{v}_2$. The order of multiplication changes the direction of $\mathbf{v}_3$ but not the orientation of the line or its magnitude.

The pole to a plane as defined by its azimuth (θ) and dip (δ) is a vector pointing in the dip direction. The orientation of this vector in terms of direction cosines is given by Eqs. (D2.15), which are:

$$l = \cos \alpha_p = -\sin \delta \sin \theta; \quad \text{(D4.8a)}$$

$$m = \cos \beta_p = -\sin \delta \cos \theta; \quad \text{(D4.8b)}$$

$$n = \cos \gamma_p = -\cos \delta. \quad \text{(D4.8c)}$$

The subscript p = the pole to a plane. Let the subscript 1 = the first domain dip, 2 = the second domain dip, and h = the hinge line, then substitute Eqs. (D4.8) into (D4.7) to obtain a vector parallel to the hinge line in terms of the direction cosines:

$$\cos \alpha = \sin \delta_1 \cos \theta_1 \cos \delta_2 - \cos \delta_1 \sin \delta_2 \cos \theta_2; \quad \text{(D4.9a)}$$

$$\cos \beta = \cos \delta_1 \sin \delta_2 \sin \theta_2 - \sin \delta_1 \sin \theta_1 \cos \delta_2; \quad \text{(D4.9b)}$$

$$\cos \gamma = \sin \delta_1 \sin \theta_1 \sin \delta_2 \cos \theta_2 - \sin \delta_1 \cos \theta_1 \sin \delta_2 \sin \theta_2. \quad \text{(D4.9c)}$$

The direction cosines must be normalized to ensure that the sum of their squares equals 1, which is done by dividing through by N_c where

$$N_c = (\cos \alpha^2 + \cos \beta^2 + \cos \gamma^2)^{1/2}. \quad \text{(D4.10)}$$

The final equation for the line of intersection between two planes is:

$$\cos \alpha_h = (\cos \alpha) / N_c; \quad \text{(D4.11a)}$$

$$\cos \beta_h = (\cos \beta) / N_c; \quad \text{(D4.11b)}$$

$$\cos \gamma_h = (\cos \gamma) / N_c. \quad \text{(D4.11c)}$$

The azimuth and dip of the line of intersection are given by Eqs. (D2.7), (D2.8) and Table D2.1.

4.9.4 Plane Bisecting Two Planes

The axial surface of a constant thickness fold hinge is a plane that bisects the angle between the two adjacent bedding planes and can be found as a vector sum. A vector sum is the diagonal of the parallelogram formed by two vectors. If the two vectors forming the sum have the same lengths, as do unit vectors, then the diagonal bisects the angle between them (Fig. D4.3). The sum of two vectors is the sum of their corresponding components:

$$\mathbf{v}_1 + \mathbf{v}_2 = (l_1 + l_2)\mathbf{i} + (m_1 + m_2)\mathbf{j} + (n_1 + n_2)\mathbf{k}. \quad \text{(D4.12)}$$

A unit vector that bisects the angle between the two vectors $\mathbf{v}_1$ and $\mathbf{v}_2$ is:

$$\mathbf{v}_3 = (1/N_b)(l_1 + l_2)\mathbf{i} + (1/N_b)(m_1 + m_2)\mathbf{j} + (1/N_b)(n_1 + n_2)\mathbf{k}, \quad \text{(D4.13)}$$

where

$$N_b = ((l_1 + l_2)^2 + (m_1 + m_2)^2 + (n_1 + n_2)^2)^{1/2} \quad \text{(D4.14)}$$

serves to normalize the length to make the resultant a unit vector. Depending on the directions of $\mathbf{v}_1$ and $\mathbf{v}_2$, this vector could bisect either the acute angle or the obtuse angle. If the two vectors are the poles to planes, the vector bisector is the pole to one of the two bisecting planes. The two bisecting planes are separated by 90°.

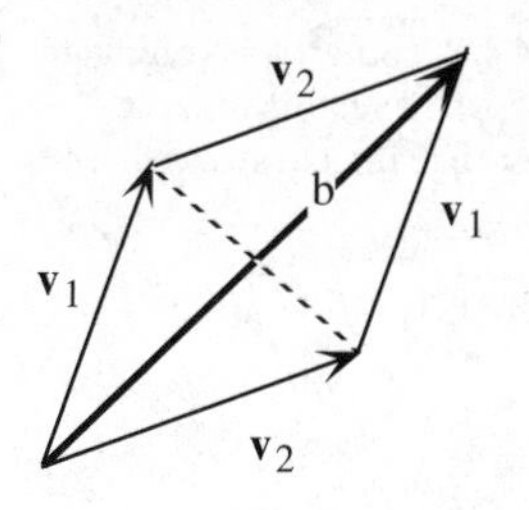

Fig. D4.3. Bisecting vector (b) of the angle between two vectors in the plane of the two vectors

To bisect a given pair of planes, begin with the poles of the planes to be bisected (Eqs. D4.8). Substitute the Eqs. (D4.8) for both planes into (D4.13) and use the same subscripting convention as above, with the subscript B = bisector, to obtain:

$$\cos \alpha_{B1} = -(1/N_B)(\sin \delta_1 \sin \theta_1 + \sin \delta_2 \sin \theta_2); \quad \text{(D4.15a)}$$

$$\cos \beta_{B1} = -(1/N_B)(\sin \delta_1 \cos \theta_1 + \sin \delta_2 \cos \theta_2); \quad \text{(D4.15b)}$$

$$\cos \gamma_{B1} = -(1/N_B)(\cos \delta_1 + \cos \delta_2), \quad \text{(D4.15c)}$$

where, from (D4.14):

$$N_B = [(\cos \delta_1 \sin \theta_1 + \cos \delta_2 \sin \theta_2)^2 + (\sin \delta_1 \cos \theta_1 + \sin \delta_2 \cos \theta_2)^2 + (\cos \delta_1 + \cos \delta_2)^2]^{1/2}. \quad \text{(D4.16)}$$

Equations (D4.15) and (D4.16) give the pole to one of the bisecting planes. Convert the pole to the plane to the dip vector of the plane by following the procedure in Eqs. (D2.21). First reverse the direction of the vector if it points upward ($\cos \gamma_{B1}$ positive) by reversing the signs of all the direction cosines. Then:

$$\cos \alpha_1 = \cos \alpha_{B1}; \quad \text{(D4.17a)}$$

$$\cos \beta_1 = \cos \beta_{B1}; \quad \text{(D4.17b)}$$

$$\cos \gamma_1 = \cos (90 + \arccos (\cos \gamma_{B1})). \quad \text{(D4.17c)}$$

The procedure given above finds one of the two bisectors of the two planes. The other bisecting plane includes the line normal to both the line of intersection of the two planes (hinge line, Eqs. D4.11) and to the first bisecting line (Eqs. D4.17). Find the pole to the second plane by substituting (D4.11) and (D4.17) into the cross product (D4.7) to obtain:

$$\cos \alpha_{b2} = (1/N_p)(\cos \beta_h \cos \gamma_{B1} - \cos \gamma_h \cos \beta_{B1}); \quad \text{(D4.18a)}$$

$$\cos \beta_{b2} = (1/N_p)(\cos \gamma_h \cos \alpha_{B1} - \cos \alpha_h \cos \gamma_{B1}); \quad \text{(D4.18b)}$$

$$\cos \gamma_{b2} = (1/N_p)(\cos \alpha_h \cos \beta_{B1} - \cos \beta_h \cos \alpha_{B1}), \quad \text{(D4.18c)}$$

where

$$N_p = [(\cos \beta_h \cos \gamma_{B1} - \cos \gamma_h \cos \beta_{B1})^2 + (\cos \gamma_h \cos \alpha_{B1} - \cos \alpha_h \cos \gamma_{B1})^2 + (\cos \alpha_h \cos \beta_{B1} - \cos \beta_h \cos \alpha_{B1})^2]^{1/2} \equiv 1. \quad \text{(D4.19)}$$

Equation (D4.19) serves as a check on the calculation: it should always equal 1 because the starting vectors are orthogonal and of unit length. Convert this pole to the bisecting plane (Eqs. D4.18) into the dip vector of the plane with Eqs. (D2.21) by first

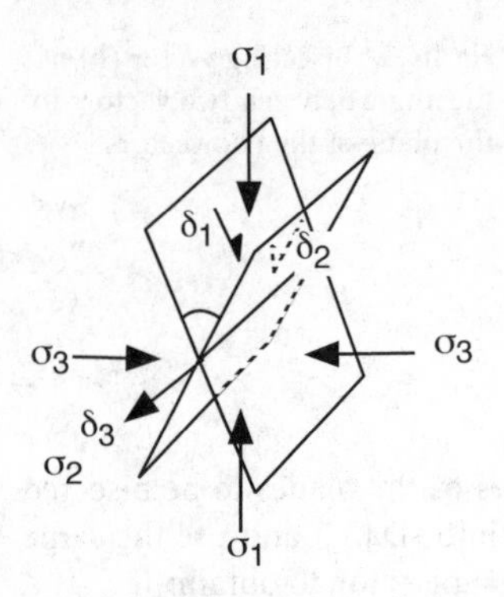

Fig. D4.4. Geometry of conjugate faults. δ_i Dip vectors of fault planes; σ_i principal stress directions

reversing the direction of the vector if it points downward (cos γ negative) by reversing the signs of all the direction cosines. Then:

$$\cos \alpha_2 = \cos \alpha_{b2}; \tag{D4.20a}$$

$$\cos \beta_2 = \cos \beta_{b2}; \tag{D4.20b}$$

$$\cos \gamma_2 = \cos (90 + \arccos (\cos \gamma_{b2})). \tag{D4.20c}$$

The azimuth and dip of the planes in Eqs. (D4.17) and (D4.20) are given by Eqs. (D2.7), (D2.8) and Table D2.1.

If the two planes being bisected are conjugate faults (Fig. D4.4), then the line of intersection of the faults is the intermediate principal compressive stress axis (σ_2), the line (pole) that bisects the acute angle is the maximum principal compressive stress axis (σ_1), and the line (pole) that bisects the obtuse angle is the least principal compressive stress axis (σ_3).

4.9.5 Axial Surface Geometry

The relationship between bed thickness and axial surface orientation is obtained as follows. Both limbs of the fold meet along the axial surface (Fig. D4.5) and have the common length h and therefore must satisfy the relationship:

$$h = t_1/ \sin \gamma_1 = t_2/ \sin \gamma_2. \tag{D4.21}$$

If $t_1 = t_2$, then $\gamma_1 = \gamma_2$ and the axial surface bisects the hinge. If the bed changes thickness across the axial surface, then the axial surface cannot bisect the hinge. The thickness ratio across the axial surface is:

$$t_1 / t_2 = \sin \gamma_1 / \sin \gamma_2. \tag{D4.22}$$

If the thickness ratio and the interlimb angle are known, then from the relationship (Fig. D4.5)

$$\gamma = \gamma_1 + \gamma_2, \tag{D4.23}$$

where γ = the interlimb angle and γ_2 = the angle between the axial surface and the dip of limb 2. Substitute $\gamma_1 = \gamma - \gamma_2$ from Eq. (D4.23) into (D4.22), apply the trigonometric identity for the sine of the difference between two angles, and solve to find γ_2 as

$$\tan \gamma_2 = t_2 \sin \gamma / (t_1 + t_2 \cos \gamma). \tag{D4.24}$$

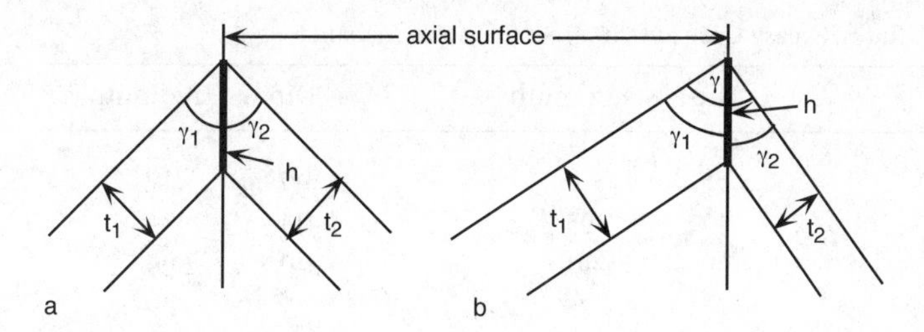

Fig. D4.5. Axial-surface geometry of a dip-domain fold hinge in cross section. **a** Constant bed thickness. **b** Variable bed thickness

4.10 Problems

4.10.1 Geometry of the Sequatchie Anticline

1. Plot the attitude data from Table 4.1 on a stereogram to find the fold style and plunge.
2. Plot the attitude data from Table 4.1 on a tangent diagram to find the fold style and plunge. How does your answer compare with the interpretation based on the stereogram?
3. Based on the results of questions 1 and 2, what is the orientation of a plunge line and the spacing of points along the line? Use your plunge line for the Sequatchie anticline to project the top of the Devonian contact along the plunge on Fig. 3.1.
4. What is the orientation of the axial surface of the first-order fold in Fig. 4.18? Describe the reason for the method you choose.

4.10.2 Geometry of the Greasy Cove Anticline

The bedding attitudes below (Table 4.2) come from the Greasy Cove anticline, a compressional structure in the southern Appalachian fold-thrust belt. Use it to answer the following:

1. What is the π axis of the fold?
2. Use a tangent diagram to find the axis and the style of the fold.
3. Plot the data from Table 4.2 on an azimuth-dip diagram (Fig. 4.51). Compare your interpretation of the fold plunge style and direction with your interpretation of the tangent diagram in the previous problem.

4.10.3 SCAT Analysis of the Sequatchie Anticline

Use the data in Table 4.1 to perform a complete SCAT analysis (Fig. 4.52) on the dip traverse across the Sequatchie anticline.

Table 4.2. Bedding attitudes, Greasy Cove anticline, northeastern Alabama

Dip	Azimuth	Dip	Azimuth	Dip	Azimuth
46	316	55	316	60	310
55	311	40	295	14	124
12	319	26	281	60	304
20	248	25	275	10	270
10	266	14	294	16	307
12	243	15	173	12	150
24	154	22	154	28	231
20	129	30	128	30	119
35	143	25	114	32	131
25	120	70	151	26	165
20	345	12	255	11	258
15	160	12	190		

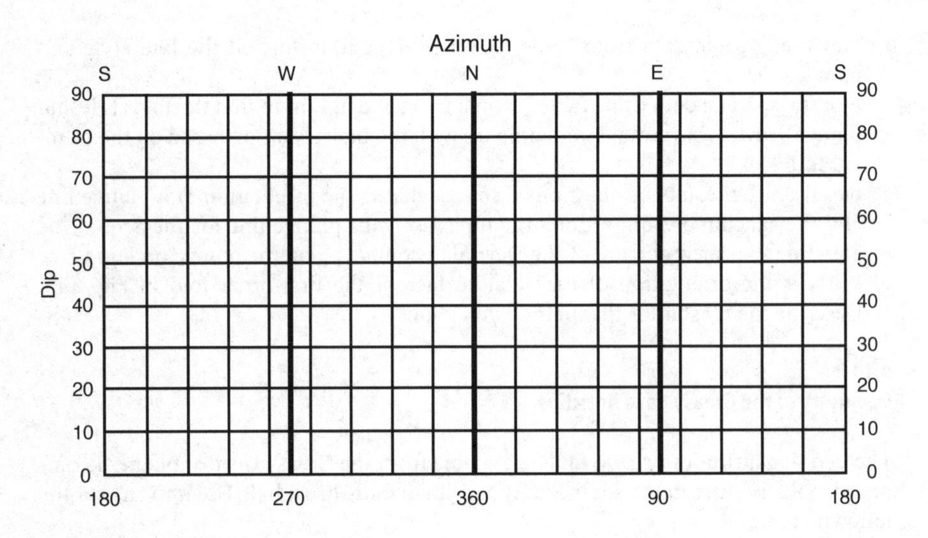

Fig. 4.51. Dip-Azimuth diagram

1. Plot the azimuth-distance and the dip-distance diagrams.
2. What are the T and L directions? Use your answers to Section 4.10.1 and the dip-azimuth diagram.
3. What are the dip components in the T and L directions? Plot them on the dip-component diagrams.

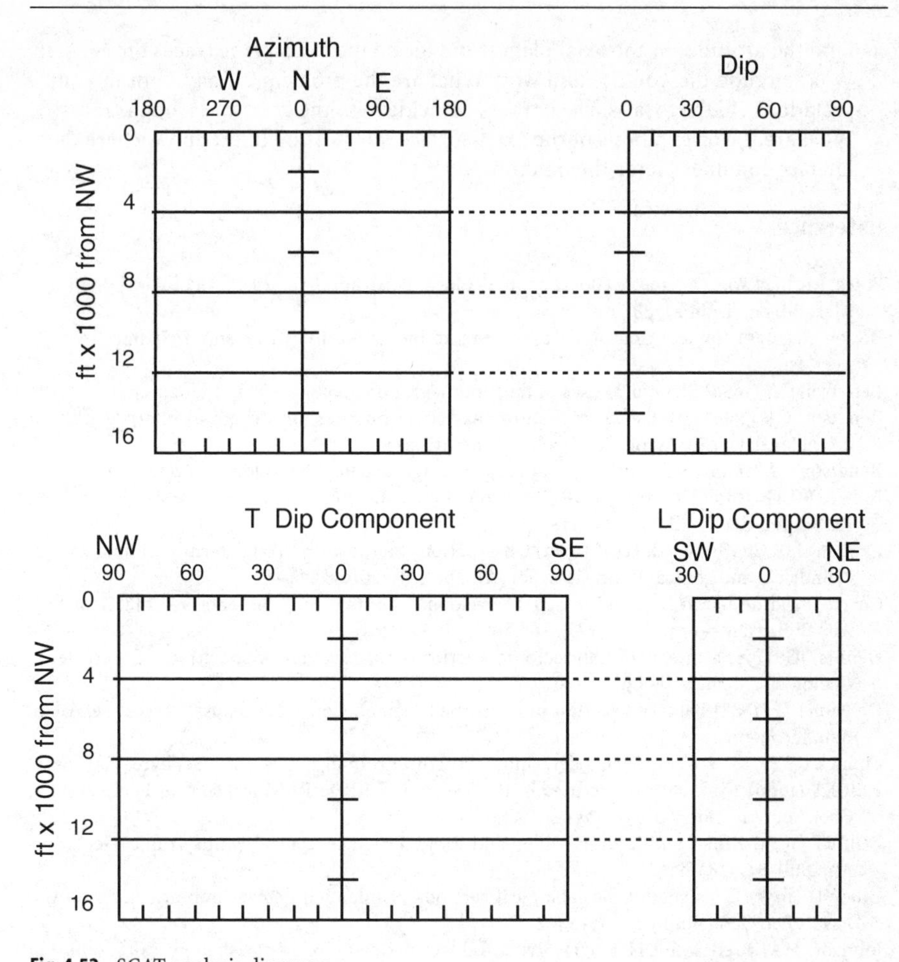

Fig. 4.52. SCAT analysis diagrams

4.10.4
Structure of a Selected Map Area

Use the map of a selected structure (for example, Fig. 3.10) to answer the following questions.

1. Measure and list all the bedding attitudes on the map. Plot the attitudes on a stereogram and a tangent diagram. What fold geometry is present? Which diagram gives the clearest result? Explain.
2. Project the position of a selected unit beyond the area where it is shown on the map using plunge lines.
3. Define the locations of the crest and trough traces from the map. Are the directions the same as given by the attitude diagrams?

4. Find the attitudes of the axial planes, and locate the axial-plane traces on the map. What method did you use and why? What are the problems, if any, with the interpretation? Do the axial-surface traces coincide with the crest and trough traces?
5. What are the orientations of the axial-surface intersection lines? Show where these intersection lines pierce the outcrop.

References

Bates RL, Jackson JA (1987) Glossary of geology, 3 rd edn. American Geological Institute, Alexandria, Virginia, 788 pp

Becker A (1995) Conical drag folds as kinematic indicators for strike slip. J. Struct. Geol. 17: 1497–1506

Bengtson CA (1980) Structural uses of tangent diagrams: Geology, v. 8, 599–602

Bengtson CA (1981a) Statistical curvature analysis techniques for structural interpretation of dipmeter data. Am. Assoc. Pet. Geol. Bull. 65: 312–332

Bengtson CA (1981b) Structural uses of tangent diagrams, Reply: Geology, v.9, 242–243

Brewer RC, Groshong RH Jr (1993) Restoration of cross sections above intrusive salt domes. Am. Assoc. Pet. Geol. Bull. 77: 1769–1780

Cruden DM, Charlesworth HAK (1972) Observations on the numerical determination of axes of cylindrical and conical folds. Geol. Soc. Am. Bull. 83: 2019–2024

Currie JB, Patnode HW, Trump RP (1962) Development of folds in sedimentary strata. Geol. Soc. Am. Bull. 73: 655–673

Dennis JG (1967) International tectonic dictionary. American Association of Petroleum Geologists, Mem. 7, 196 pp

De Paor DG (1988) Balanced section in thrust belts. Part I: Construction. Am. Assoc. Pet. Geol. Bull. 72: 73–90

Elphick RY (1988) Program helps determine structure from dipmeter data. Geobyte, 3: 57–62

Faill RT (1969) Kink band structures in the Valley and Ridge Province, central Pennsylvania. Geol. Soc. Am. Bull. 80: 2539–2550

Faill RT (1973) Kink band folding, Valley and Ridge Province, central Pennsylvania. Geol. Soc. Am. Bull. 84: 1289–1314

Gill WD (1953) Construction of geological sections of folds with steep-limb attenuation. Am. Assoc. Pet. Geol. Bull. 37: 2389–2406

Johnson HA, Bredeson DH (1971) Structural development of some shallow salt domes in Louisiana Miocene productive belt. Am. Assoc. Pet. Geol. Bull. 55: 204–226

Ramsay JG (1967) Folding and fracturing of rocks. McGraw-Hill, New York, 568 pp

Rowland SM, Duebendorfer EM (1994) Structural analysis and synthesis, 2nd edn. Blackwell, Oxford, 279 pp

Stauffer MR (1973) New method for mapping fold axial surfaces. Geol. Soc. Am. Bull. 84: 2307–2318

Stockmal GS, Spang JH (1982) A method for the distinction of circular conical from cylindrical folds. Can. J. Earth Sci. 19: 1101–1105

Stockwell CH (1950) The use of plunge in the construction of cross sections of folds. Proc. Geol. Assoc. Can. 3: 97–121

Thomas GB (1960) Calculus and analytic geometry. Addison-Wesley, London, 1010 pp

Thorsen CE (1963) Age of growth faulting in southeast Louisiana. Trans. Gulf Coast Assoc. Geol. Soc. 13: 103–110

Wilson G (1967) The geometry of cylindrical and conical folds. Proc. Geol. Assoc. 78: 178–210

Chapter 5
Faults and Unconformities

5.1
Introduction

This chapter characterizes the geometrical properties of faults, fault displacements, and unconformities. It begins with a discussion of how to recognize a fault on a geologic map, seismic line, electric log, caliper log, dip traverse or dipmeter, and from the rock type. This chapter gives the techniques for determining fault displacement, slip, separation, throw and heave. Throw and heave are the components shown on structure contour maps. The effect of stratigraphic growth across faults on the throw and heave calculation is given and growth is quantified with the expansion index. Unconformities are treated here because, by truncating map units at the map scale, they share characteristics with faults and need to be distinguished from faults.

5.2
Recognition of Faults

Faults are recognized where they cause discontinuities in the traces of marker horizons on maps and cross sections, discontinuities in the stratigraphic sequence, and anomalies in the thicknesses. Faults may also be recognized from the diagnostic shape of drag folds adjacent to the fault and by the distinctive rock types caused by faulting.

5.2.1
Discontinuities in Geological Map Pattern

At the map scale, a fault is inferred where it causes a break in the continuity of the stratigraphy or the structure on a geologic map. A seismic-reflection time slice is similar to a geologic map in a region of low topographic relief and will also show faults as discontinuities. As an example, the coal seams in the north half of Fig. 5.1 are abruptly truncated along strike where they intersect faults (points A). Fault dips and hence fault separations (Sect. 5.5.2) cannot be determined from the horizontal map surface, but these are known to be normal-separation faults. The truncation of units by a contact (B, Fig. 5.1) indicates that the contact is a fault or that the beds are cut by an unconformity. The base of the continuous bed that crosses the truncated beds (C, Fig. 5.1) is either a fault contact or an unconformity. The contact indicated by B and C in Fig. 5.1 is a fault because the truncated beds to the north (bc, ml) are younger than the cross-cutting unit to the south (by) and so should be above the unconformity, not below it. At D, E, and F, the contacts are parallel and so provide no direct evidence of faulting, although the absence of stratigraphic units at the contacts suggests the presence of faults. The contact at D can be traced into a location (C) where it is faulted;

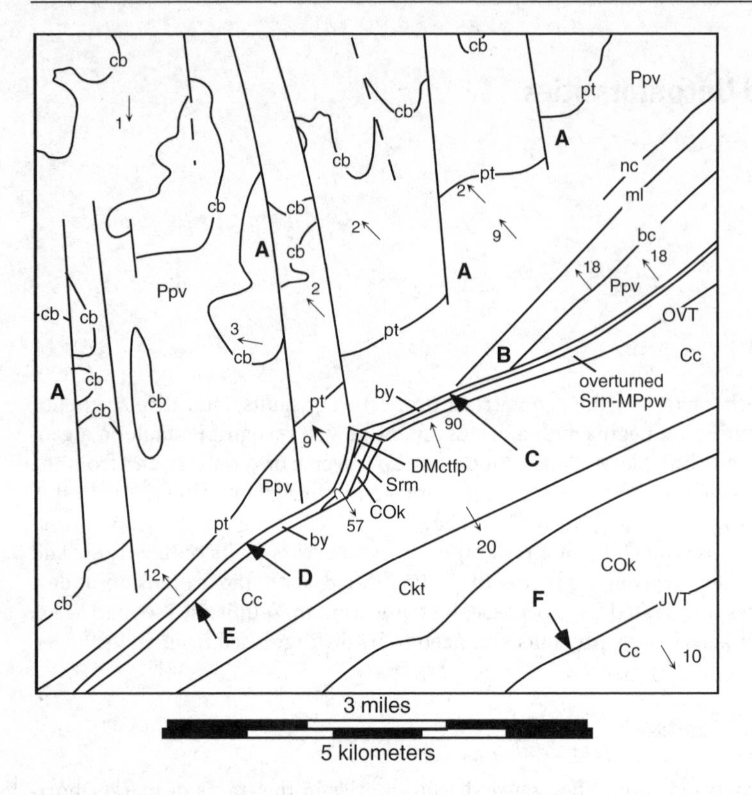

Fig. 5.1. Geologic map of the Ensley area, Alabama The topography is nearly flat, making the map close to a horizontal section. Units (oldest to youngest): Cambrian, *Cc* Conasauga Ls., *Ckt* Ketona dolomite; Cambro-Ordovician, *COk* Knox dolomite; Silurian, *Srm* Red Mountain Formation; Devonian-Mississippian, *DMctfp* Chattanooga Shale, Tuscumbia Limestone, Fort Payne Chert; Mississippian-Pennsylvanian, *MPpw* Parkwood Formation; Pennsylvanian, *Ppv* Pottsville, containing the following marker units (oldest to youngest): *by* Boyles sandstone, *bc* Black Creek coal, *ml* Mary Lee coal, *nc* Newcastle (upper Mary Lee) coal, *pt* Pratt (American) coal, *cb* Cobb coal. *OVT* Opossum Valley thrust, *JVT* Jones Valley thrust. Contact relationships: *A* offset of strike traces, *B* truncation of units at contact, *C* unit crosses contacts, *D* missing section, *E* missing section, *F* older over younger. (After Butts 1910; Kidd 1979)

hence it is probably a fault. The contact at F is a reverse fault because the dips of bedding on both sides of the contact show that the older units to the south lie on top of the younger units to the north. The contact indicated by E places upright lower Cambrian rocks adjacent to overturned younger units to the north, suggesting reverse fault drag (Sect. 5.2.7). The contact indicated by C and D is either a reverse fault or the folded lower detachment of a normal fault. Once a contact has been identified as a fault, it is usually shown as a heavier line on the map.

The juxtaposition of different units across a fault favors some form of topographic expression for the fault because of the likelihood that the two units will not weather and erode identically. The boundary between the juxtaposed units is likely to form a topographic lineament. The fault-zone material (Sect. 5.2.6) may erode differently from any of the surrounding country rocks and thus cause a topographic valley or ridge along the fault.

5.2.2 Discontinuities on Reflection Profile

The primary criterion for recognizing a fault on a seismic- or radar-reflection profile is as a break in the continuity of a reflector (Fig. 5.2a, location A). Reflections are recorded at locations that depend on the velocity distribution and the dips of the reflectors. The abrupt termination of a reflecting horizon at a fault provides a point source of diffractions, arcuate reflectors that emanate from the fault and cross other reflectors. Time migration of the seismic profile is designed to restore the reflectors to their correct relative locations and dips and to remove the diffractions. Some diffractions, however, may remain in time-migrated profiles, as seen in the region between B and C in Fig. 5.2a.

Features other than faults may resemble faults. Reflectors are also truncated at an unconformity (Fig. 5.2a, between the arrows labeled D). The presence of parallel reflectors above the unconformity supports the interpretation that the truncation is indeed an unconformity, not a fault. Reflections from steeply dipping beds may fail to be recorded or may not be correctly migrated, leading to zones of disturbed reflectors that can easily be mistaken for fault zones. The lack of reflector continuity around and above location E (Fig. 5.2a) is due to the steep limb of a fold (Fig. 5.2b), not a fault. In some areas the faults themselves produce reflections as, for example, along the large normal faults in the Gulf of Mexico (Fig. 5.3).

Regions of complex structure are commonly associated with faults and may have steep dips and large lateral velocity changes, either of which can create discontinuities at deeper levels on the seismic profile (Fig. 5.4). Discontinuities in the otherwise continuous reflectors below 2-s two-way travel time (Fig. 5.4a) might easily be interpreted as being normal faults (Fig. 5.4b), although the interpreters of the line correctly did not do so (Fig. 5.4c). This region is below a major thrust fault that places rocks as old as Devonian on top of Cretaceous units. The older rocks have significantly higher seismic velocities than the Cretaceous rocks. The geological interpretation (Fig. 5.4d), based on well control and the seismic profile, indicates no offset of the Cretaceous and older units below the thrust, but rather a continuous westward regional dip of the subthrust units. The apparent eastward dip of the subthrust units is a velocity pull up caused by the high velocities of the rocks in the thrust sheet. The apparent normal faults in the subthrust sequence are probably caused by the rapid lateral velocity gradients associated with fault slices within the thrust sheet.

Reflection profiles can be one of the most powerful tools for structural interpretation. Because the profile itself is an interpretation based on the inferred velocity structure of the region, the geological interpretations must be tested with all of the same techniques that are applied to geological data.

5.2.3 Discontinuities on Structure Contour Map

A linear trend of closely spaced structure contours that form a monoclinal fold may represent an unrecognized fault (F, Fig. 5.5). The monocline could be replaced by a

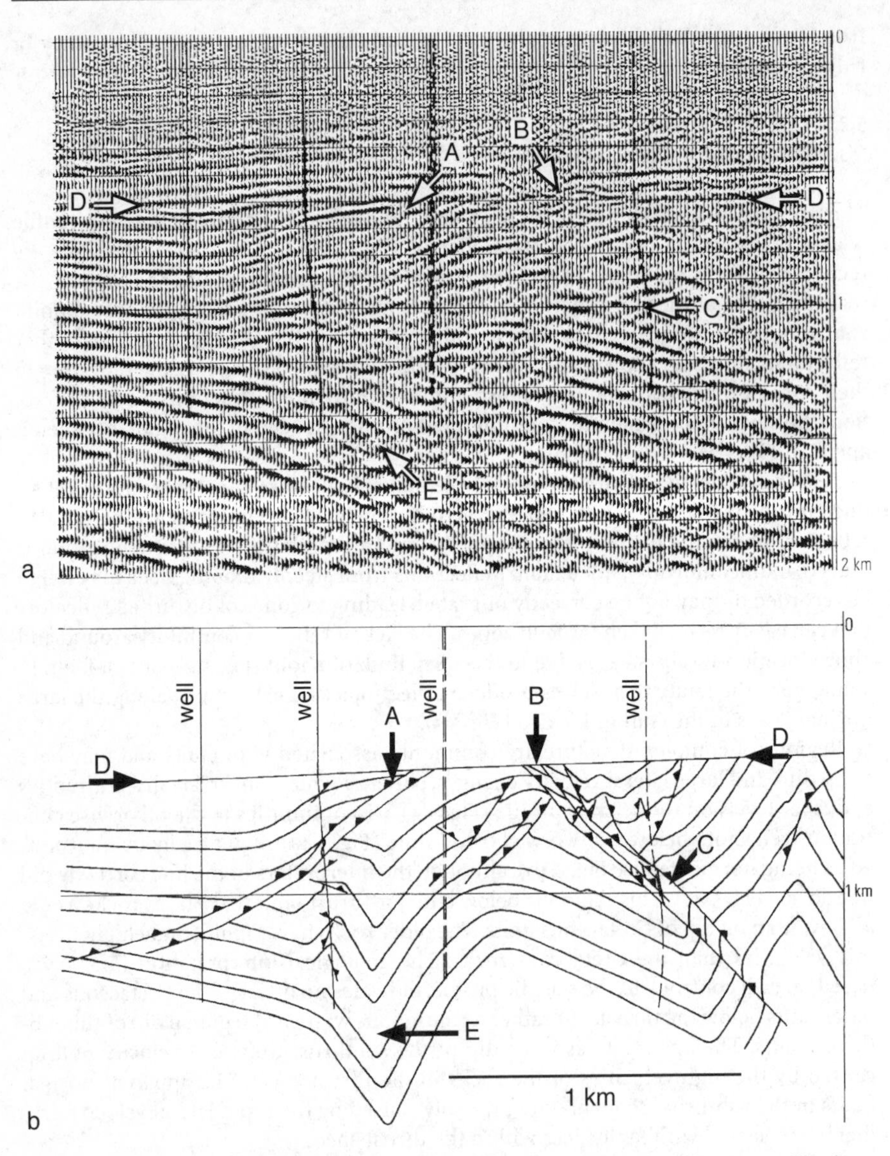

Fig. 5.2. Faults in a portion of the Ruhr coal district, Germany. **a** Seismic reflection profile: dynamite source, 12-fold common-depth-point stack, time migrated. *A* reflector discontinuity at a large thrust fault; *B* reflector discontinuity at a small, complex, reverse fault zone; *C* crossing reflectors in a fault zone; *D* reflector discontinuity at an angular unconformity; *E* zone of disturbed reflectors on the steep limb of a fold. **b** Geological cross section based on the seismic profile and the wells shown. Letters designate the same features as in **a**. (adapted from Drozdzewski 1983)

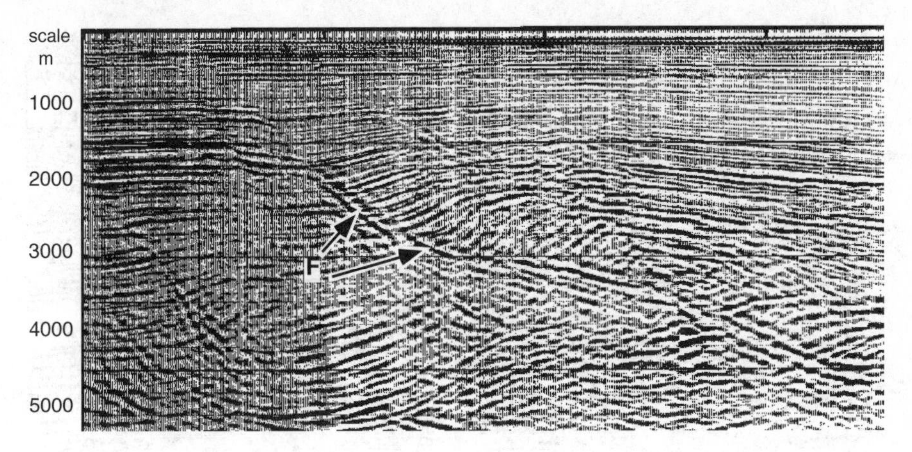

Fig. 5.3. Corsair fault, a thin skinned growth fault on the Texas continental shelf. 48-fold, depth-converted seismic line. *F* fault reflectors. (After Christensen 1983)

fault. Linear fold trends are, of course, perfectly reasonable and so independent evidence for faulting is desirable before the fold is reinterpreted as a fault as, for example, a direct observation of the fault somewhere along the trend. In some areas, especially in extensional terrains, folds with steep limb dips are rare or do not occur at all and so the closely spaced contours could be replaced by a fault on the basis of consistency with the local structural style.

5.2.4 Stratigraphic Thickness Anomaly

A linear thickness anomaly in a generally uniform stratigraphic unit may be caused by a fault. The anomaly can be caused by the stratigraphic section missing or repeated by an unrecognized fault. A fault appearing as a thickness anomaly is a very likely occurrence where the stratigraphic separation on the fault is less than the stratigraphic resolution. A unit penetrated by three wells (Fig. 5.6a) might be interpreted as having a stratigraphic thin spot in the middle well. The thin spot might also represent a normal fault cut (Fig. 5.6b). Where a thickness anomaly is recognized as a possible fault, the data should be re-examined for other types of evidence of a fault cut. A cross section involving multiple units, rather than just one unit as in Fig. 5.6, should be constructed. Thickness anomalies that line up on a cross section with a dip appropriate for a fault probably represents a fault.

On an isopach map a thickness anomaly will follow the trace of the fault (Fig. 5.7). A fault with normal separation will cause the unit to be thinner and a reverse separation will cause the unit to be thicker. The greater the stratigraphic separation of the fault, the greater the thickness change from the regional value. The maximum thinning is to zero thickness but the maximum thickening can be any amount, depending on the number of repetitions that occur within the boundaries of the formation. The example shows an east-west trending fault that is losing displacement to the west

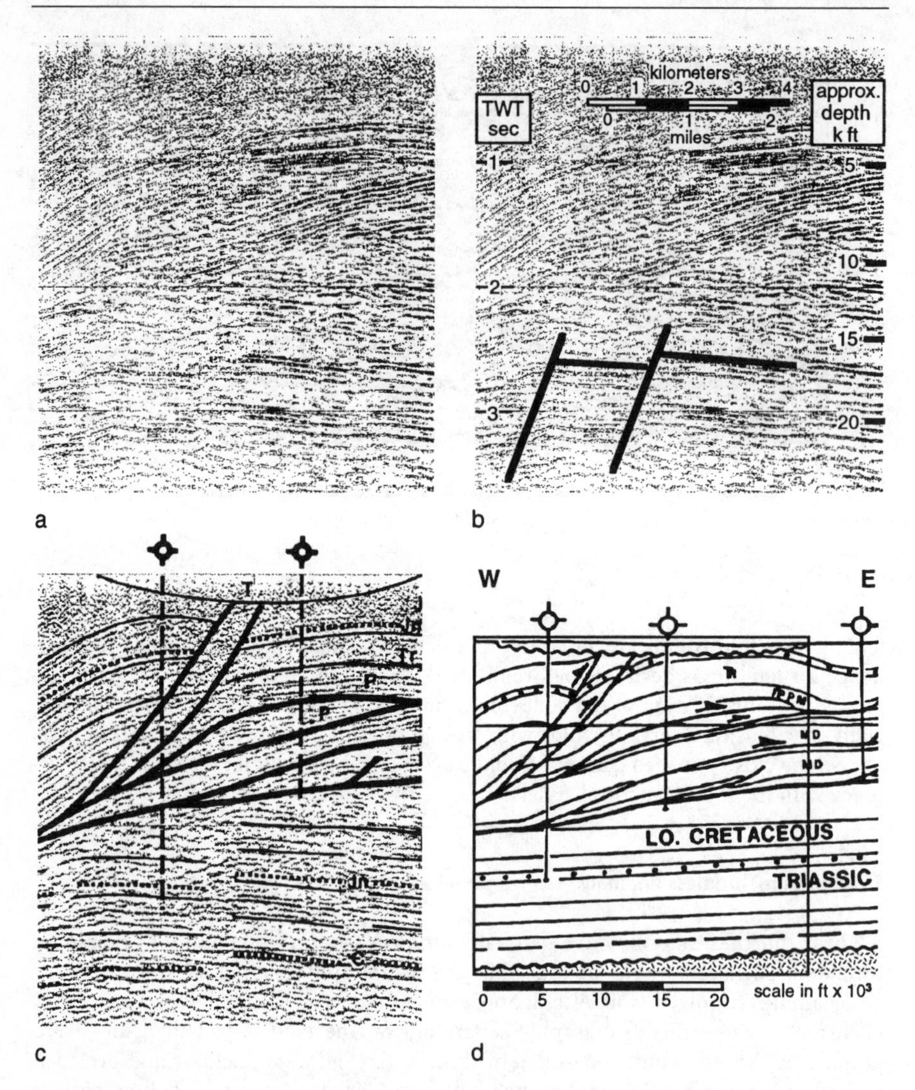

Fig. 5.4. Velocity discontinuities create features that look like faults on seismic profiles. **a** Segment of a seismic line across Wyoming thrust belt (dynamite source, eight-fold common-depth-point stack, migrated time section, approximate vertical exaggeration 1.3 at 2.7 s; Williams and Dixon 1985). **b** Discontinuities in seismic reflectors that might be normal faults. **c** Interpretation by Williams and Dixon (1985). **d** Geological cross section using well control and the seismic line; no vertical exaggeration (Williams and Dixon 1985). The box outlines the area of the seismic line. No normal faults are present. *TWT* two-way traveltime (seconds); C Cambrian; *MD* Mississippian-Devonian; *IP PM* Pennsylvanian, Permian, Mississippian undifferentiated; *P* Permian; *Tr, ₮* Triassic, *J* Jurassic undifferentiated; *Jn* Jurassic Nugget sandstone; *T* Tertiary

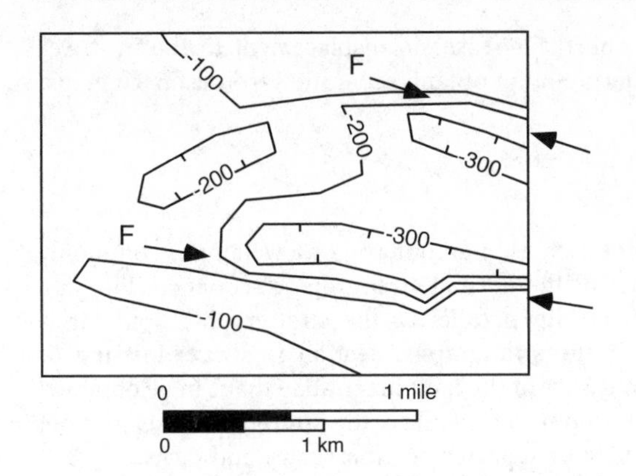

Fig. 5.5. Structure contour map showing the possible locations of faults (*F*) between the *arrows*. Contours are in feet

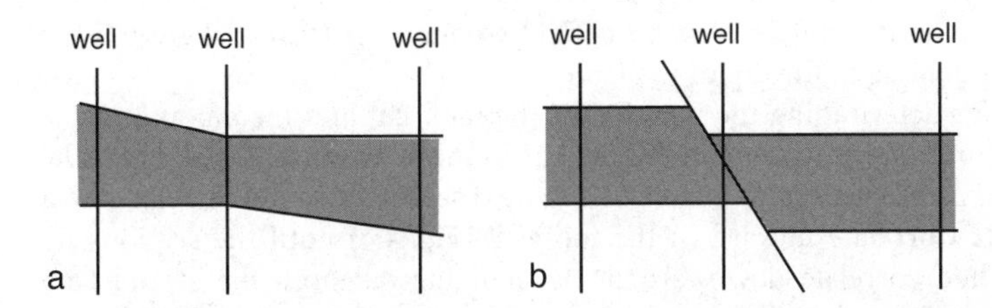

Fig. 5.6. Cross sections showing alternative interpretations of a stratigraphic thickness anomaly in the *shaded* unit of the center well. **a** Thinning interpreted as due to stratigraphic change. **b** Thinning interpreted as due to a normal fault

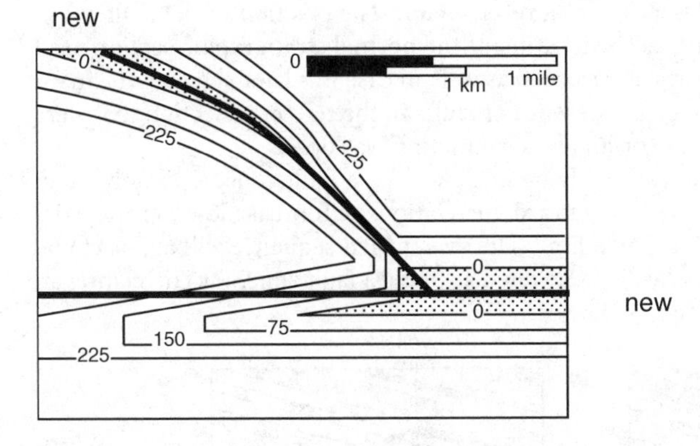

Fig. 5.7. Isopach map showing the effect of two intersecting normal faults on the thickness of a formation that is regionally 225 or more units thick. Contours are of thickness, the formation is *absent* in the shaded areas. *Heavy lines* mark the fault trends; because the faults are dipping planes, the lines do not represent the exact map traces of the faults

while an intersecting fault to the north increases in displacement to the west. See Section 6.7 for the quantitative interpretation of fault separation from isopach maps.

5.2.5 Discontinuity in Stratigraphic Sequence

The recognition of a fault at a point such as in an outcrop or a well log is commonly based on a break in the continuity of the normal stratigraphic sequence (Fig. 5.8). Except for the special case of fault slip parallel to the stratigraphic boundaries (Redmond 1972), some amount of the stratigraphic section is always missing or repeated at the fault contact. The measure of the fault magnitude that can be obtained from the information available at a point on a fault is the stratigraphic separation, equal to the thickness of the missing or repeated section (Bates and Jackson 1987). Only the stratigraphic separation (Sect. 5.5.2) can be determined at a point on the fault, not the slip. Determination of the slip requires additional information (Sect. 5.5.1). The true slip on the faults in Fig. 5.8 could be oblique or even strike slip if the dip of bedding is oblique to the dip of the fault.

The method for determining the position of the fault cut and the amount of the stratigraphic separation is the same in the field as in the subsurface. A well log is the same as a section measured across a fault. The faulted section must be correlated to a reference section. Correlate upward in the footwall (Fig. 5.9) until the sections no longer match. Then correlate downward in the hangingwall until the sections no longer match. The mismatch will occur at the same place in the faulted section if a single fault is present. A mismatch in upward and downward correlation that fails to occur at the same place in the faulted section could be caused by the presence of more than one fault or by a fault zone of finite thickness. In the former situation, the process should be repeated on a finer scale until all the faults are located. An inverted stratigraphic section may be present below a reverse fault. The position of the fault cut in the faulted section is recorded. The location of the normal stratigraphic section used as the reference should always be recorded as well, in case it is later shown to be inappropriate due to, for example, the presence of faults in the reference section that were not recognized at the time the original interpretation was done.

An overturned fold limb can easily be mistaken for a fault zone, especially in an isoclinal fold. The upward and downward correlations will break down at the axial surfaces that bound the overturned limb. The overturned sequence will appear to be an unfamiliar unit and so is easily interpreted as being a fault zone. Try correlating an

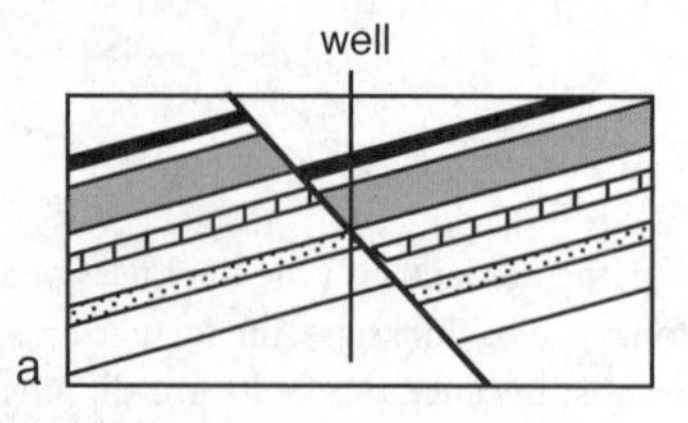

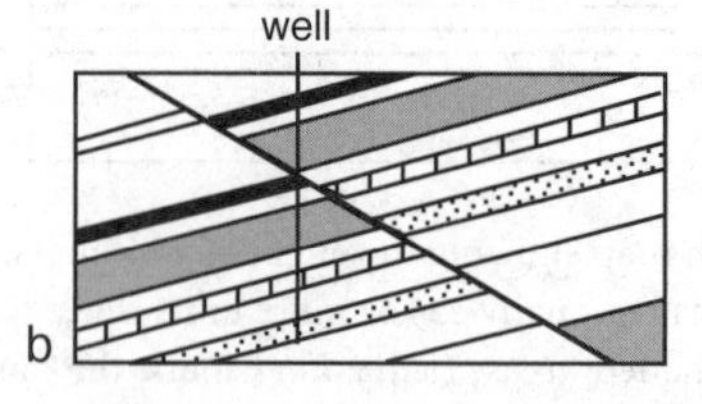

Fig. 5.8. Cross sections of faults recognized from a stratigraphic discontinuity where the well cuts a fault. **a** Normal separation. Part of the normal stratigraphic sequence is missing in a vertical well. **b** Reverse separation. Part of the normal stratigraphic sequence is repeated in a vertical well

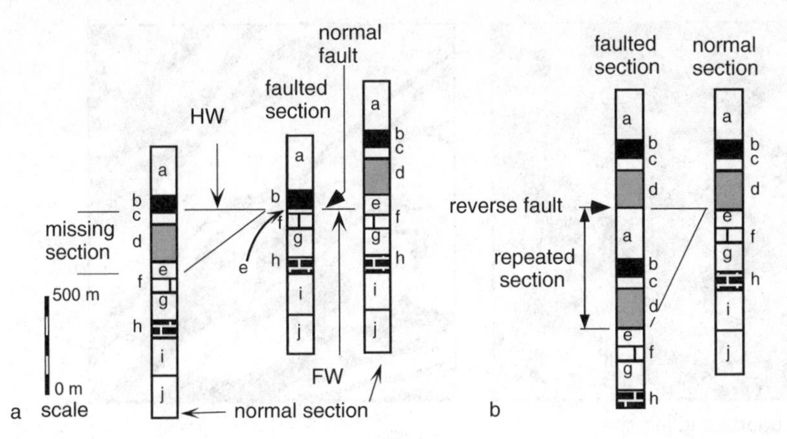

Fig. 5.9. Location and amount of stratigraphic separation determined from the missing or repeated section across the fault. **a** About 300 m of section is missing across a normal fault. **b** About 600 m of section is repeated across a reverse fault

upside-down stratigraphic column to the possible fault zone to see if an overturned section is present. Beds near the fault may be steeply dipping, causing a great exaggeration of the bed thicknesses in a vertical well, representing another factor to consider when correlating to the type section.

The precision of the determination of stratigraphic separation depends on the level of detail to which the stratigraphy is known. For example, if the fault cut in Fig. 5.9a was at the top of unit e rather than near the base, the missing section would be significantly (50–75 m) greater. The error in the amount of missing section is the sum of the stratigraphic uncertainties in the units on both sides of the fault. Lack of stratigraphic resolution means that small faults may not be detectable from correlation data alone.

The *fault cut* refers to the position of the fault in a well and the to the amount of the missing or repeated section compared to the reference section (Tearpock and Bischke 1991). The location of the fault cut, the amount of the fault cut and the reference section should be recorded for each fault cut. It is not unusual to find that after preliminary mapping has been completed, the original stratigraphic thicknesses were inappropriate. For example, the units in a reference well may be shown by mapping to have a significant dip and their thicknesses therefore will have been exaggerated. Mapping may also reveal that the reference section is faulted. Any change in thickness in the reference section requires changes in the magnitudes of all the fault cuts determined from it.

5.2.6
Rock Type

Rock types diagnostic of faulting (Fig. 5.10) include the cataclasite suite, produced primarily by fragmentation, and the mylonite suite, produced by large crystal-plastic strains and/or recrystallization (Sibson 1977; Ramsay and Huber 1987). A cataclastic fault surface, or slickenside, is usually grooved, scratched, or streaked by mineral

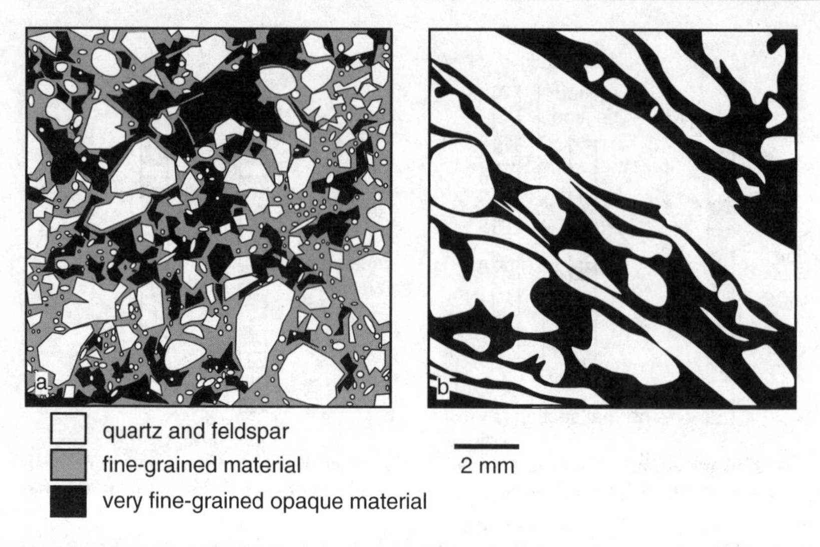

Fig. 5.10. Representative fault-rock textures. The scale is approximate. **a** Cataclasite. **b** Mylonite. Drawn from thin section photographs in Ramsay and Huber (1987)

fibers or may be highly polished. The parallel striations or mineral elongation directions are slickenlines which indicate the slip direction at some stage in the movement history. A mylonite typically has a thinly laminated compositional foliation and a strong penetrative mineral lineation which is elongated in the displacement direction.

A fault zone may be recognizable in well logs. A caliper log is produced by a tool that measures the size of the well bore and a fault may be recorded as an expansion in the size of the borehole. An uncemented cataclastic fault zone is mechanically weak and very friable. This is the reason why an outcropping fault zone may erode to a valley. It is also the reason for a characteristic response on a caliper log. Where the well encounters a mechanically weak unit, such as an uncemented fault zone, an enlargement of the well-bore diameter occurs. Well enlargement may occur in any mechanically weak unit such as coal, uncemented clay or sand, salt, etc. If the normal stratigraphic reasons for well enlargement can be eliminated, however, then the presence of a fault zone is a good possibility. The latest generation of extremely high resolution dipmeters provides a log that resembles a photograph of the wall of the borehole. A fault zone may be inferred from the same observational criteria that would apply to an outcrop, in particular, a high concentration of fractures or an abrupt change in lithology not part of the normal stratigraphic sequence.

5.2.7
Fault Drag

The systematic variation of the dip of bedding adjacent to the fault shown in Fig. 5.11b is known as "normal" drag, or simply *drag*, as the term will be used here. A drag fold gives the sense of slip on the fault. This geometry is probably best interpreted as being the result of the permanent strain that occurs prior to faulting (Fig. 5.11a), a sequence of formation well known from experimental rock mechanics. The term "drag" is thus

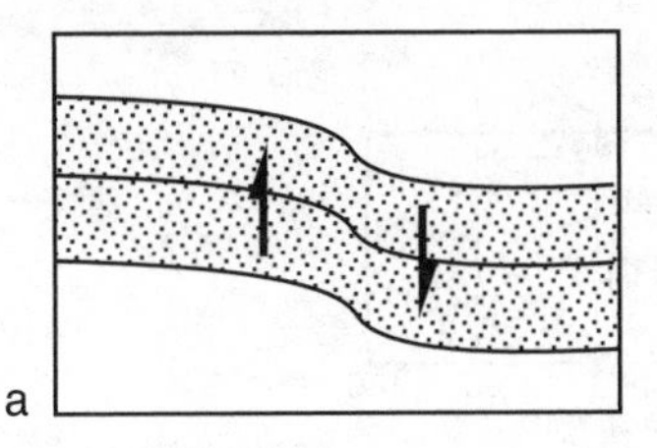

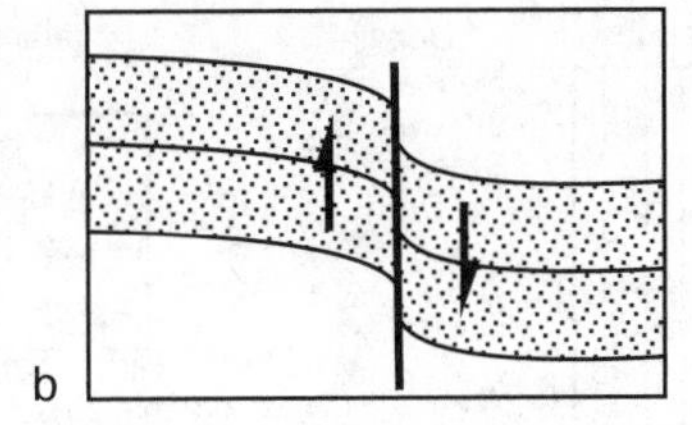

Fig. 5.11. Drag folds caused by permanent bending strain before faulting. **a** Fold before the units reach their ductile limit. **b** A fault forms after the ductile limit is exceeded, and the permanent strain remains as a drag fold

a misnomer because the fold forms prior to faulting, not as the result of frictional drag along a pre-existing fault. In fact, the strain associated with slip on a pre-existing fault tends to produce bending in the opposite direction relative to the sense of slip (Reches and Eidelman 1995) called "reverse" drag. The reverse drag described by Reches and Eidelman (1995) is significantly smaller in magnitude than the normal drag produced before faulting and might not be noticeable at the map scale. Map-scale reverse drag is also well known as a consequence of slip on a downward-flattening fault (Hamblin 1965). Because the term drag is long established for the folds close to a fault and because the curvature of drag folds gives the correct interpretation of the sense of shear, use of the term is continued here. The drag-fold geometry usually extends no more than a few to tens of meters away from the fault zone. Although the drag geometry is common, not all faults have associated drag folds.

5.3 Dip Sequence Analysis

The drag geometry (Sect. 5.2.7) provides the basis for fault recognition by dip sequence (SCAT) analysis, which is performed on dip data collected along a traverse in the field or with a dipmeter in a well. See Section 4.6 for an introduction to dip sequence analysis.

5.3.1 Fault Interpretation

The presence of a fault is recognized from the distinctive cusp pattern on the transverse dip component plot (Figs. 5.12–5.15). A cusp is also present on the dip vs. depth plot but may not be as clearly formed. The cusp is caused by the dips in the drag fold adjacent to the fault and is expected to occur within a distance of meters to tens of meters from the fault cut. A traverse perpendicular to the fault plane will show the minimum affected width, whereas a traverse at a low angle to the fault plane, such as a vertical well drilled through a normal fault, will show the maximum width. The fault cut is at the depth indicated by the point of the cusp. The azimuth vs. depth plots distinguish between steepening drag that occurs where the faults dip in the direction of the regional dip of bedding (Fig. 5.12) and flattening drag that occurs where the fault dip is opposite to the regional dip of bedding (Fig. 5.13). Steepening drag maintains a constant dip direction whereas flattening drag may produce a reversal in the dip direction. A drag-

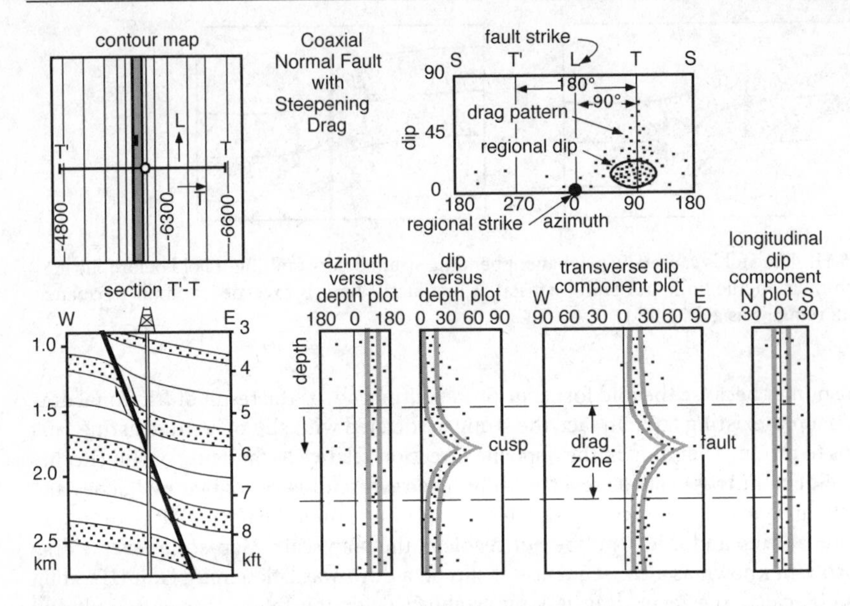

Fig. 5.12. Structure contour map, cross section, and SCAT plots for a normal fault whose drag steepens the regional dip. Fault strike is parallel to the regional strike. *L* regional strike; *T* regional down-dip direction; *T′* regional up-dip direction. (After Bengtson 1981)

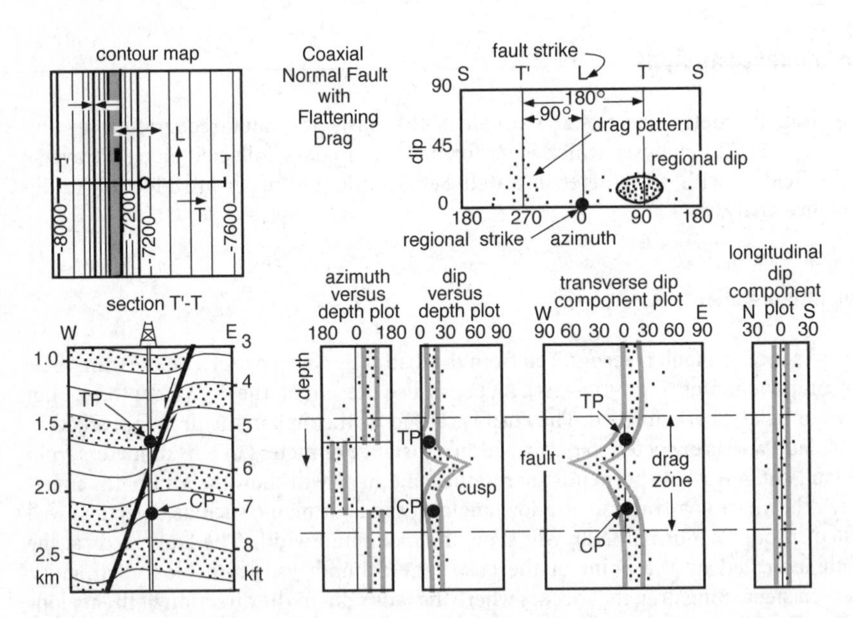

Fig. 5.13. Structure contour map, cross section, and SCAT plot for a normal fault whose drag flattens the regional dip. Fault strike is parallel to regional strike. *L* regional strike; *T* regional down-dip direction; *T′* regional up-dip direction; *CP* crestal plane; *TP* trough plane. (After Bengtson 1981)

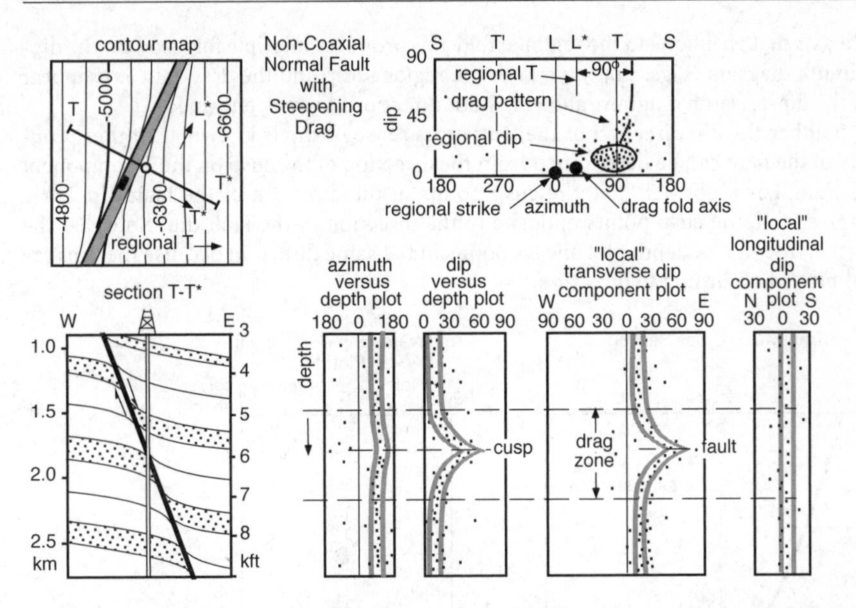

Fig. 5.14. Structure contour map, cross section, and SCAT plot for a normal fault striking oblique to regional dip with drag that steepens the regional dip. *L* regional strike; *T* regional down-dip direction; *T′* regional up-dip direction; *L** fault strike; *T** normal to fault strike. (After Bengtson 1981)

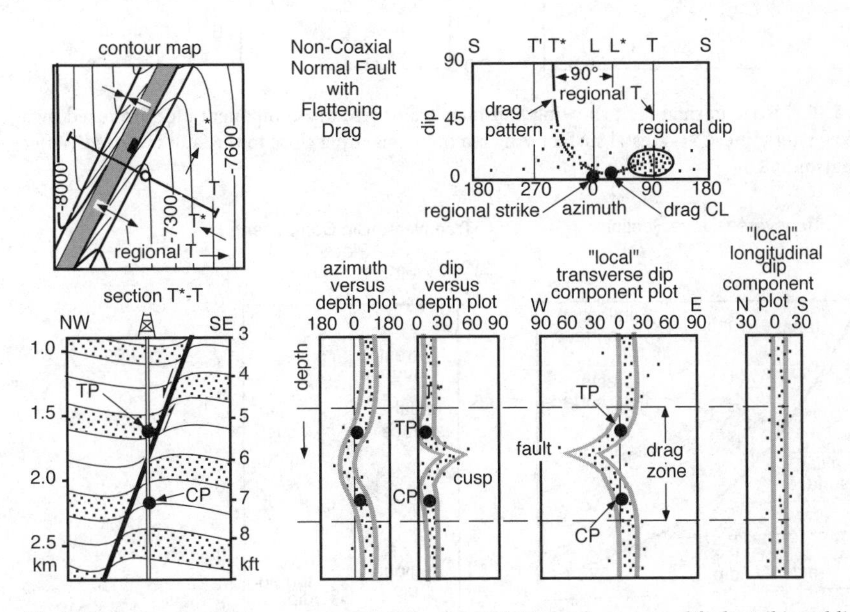

Fig. 5.15. Structure contour map, cross section, and SCAT plot for a normal fault striking oblique to regional dip with drag that flattens the regional dip. *L* regional strike; *T* regional down-dip direction; *T′* regional up-dip direction; *L** fault strike; *T** normal to fault strike; *CP* crestal plane; *TP* trough plane. (After Bengtson 1981)

fold axis that is oblique to the regional fold axis produces multiple fold axes on the dip-azimuth diagram (Figs. 5.14, 5.15). Both the regional dip and the drag-fold axis appear on the dip-azimuth diagram, allowing both directions to be determined.

If either the dip direction of the fault or its sense of slip is known, the other property of the fault can be determined from the direction of the cusp on the T component diagram. For a normal fault, the cusp points in the direction of the fault dip. For a reverse fault, the cusp points opposite to the direction of the fault dip. Note that the cusp on the dip vs. depth plot always points in the same direction because the dips are not plotted according to direction.

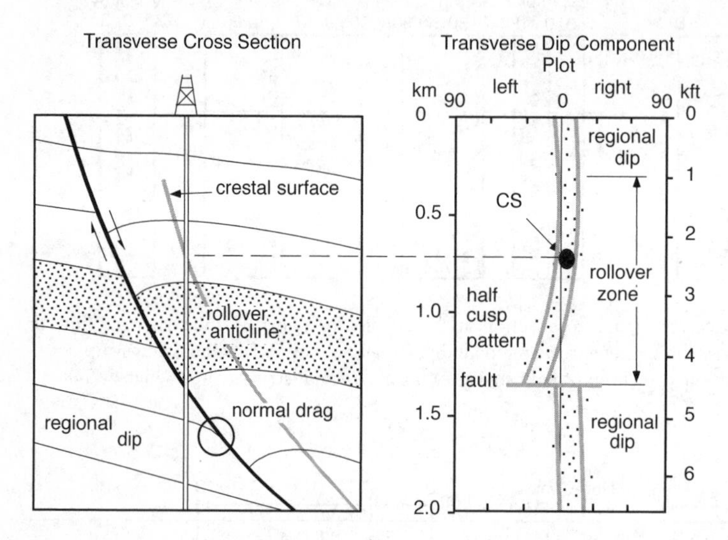

Fig. 5.16. Listric normal fault showing half-cusp transverse dip component plot produced by a rollover anticline. *CS* = crestal surface. An area of normal drag close to the fault is *circled*. (After Bengtson 1981)

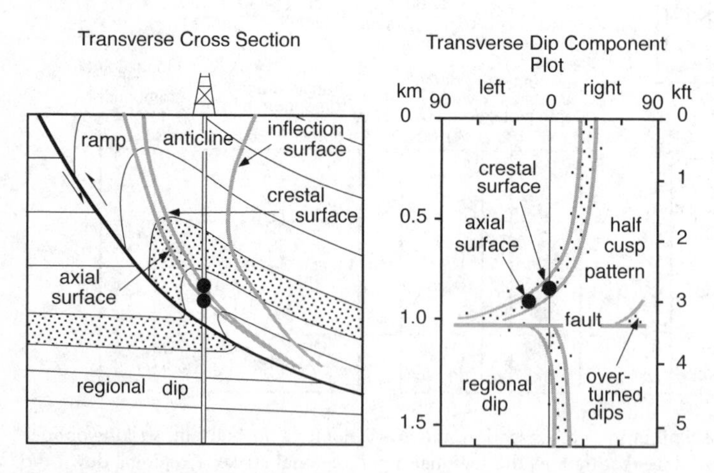

Fig. 5.17. Thrust-ramp anticline showing half-cusp transverse dip component plot. (After Bengtson 1981).

At both the local scale and the map scale, a fold may be present on one side of the fault but absent on the other, resulting in a half-cusp pattern. The examples in Figs. 5.16 and 5.17 are ramp-related anticlines, caused by displacement on curved faults. Folds of this type produce a large-scale half-cusp pattern on the transverse dip component plot (Figs. 5.16, 5.17). Fault displacement large enough to transport a ramp anticline from a ramp onto a flat will also produce a half-cusp pattern. Along a single fault, certain (ductile) units may form drag folds whereas other (brittle) units may break without the formation of drag folds. After displacement, units with drag folds may be juxtaposed against units without folds (Fig. 5.16), resulting in local half-cusp geometries.

5.3.2. Example

A portion of the northwest dipping forelimb of the major Greasy Cove–Wills Valley anticline in the southern Appalachian fold-thrust belt serves as an example of the dip-sequence analysis for finding and interpreting a fault (Fig. 5.18). The regional dip of the forelimb of around 30° northwest is seen outside the map area. On the azimuth-dip diagram (Fig. 5.19), the dips from the map area show a vertical scatter trend along the dip direction of the forelimb, as expected for a fault parallel to regional strike (compare with Fig. 5.12). The attitudes of bedding in the map area give the same T and L directions (Fig. 5.19) as determined with a tangent diagram of data from the entire southwest end of the anticline. The map (Fig. 5.18) shows the line of the dip traverse. Attitudes are projected onto the traverse according to their distances from the northwest end point (Table 5.1).

The azimuth distance diagram (Fig. 5.20a) shows the regional northwesterly dip with one aberrant point that dips to the southeast. The dip-distance diagram (Fig. 5.20b) shows a high degree of scatter that does not suggest any interpretation. The component plots significantly clarify the interpretation. Dips over about 80° are not practical to break into components because of their extreme magnitude, and so the 90° dip is shown on the component plots as an open circle at the edge of the diagrams. The T

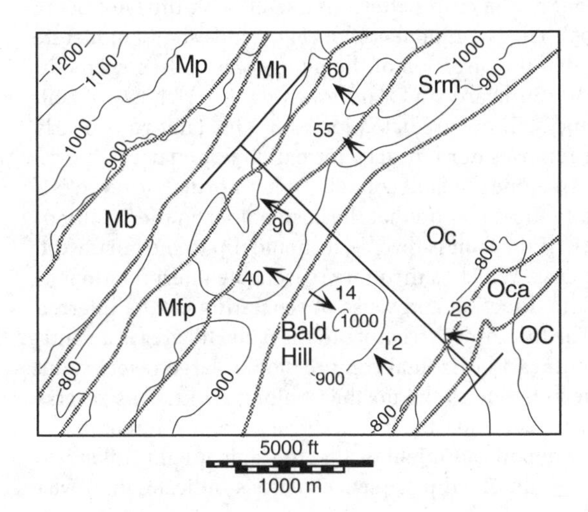

Fig. 5.18. Preliminary geologic map of the Bald Hill area, northwest limb of the Greasy Cove anticline, Alabama. Topographic contours (in feet) are *thin lines*, geologic contacts are *wide gray lines*. The fault contacts are not yet identified. Location of the northwest-southeast dip traverse is shown. (Modified from Burchard and Andrews 1947)

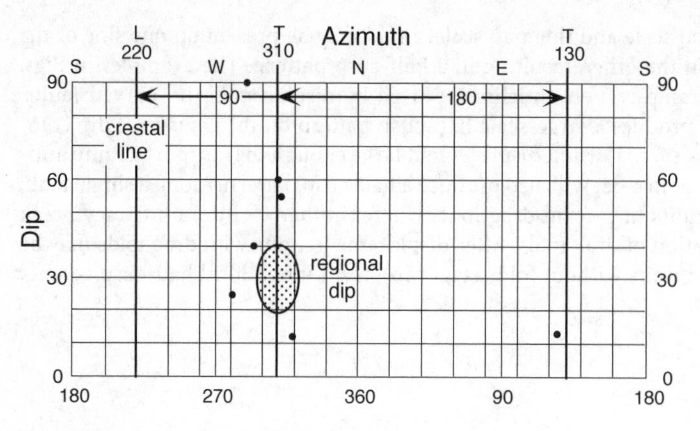

Fig. 5.19. Dip-Azimuth diagram for the Bald Hill map area

Table 5.1. Dip traverse data from the Bald Hill map area. Dips of 90° are at infinity on the tangent diagram

Distance from the northwest (km)	Attitude Dip, azimuth	T component	L component
0.54	60, 310	60 NW	0
0.70	90, 289	∞ NW	∞ SW
0.90	55, 311	55 NW	2 NE
1.10	40, 295	38 NW	12 SW
1.30	14, 124	13 SE	4 NE
2.38	12, 319	12 NW	3 NE
2.68	26, 281	23 NW	15 SW

component plot (Fig. 5.20c) suggests the cusp pattern of a fault, with the fault being close to the fourth point (40, 295). The L component plot (Fig. 5.20d) shows moderate scatter around zero, indicating that the direction of the plunge was chosen correctly. The map as originally given by Burchard and Andrews (Fig. 5.21) shows a fault between dip measurements 4 and 5, in the predicted location. This fault runs nearly parallel to bedding and neither removes nor repeats a formation boundary.

From the dip sequence analysis alone the fault could be either normal or reverse. If the fault is normal it should dip in the direction that the cusp points on the T component plot, that is, to the southeast. If the fault is reverse, it should dip to the northwest, opposite to the direction of the cusp. Which is the more reasonable interpretation?

The best choice of the fault dip direction and sense of separation can be inferred from the local structural style. The first-order structure in the Bald Hill area is an anticline produced by northwest-southeast horizontal compression. A large reverse fault crops out in the forelimb of the anticline to the northeast along strike. This reverse fault dips to the southeast, as is usual for the major thrusts in the forelimbs of Appalachian folds. A preliminary hypothesis might be that the fault at Bald Hill is also a southeastward dipping reverse fault. The dip sequence analysis indicates, however,

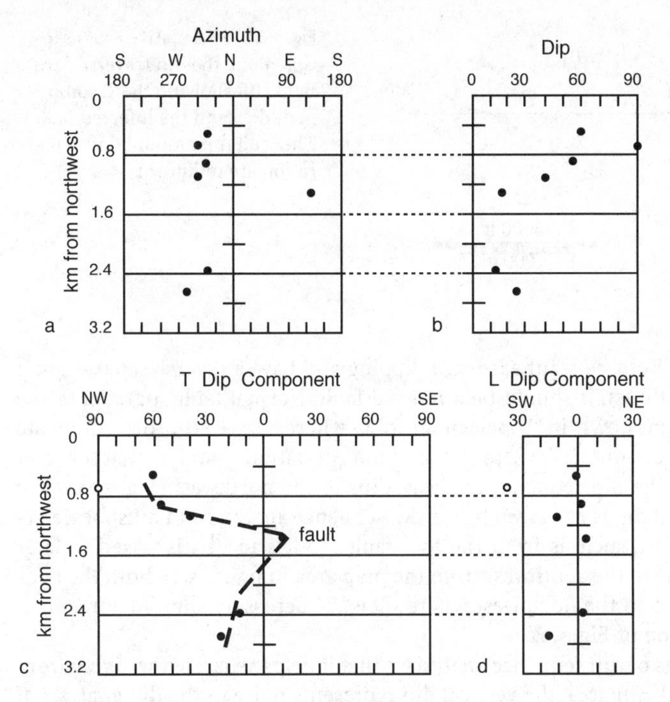

Fig. 5.20. Dip sequence analysis of a portion of the Greasy Cove anticline. **a** Azimuth-distance diagram. **b** Dip-distance diagram. **c** T component dip-distance diagram. *Open circle* is the 90° dip which plots off scale. **d** L component dip-distance diagram. *Open circle* is the 90° dip which plots off scale

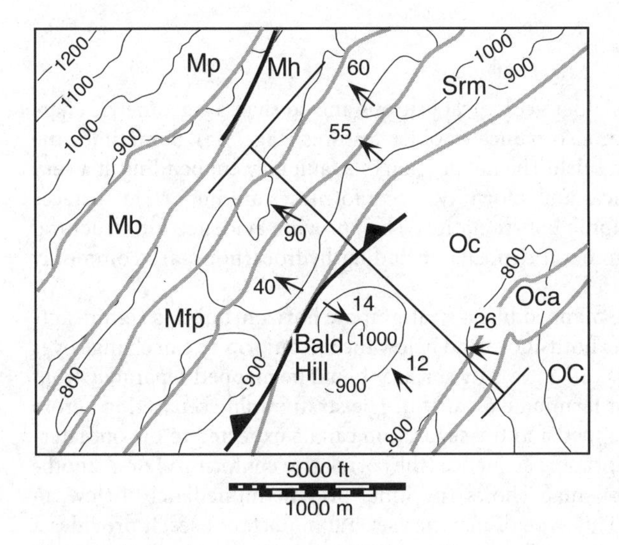

Fig. 5.21. Geologic map of the Bald Hill area. Topographic contours (in feet) are *thin lines*; geologic contacts are *wide gray lines*; faults are heavy solid lines. The fault at Bald Hill is interpreted to be a thrust that dips to the northwest. (After Burchard and Andrews 1947)

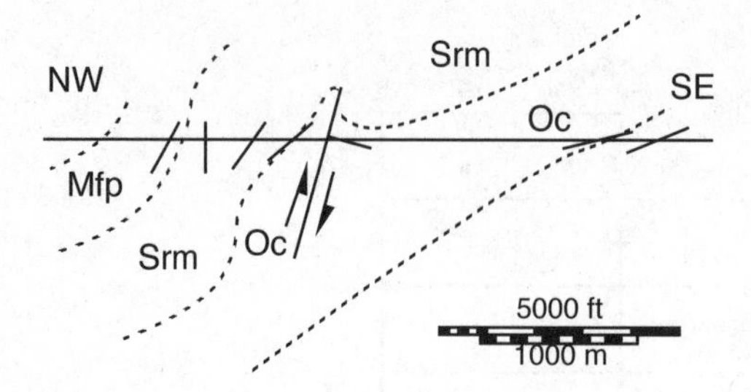

Fig. 5.22. Speculative cross section along the dip traverse across Bald Hill showing the T component dips and the inferred fault. The fault dip amount and separation are unknown

that the fault cannot both be southeastward dipping and have a reverse separation. If the dip is to the southeast, it should be a normal fault. Normal faults parallel to the strike are virtually unknown in Appalachian folds whereas second-order conjugate thrusts are relatively common. Thus, the favored interpretation would be that the fault is reverse and, from the dip sequence analysis, dips to the northwest (Fig. 5.22). The magnitude of the fault dip is not given by the dip sequence analysis and must be determined by other means, such as from the fold-fault models to be discussed in later chapters. Along strike to the southwest from the map area in figure 5.21, both the bedding and the fault dip to the northwest where they "V" across a valley, in agreement with the interpretation of Fig. 5.22.

One of the benefits of dip sequence analysis is that it helps separate the signal from the noise. In the Bald Hill area, the vertical dip represents noise in the dip analysis of the fault (Fig. 5.20). All the other dips are clearly consistent with the fault interpretation. The vertical dip does not contribute to the fault interpretation but can be included as part of a second-order fold on the speculative interpretation (Fig. 5.22).

5.4 Unconformities

An unconformity truncates older geological features and so shares a geometric characteristic with a fault which also truncates older features (Fig. 5.23). Sometimes the two can be difficult to distinguish. The flat portion of a fault may cut bedding at a very low angle, just like the typical unconformity. Unconformities are important surfaces for structural and stratigraphic interpretation. They provide evidence for structural and sea-level events, and mineral deposits, including hydrocarbons, are commonly trapped below unconformities.

An unconformity can be mapped like a stratigraphic horizon, but it is significantly different because the units both above and below the unconformity can change over the map area (Fig. 5.24a). An unconformity surface should be mapped separately from the units that it cuts or that terminate against it. The stratigraphic separation across an unconformity can be mapped and the separation can be expected to die out laterally into a continuous stratigraphic sequence (the correlative conformity) or at another unconformity. A subcrop map shows the units present immediately below an unconformity (Fig. 5.24b). This type of map has several important uses. It provides a guide to the paleogeology, can indicate the trends of important units, for example hydrocarbon reservoirs, and can suggest the paleostructure. An analogous map type is a subcrop map of the units below a thrust fault.

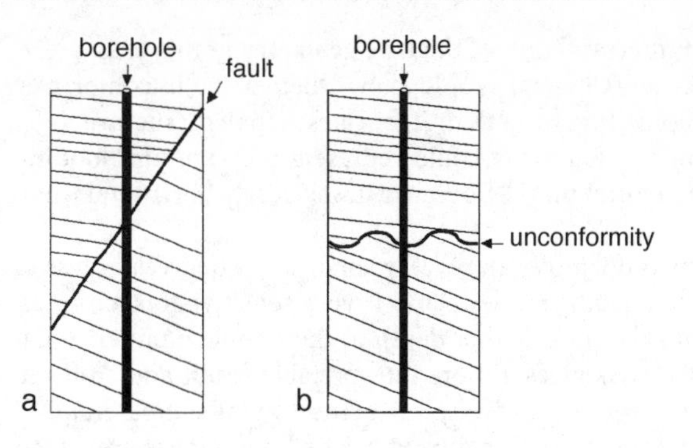

Fig. 5.23. A dip change across a stratigraphic discontinuity may be either a fault or an unconformity. **a** Fault. **b** Unconformity. (After Hurley 1994)

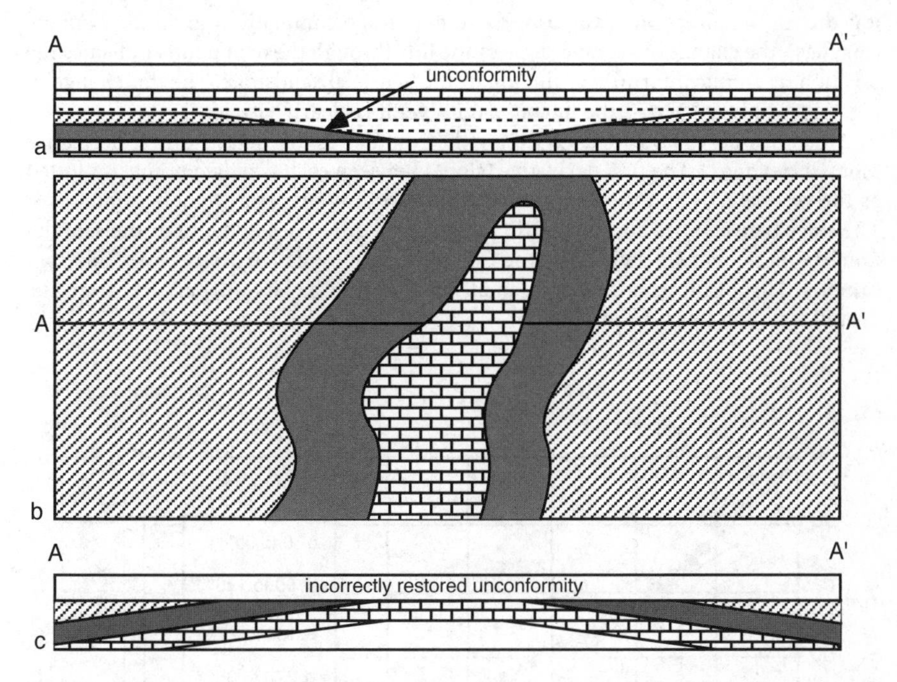

Fig. 5.24. The geometry of a low-angle unconformity. **a** Cross section showing a filled paleo-valley. Vertical exaggeration is about two to three times. **b** Subcrop map of units below the unconformity. **c** Paleo-valley converted into a paleo-arch by incorrectly restoring the unconformity to horizontal. (After Calvert 1974)

As map horizons for structural interpretation, unconformities contain an important pitfall: they were not usually horizontal when they formed (Calvert, 1974). The original topography on an unconformity surface (Fig. 5.24a) will be converted into a false structure by restoration of the unconformity to horizontal (Fig. 5.24c). An analogous problem occurs if the thickness variation of the unit directly above an unconformity is interpreted as being a paleostructure when it, in fact, represents paleotopography. The thickest shale overlying the unconformity in Fig. 5.24a might be interpreted as representing a growth syncline and the adjacent thinner areas as being growth anticlines, an incorrect interpretation of the true geometry in this example. A carefully constructed cross section (Chap. 7) should show whether the unconformity is developed on paleotopography (Fig. 5.24a) or truncates a paleostructure (Fig. 5.24c). The upper limestone unit in Fig. 5.24a could represent a maximum flooding surface and hence be a good structural marker, because it was nearly horizontal when it was deposited.

The dip change across an unconformity can be very small, a few degrees or less. A plot of the cumulative dip versus depth (Hurley 1994) is very sensitive to such small changes (Fig. 5.25). The vertical axis can be either depth or the sample number. Equal spacing of points on the vertical axis gives a more interpretable result and so if the bedding planes are sampled at irregular intervals, it is better to plot sample number than true depth or distance (Hurley 1994). The dips of the bedding planes are added together in the direction of the traverse to obtain the cumulative dip. In the example (Fig. 5.25), the change in average slope of the line through the data points indicates the position of an unconformity at the top of the Tensleep sandstone. The dip change in Fig. 5.25 is only 1.3° across the unconformity, yet it is clearly visible.

An additional analysis technique for the cumulative dip data is a first derivative plot (Hurley 1994). The first derivative (slope) between each two data points is plotted as the vertical axis and the cumulative dip of bedding planes as the horizontal axis. The slope between two successive dip points is the difference in depth or the sample-number increment (1) divided by the difference in cumulative dip between adjacent samples. Abrupt dip changes, even small ones, appear as large departures from the overall trend on this type of plot.

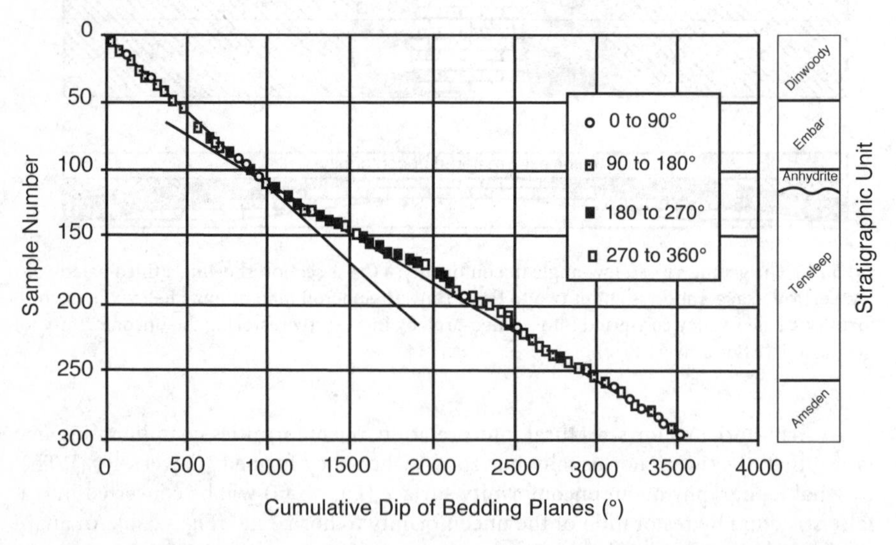

Fig. 5.25. Cumulative dip versus sample number from a well in the Bighorn basin. Samples are *numbered* from the top of the well. Bedding planes are derived from a high-resolution dipmeter log. Points falling into different quadrants of the compass are coded with different *symbols*. (After Hurley 1994)

5.5 Displacement

Displacement is the general term for the relative movement of the two sides of a fault, measured in any direction (Bates and Jackson 1987). The components of displacement depend on five measurable attributes: the attitude of the fault, the magnitude of the stratigraphic separation, the attitudes of the beds on both sides of the fault, and direction of the slip vector in the fault plane. Different sets of these attributes are used to define the displacement components of slip, separation, throw and heave, as will be discussed next. Throw and heave are the components usually most useful in map construction.

5.5.1 Slip

Slip (or net slip) is the relative displacement of formerly adjacent points on opposite sides of the fault, measured along the fault surface (Bates and Jackson 1987). Slip is commonly described in terms of its components in the plane of the fault, dip slip and strike slip (Fig. 5.26a). Dip slip components are normal or reverse and strike slip components are right lateral or left lateral. The offset of a marker horizon, as recorded on a map, can be produced by slip in a variety of directions. The geometry of the fault and the offset horizon are identical in Figs. 5.26a,b but the fault in 5.26a is oblique slip and in 5.26b is pure dip slip.

The direction of the net slip may be given by the direction of the grooves or slickenlines observed on the fault surface or by the trend of the lineation in the fault-zone rock. The slip direction is usually perpendicular to the axis of the drag fold at the border of the fault zone, which can be observed in outcrop or be inferred by dip sequence analysis (Sect. 5.3, see also Becker 1995). The axes of drag folds within a fault zone (Sect. 4.7) are approximately normal to the slip direction, but may be arcuate, in which case the slip direction bisects the arc (Hansen 1971). Slip vector determination contains inherent ambiguities that must be considered, however. Different slip directions may be overprinted with only the last increment being observed or the slip trajectories may be curved. An unambiguous slip direction is the orientation of the line joining two points that can be correlated across the fault (Fig. 5.26). The orientation of the slip vector may be recorded as the bearing and plunge of the lineation, or as the rake, which is the angle from horizontal in the plane of the fault.

There are two general approaches to finding the amount of the net slip on a fault. The most direct is to find the offset of a geological line that can be correlated across the fault. A geological line may be a geographic or a paleogeographic feature such as a stream channel or a facies boundary. The most complete method for the determination of the slip from the offset of geological features that intersect the fault surface is based on fault-surface cross sections (see Sect. 7.7.2). An alternative approach is based on the distance in the slip direction between correlative planar surfaces. In this method the slip direction must be known, for example, from the slickenline direction. To determine the fault slip, draw the slip vector on the structure contour map of the fault so that it connects the correlated points on the hangingwall and the footwall as in Fig. 5.26. If the end points of the slip vector are specified by their xyz coordinates, the length of the slip vector in the plane of the fault is (from Eq. 2.2)

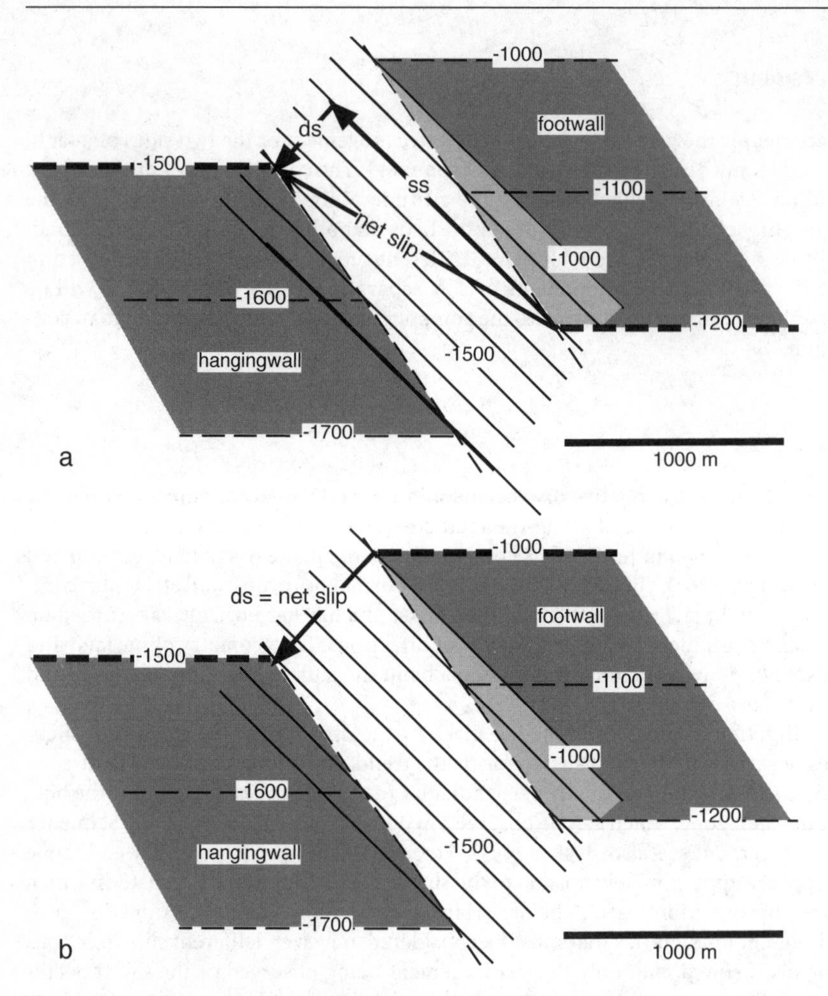

Fig. 5.26. Displacement of a marker horizon (*shaded*, with *dashed contours*) across a fault (*unshaded*, with *solid contours*). The final geometries of **a** and **b** are identical although the slip is different. *Wide contour lines* represent correlative linear features displaced by the fault. *ds* dip slip; *ss* strike slip. **a** Right-lateral, normal oblique slip. **b** Pure normal dip slip

$$L = [(x_2 - x_1)^2 + (y_2 - y_1)^2 + (z_2 - z_1)^2]^{1/2}, \tag{5.1}$$

where L = the slip and the subscripts 1 and 2 = the coordinates of the opposite ends of the slip vector. If the end points are specified by the horizontal and vertical distance between them (from Eq. 2.28)

$$L = (v^2 + h^2)^{1/2}, \tag{5.2}$$

where L = the slip, v = vertical distance between end points, and h = the horizontal distance between end points.

Without the information provided by the correlation of formerly adjacent points across the fault or knowledge of the slip vector, it is impossible to determine whether a fault is caused by strike slip, dip slip, or oblique slip. Map interpretation must frequently be done without knowledge of the slip amount or direction, a situation for which the fault separation, given next, provides the framework for the interpretation.

5.5.2 Separation

Separation is the distance between any two index planes disrupted by a fault (Dennis 1967; Bates and Jackson 1987). Several different separation components are commonly used to describe the magnitude of the fault displacement. Strike and dip separations (Fig. 5.26) are, respectively, the separations measured along the strike and directly down the dip of the fault (Billings 1972). The stratigraphic separation (Fig. 5.27) is the

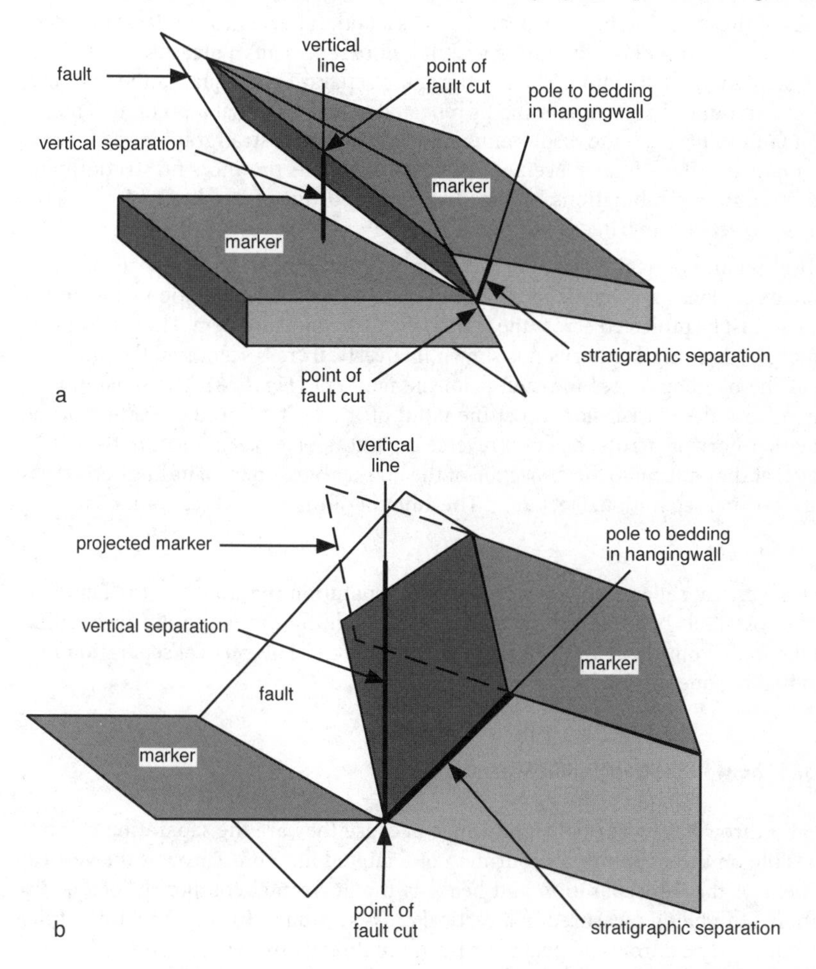

Fig. 5.27. Vertical and stratigraphic fault separation. **a** Reverse fault. **b** Normal fault: the marker horizon must be projected across the fault in order to measure the vertical separation

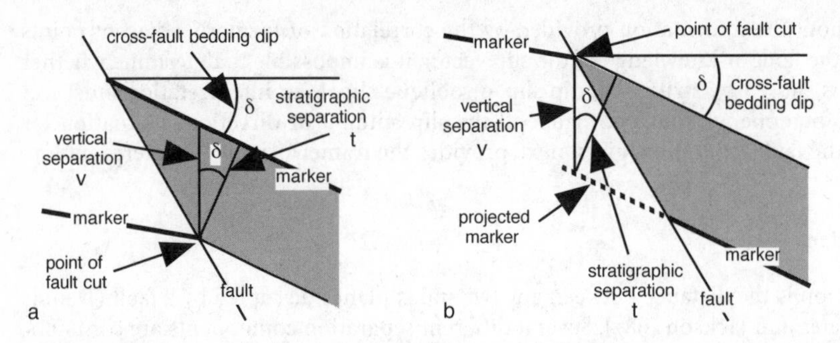

Fig. 5.28. Vertical separation calculated in a vertical cross section in the direction of the cross-fault bedding dip. **a** Reverse separation fault. **b** Normal separation fault.

thickness of the beds missing or repeated across a fault (after Bates and Jackson 1987). Stratigraphic separation is a thickness, and therefore represents a measurement direction perpendicular to bedding. The stratigraphic separation is equal to the fault cut, that is, the amount of section missing or repeated across a fault at a point. It is possible for a fault to have a large displacement and yet show no stratigraphic separation. For example, a strike-slip displacement of horizontal beds produces no stratigraphic separation. Other combinations of slip direction and dip of marker beds can also result in zero separation (Redmond 1972).

Vertical separation is the distance, measured vertically, between two parts of a displaced marker (Fig. 5.27; Dennis 1967). In the case of a normal-separation fault, one of the planes must be projected across the fault to make the measurement. The vertical separation of an offset marker horizon is shown in a vertical cross section in the direction of dip of the bedding across the fault from the fault cut (Fig. 5.28). The separation is measured from the marker horizon at the point of the fault cut to the position of the same marker horizon across the fault (reverse separation; Fig. 5.28a) or from the marker horizon at the fault cut to the *projection* of the marker horizon from its location across the fault (normal separation, Fig. 5.28b). The amount of the vertical separation is

$$v = t / \cos \delta, \tag{5.3}$$

where v = vertical separation, t = stratigraphic separation = amount of the fault cut, and δ = cross-fault bedding dip. The cross-fault bedding dip is the dip of bedding across the fault from the marker horizon at the fault cut. The vertical separation of a vertical bed is undefined.

5.5.3 Heave and Throw from Stratigraphic Separation

Heave and throw have a special significance because they are the separation components visible on the structure contour map of a faulted horizon. Throw is the vertical component of the dip separation and heave is the horizontal component of the dip separation, both being measured in a vertical cross section in the dip direction of the fault (Dennis 1967). Throw and heave can be found directly from the stratigraphic separation. The following discussion refers to the geometry shown in Fig. 5.29. Let point P_1 be the location of the fault cut in a well or exposed in outcrop. The marker horizon

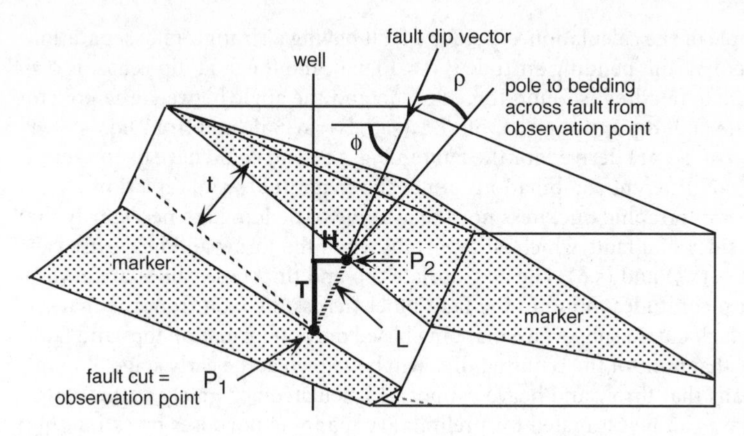

Fig. 5.29. Fault heave (*H*) and throw (*T*) at a fault cut (point P_1). *L* Dip separation; $H = L \cos \phi$; $T = L \sin \phi$; *t* stratigraphic separation = fault cut; ϕ dip of the fault; ρ angle between the fault dip and the pole to the cross-fault bedding attitude at P_2. The marker horizon is *shaded*

is shaded. P_2 is the location of the marker horizon in a vertical plane oriented in the direction of fault dip. The calculation of throw and heave is a projection across the fault from the control location (P_1) to a predicted location (P_2). The stratigraphic separation at the point of the fault cut is the thickness of the missing or repeated section, t. The dip separation, L, is equal to the apparent thickness of the missing or repeated section in the direction of the fault dip (dip vector) between points P_1 and P_2. The dip separation can be found from Eq. (2.24) as

$$L = t / \cos \rho, \tag{5.4}$$

where t = the stratigraphic separation and ρ = the angle between the pole to the cross-fault bedding attitude and the dip vector of the fault. The dip of bedding used is always that belonging the side of the fault to which the projection is being made (P_2), that is, across the fault from the fault cut on the marker horizon. The value of ρ can be found with a stereonet (Fig. 2.28) or from Eqs. (2.25) and (2.26) or (2.27) and (2.28).

The throw and heave are determined from the stratigraphic separation with the following equations based on the geometry of Fig. 5.29 and using the value of L from Eq. (5.4):

$$H = t \cos \phi / \cos \rho; \tag{5.5}$$

$$T = t \sin \phi / \cos \rho, \tag{5.6}$$

where H = heave, T = throw, t = stratigraphic separation, ϕ = the dip of the fault, and ρ = the angle between the cross-fault bedding attitude and the dip vector of the fault. If bedding is horizontal, its pole is vertical, causing ρ to be equal to 90 - ϕ, and Eq. (5.5) and (5.6) reduce to:

$$H = t / \tan \phi; \tag{5.7}$$

$$T = t. \tag{5.8}$$

In the case of zero dip of bedding, and only in the case of zero dip of bedding, the stratigraphic separation is equal to the fault throw and equal to the vertical separation.

As an example of the calculation, consider a fault having a stratigraphic separation of 406 m, the cross-fault bedding attitude is $\delta = 10, 180$, and the fault dip vector is $\phi = 37, 220$. The pole to bedding is found to be 80, 360 and the angle between the pole to bedding and the fault dip is $\rho = 60°$. From Eq. (5.5), H = 648 m, and from Eq. (5.6), T = 489 m. These values are derived for the fault in Fig. 5.26 and could have been caused by any number of different combinations of net slip magnitudes and directions.

Neither the stratigraphic thickness nor the attitude of bedding are necessarily the same on both sides of a fault, which is important when the throw and heave are calculated from Eq. (5.5) and (5.6). The appropriate dip and thickness values are always the cross-fault magnitudes found in the fault block across the fault from the marker horizon at the fault cut. In subsurface mapping based only on formation tops and fault cuts, it is likely that none of the required dips will be known at the early stage of mapping. This means that throw and heave cannot be calculated accurately at this stage. Throw and heave can be estimated for preliminary mapping purposes by estimating the dip of the fault and the dip of cross-fault bedding. If no other information is available, normal-separation faults can be estimated to dip 60°, reverse separation faults to dip 30°, and bedding can be assumed to be horizontal. As the required dips are determined by mapping, the throw and heave magnitudes can be corrected.

Equations (5.5) and (5.6) can be solved to calculate the stratigraphic separation from either the throw or the heave,

$$t = H \cos \rho \,/\, \cos \phi; \tag{5.9}$$

$$t = T \cos \rho \,/\, \sin \phi. \tag{5.10}$$

5.6 Geometric Properties of Faults

In this section the typical characteristics of an individual fault and its displacement distribution are described.

5.6.1 Surface Shape

A fault surface is usually planar to smoothly curved or gently undulating (Fig. 5.30). The structure contour map of a fault surface should usually be smooth. Viewed over their entire surface, many faults are smoothly curved into a spoon-like shape. Primary surface undulations, if present, are typically aligned in the slip direction and provide a good criterion for the slip direction (Thibaut et al. 1996).

5.6.2 Displacement Distribution

The displacement on a fault dies out at a tip line which is the trace in space of the termination of a fault (Boyer and Elliott 1982). If the displacement dies out in all directions (Barnett et al. 1987), the fault is surrounded by a tip line (Fig. 5.31) like a dislocation in the theory of crystal plasticity (Nicolas and Poirier 1976). Faults are typically significantly longer in the strike direction than in the dip direction. Faults always die out along strike (Figs. 5.31, 5.32a) or transfer displacement to some other fault or to a

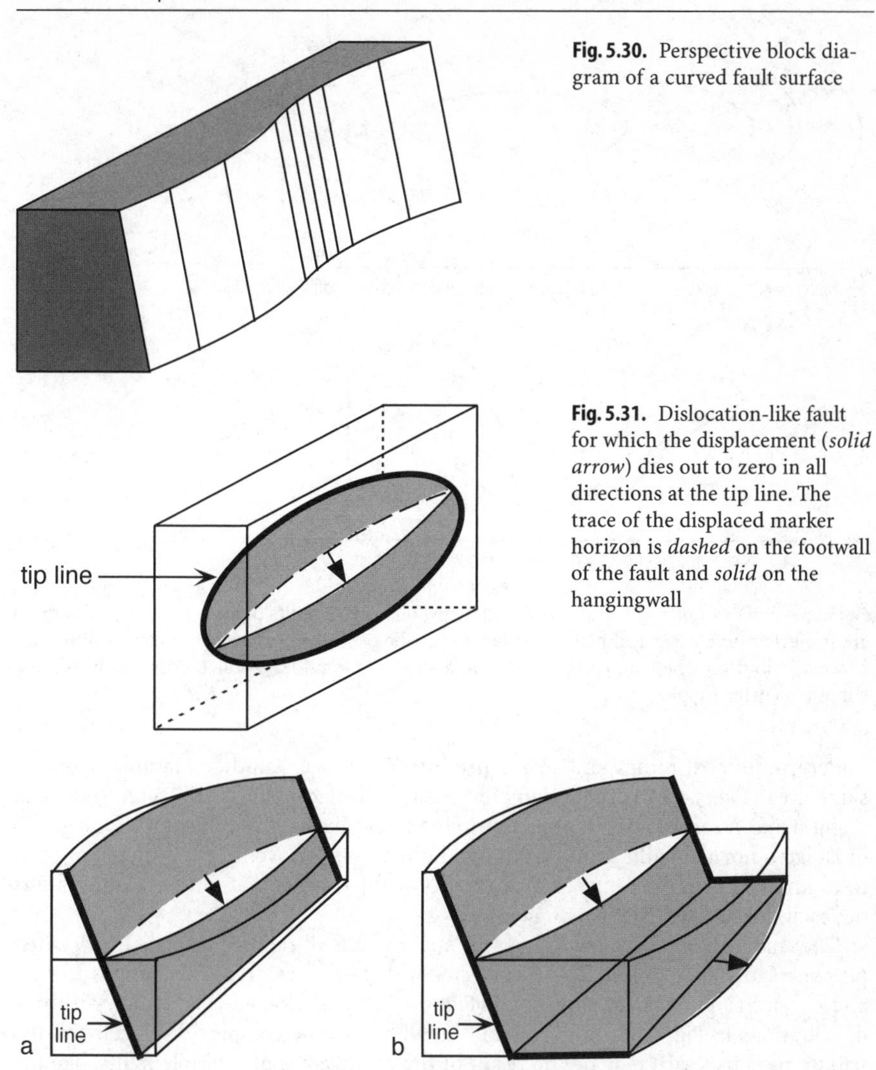

Fig. 5.30. Perspective block diagram of a curved fault surface

Fig. 5.31. Dislocation-like fault for which the displacement (*solid arrow*) dies out to zero in all directions at the tip line. The trace of the displaced marker horizon is *dashed* on the footwall of the fault and *solid* on the hangingwall

Fig. 5.32. Faults for which the displacement does not die out in the dip direction. The displacement (*solid arrows*) dies out along strike at the tip lines marked by *heavy lines*. The trace of the displaced marker horizon is *dashed* on the footwall of the fault and *solid* on the hangingwall. **a** Fault of unspecified extent in the dip direction. **b** Fault that bends into a planar lower detachment without losing displacement

fold. Displacement in the dip direction may or may not go to zero. A fault may flatten into a detachment without losing displacement (Fig. 5.32b; Gibbs 1989). The variation in displacement in the dip direction is a function of the structural style and a variety of relationships are possible.

An accurately mapped example from the Westphalian coal measures in the United Kingdom (Fig. 5.33a) has a displacement distribution nearly as simple as that in Fig. 5.31. The fault-displacement maps are derived from mapping at five different levels in

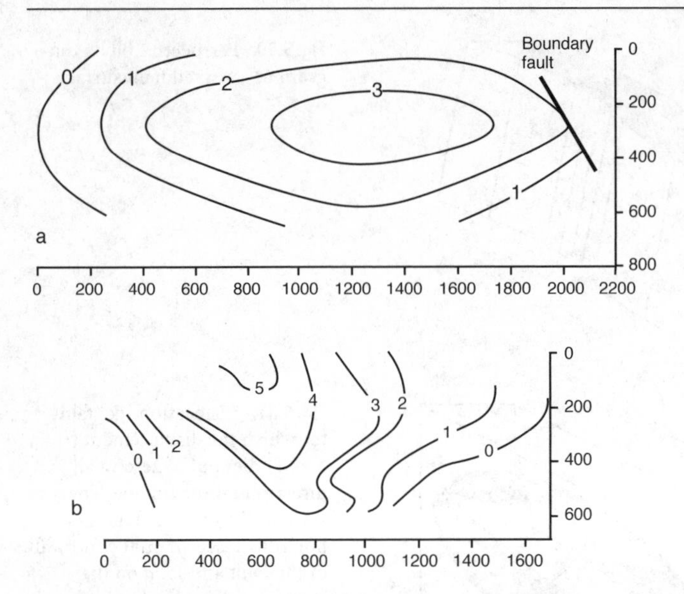

Fig. 5.33. Two examples of contours of fault throw on normal faults dipping about 65°. Contours are projected onto a vertical plane that has the strike of the fault. Scales in meters. **a** Elliptical heave distribution. The fault ends to the right against the boundary fault. **b** Complex heave distribution. (After Rippon 1985)

underground coal mines and so require little inference. Another example from the same area (Fig. 5.33b) is more complex but the displacement is still seen to die out along strike. A representative aspect ratio is 2.15 (strike length/dip length) for a group of isolated normal faults that die out in all directions for over a length range of 10 m to 10 km, and is in the range of 0.5 to 8.4 for normal faults that intersect other faults or reach the surface (Nicol et al. 1996).

The change in displacement along the strike of a fault can be illustrated with a displacement-distance graph. The displacement-distace curve typically ranges from a circular arc (Fig. 5.34a) to sinusoidal (Fig. 5.34b) to "D-shaped (Fig. 5.34c). All three distributions in Fig. 5.34 come from one small area. More complex displacement distributions (Fig. 5.33b) may be the result of the coalescence of multiple faults, like the two faults in Fig. 5.34c.

The smooth variation of displacement along a fault in the style of Figs. 5.33a and 5.34 has been described as the bow and arrow rule (Elliott, 1976). The trace of a faulted horizon on one side of the fault (fault cutoff line) forms the bow and a straight line joining the tips of the fault is the bow string. A line perpendicular to the bow string at its center is the arrow. The distance along the arrow between the bow and the bow string is an estimate of the displacement amount and the direction of the arrow is an estimate of the displacement direction. These are reasonable first approximations for dip slip faults although they should be used with caution. If the displacement distribution on the fault resembles that in Fig. 5.31, then the displacement with respect to the correlative horizon across the fault is twice that given by the bow and arrow rule because both hangingwall and footwall are displaced from their original positions (the position of the bow string). The rule does not apply to strike-slip faults.

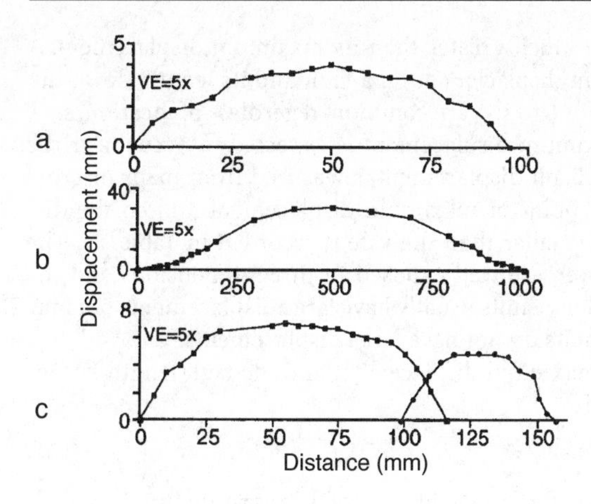

Fig. 5.34. Total displacement versus distance plots for completely exposed small normal faults. **a** Approximately circular-arc distribution. **b** Approximately sinusoidal distribution. **c** Two overlapping faults with D-shaped distributions. *VE* vertical exaggeration. (Schlische et al. 1996)

Table 5.2. Ratio of maximum displacement to length for a variety of faults that die out in tip lines in the direction of the measurement. The first two measurements from Rippon (1985) are from Fig. 5.33a, the third is from Fig. 5.33b

Max. displ./ length	Direction measured	Size range	Fault type	Location	Reference
1/10 to 1/20	Dip	40–450 cm	Normal	Japan	Muraoka and Kamata (1983)
1/10 to 1/20	Strike	10–400 km	Thrust	Canadian Rocky Mts.	Elliott (1976)
1/8	Strike	1.5–21 km	Strike slip	Iran	Freund (1970)
1/82	Strike	2.5 km	Normal	Alabama	Ch. 7, Fig. 7.46
1/700	Strike	2 km	Normal	Derbyshire, UK	Rippon (1985)
1/450	Dip	1 km	Normal	Derbyshire, UK	Rippon (1985)
1/340	Strike	1.5 km	Normal	Derbyshire, UK	Rippon (1985)
1/90	Strike	10–200 m	Normal	Western USA	Dawers et al. (1993)
1/125	Strike	0.2–10 km	Normal	western USA	Dawers et al. (1993)
1/30 to 1/50	Strike	0.2–10 km	Normal	Timor Sea	Nicol et al. (1996)
1/33	Strike	1–123 cm	Normal	Eastern USA	Schlische et al. (1996)

The length of a fault is usually much greater than its maximum displacement. A comparison between the maximum displacement on a fault and its length, down dip or along strike, shows that ratios of 1:8 to 1:33 are common, regardless of location, size, or fault type (Table 5.2). The maximum displacement is expected to occur near the center of the fault as in Fig. 5.33a. Fault displacements measured from maps or cross sections may not go through the point of maximum displacement and so the displacement/length ratios could be smaller than the values recorded in Table 5.2. The examples of Rippon (1985), however, are well exposed in three dimensions and the ratios are very large. In summary, long faults usually have large displacements but may have small displacements; short faults do not have large displacements.

The relationship between the maximum displacement and the fault length is generally considered to have the form

$$D = \gamma L^C, \tag{5.11}$$

where D = maximum displacement, γ = a constant of proportionality, L = fault length, and C is between 1 and 2 (Watterson 1986; Marrett and Allmendinger 1992; Dawers et al. 1993). The exact relationship appears to depend on many factors including the mechanical stratigraphy and the nature of interactions between overlapping faults (for example, Cowie and Scholz 1992). For the practical estimation of the displacement–length relationship in map interpretation, it appears satisfactory to let C = 1 and recognize that the value of γ may change when the size of the fault changes by an order of magnitude and will be different in different locations (Cowie and Scholz 1992; Dawers et al. 1993; Schlische et al. 1996). With C = 1, the value of γ is the D/L ratio given in Table 5.2. This relationship can be used to estimate the length of a fault from its maximum displacement or estimate the maximum displacement from the length.

5.7 Growth Faults

A growth fault is a fault that moved during deposition and controlled the thickness of the deposits on both sides if the fault. The classic Gulf of Mexico thin-skinned extensional growth fault, from the region where the concept was developed (Fig. 5.3), is depositional on both sides and the sediments thicken across the fault. A growth fault in a rifted environment typically shows footwall uplift and erosion concurrent with deposition on the hangingwall. Growth faults can be normal (Fig. 5.35) or reverse (Fig. 5.36); in fact, any type of fault can record growth, including strike-slip faults.

5.7.1 Effect on Heave and Throw

Because the stratigraphic thicknesses are different on opposite sides of a growth fault, the heave and throw determined from the fault cut depend on whether the marker being mapped is in the hangingwall or footwall of the fault. The appropriate thickness is that belonging to the section across the fault from the fault cut in the marker horizon. For a marker on the hangingwall of a normal fault (Fig. 5.35a), the appropriate fault-cut thickness is that of the footwall stratigraphic section. To map a footwall marker in the same well (Fig. 5.35b), the appropriate thickness is that of the hanging-

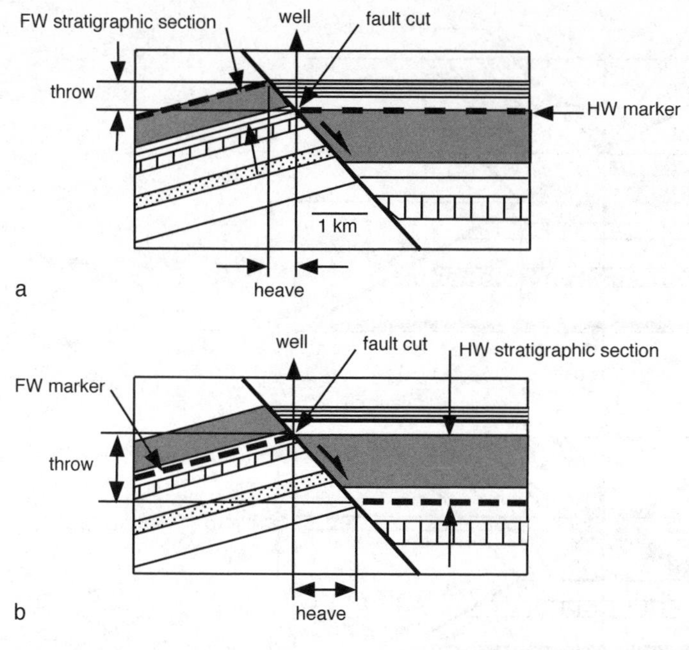

Fig. 5.35. Identical cross sections across a growth normal fault showing effect of growth on throw and heave. The section is vertical and in the direction of fault dip. The fault cut is at the same point in both sections. **a** The fault cutoff of a hangingwall marker is extrapolated to the footwall cutoff of the same marker using dip and thickness values from the footwall. **b** The fault cutoff of a footwall marker is extrapolated to the hangingwall cutoff of the same marker using dip and thickness values from the hangingwall

wall stratigraphic section. The same considerations apply to the correct choice of thickness for growth reverse faults (Fig. 5.36).

5.7.2 Expansion Index

The growth history of a fault can be illustrated quantitatively with the expansion index (E) of Thorsen (Fig. 5.37):

$$E = t_d / t_u, \tag{5.12}$$

where t_d = the downthrown thickness (hangingwall) and t_u = the upthrown thickness (footwall). The thicknesses should be measured perpendicular to bedding so as not to confuse dip changes with thickness changes, and as close to the fault as possible because that is where the maximum thickness changes occur.

The simplest possible growth fault is one that starts, increases in growth rate, then slows and stops (Fig. 5.37). In the pre-growth interval, E = 1.0 (units j-k); as the growth rate increases, so does the expansion index, to a maximum of 2.1 in figure 5.37. The growth of the fault slows and eventually stops and E returns to 1.0 (units a-b). The plot of stratigraphic interval versus expansion index gives a visual picture of the growth

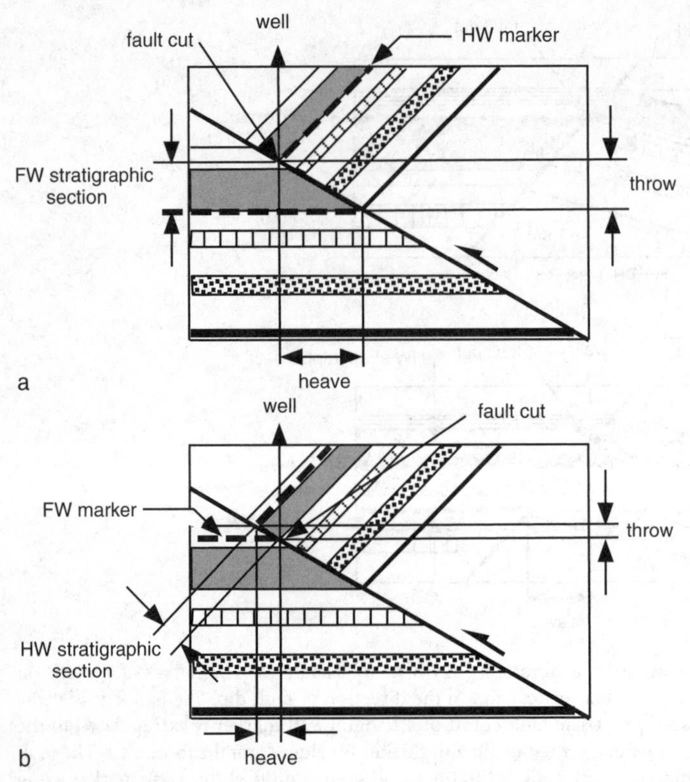

Fig. 5.36. Identical cross sections of a growth reverse fault showing effect of growth on throw and heave. The section is vertical and in the direction of fault dip. The fault cut is at the same point in both sections. **a** The fault cutoff of a hangingwall marker is extrapolated to the footwall cutoff of the same marker using dip and thickness values from the footwall. **b** The fault cutoff of a footwall marker is extrapolated to the hangingwall cutoff of the same marker using dip and thickness values from the hangingwall

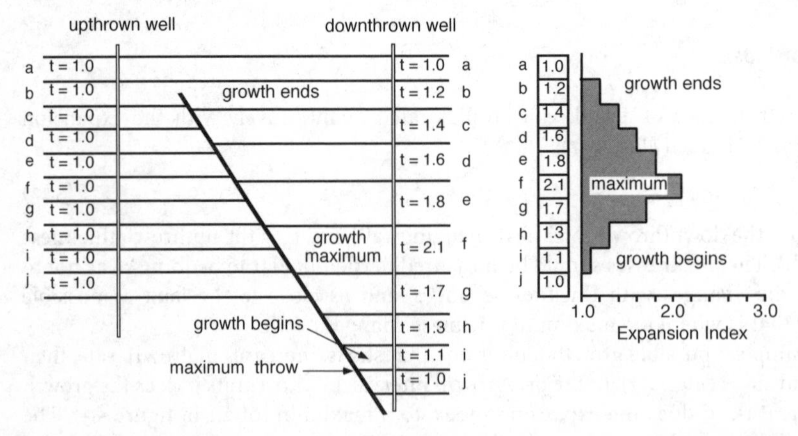

Fig. 5.37. Expansion index for a fault that begins to move, reaches a growth maximum and then stops. (After Thorsen 1963).

history. If the relative offset is large and if shale units are being compared at less than 3000 ft of burial, it might be important to correct E for compaction. At greater depths, the relative change in thickness will be small and have little effect on E.

The expansion index plot characterizes the growth history of a fault and might be correlated to other time-dependent features such as the sand/shale ratio or the time of hydrocarbon migration. The use of E eliminates the effect of absolute interval thickness, allowing the growth rates of a generally thin interval to be directly compared to those of a generally thick interval. The use of the similarity between expansion index plots at different fault cuts as an aid to fault correlation will be illustrated in Section 6.2.2.

5.8 Problems

5.8.1 Fault Recognition on a Map

Use the partially complete geologic map of Fig. 5.38, to do the following:

1. Mark all the faults. Explain the reason for each fault.
2. Indicate the sense of displacement on each fault.

5.8.2 Fault Recognition on a Seismic Line

Interpret the faults and unconformity in Fig. 5.39.

5.8.3 Faults Cuts

Locate the position of the faults and the amounts of the fault cuts on the logs in Fig. 5.40.

5.8.4 Dip-sequence analysis

1. Bald Hill. Perform a dip sequence analysis on the data in Table 5.3 to see if a fault is present. Plot the data on the graphs in Figs. 5.41 and 5.42. See Fig. 4.35 for an illustration of how to find the dip components from the true dip.
2. Southeastern Greasy Cove anticline. Perform a dip sequence analysis on the data in the Table 5.4. Consider both fold and fault geometry. Discuss your interpretation.

5.8.5 Fault Offset

Find the heave, throw, and vertical separation for the faults listed in Table 5.5.

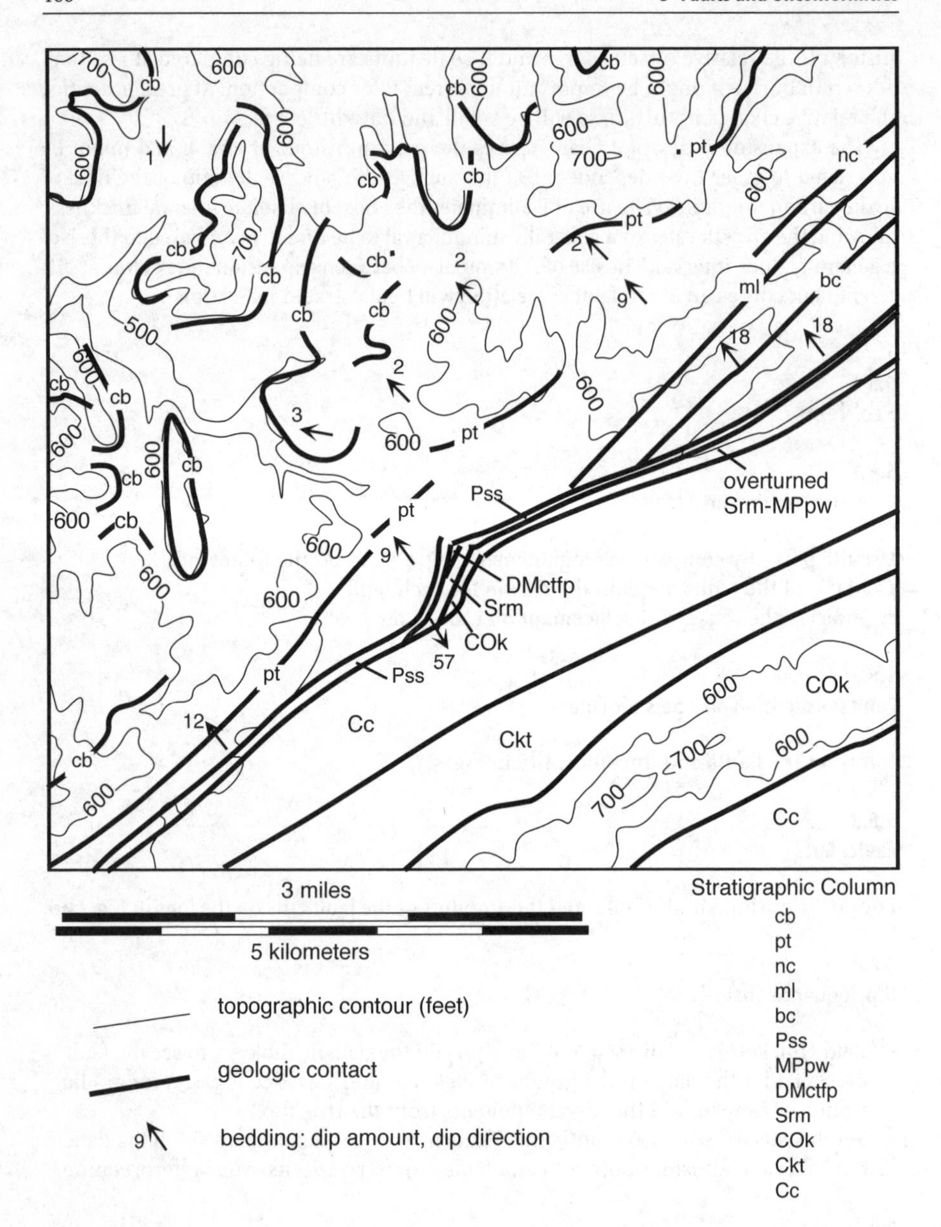

Fig. 5.38. Geologic map of the Ensley area, Alabama, with fault contacts not marked. (after Butts 1910; Kidd 1979)

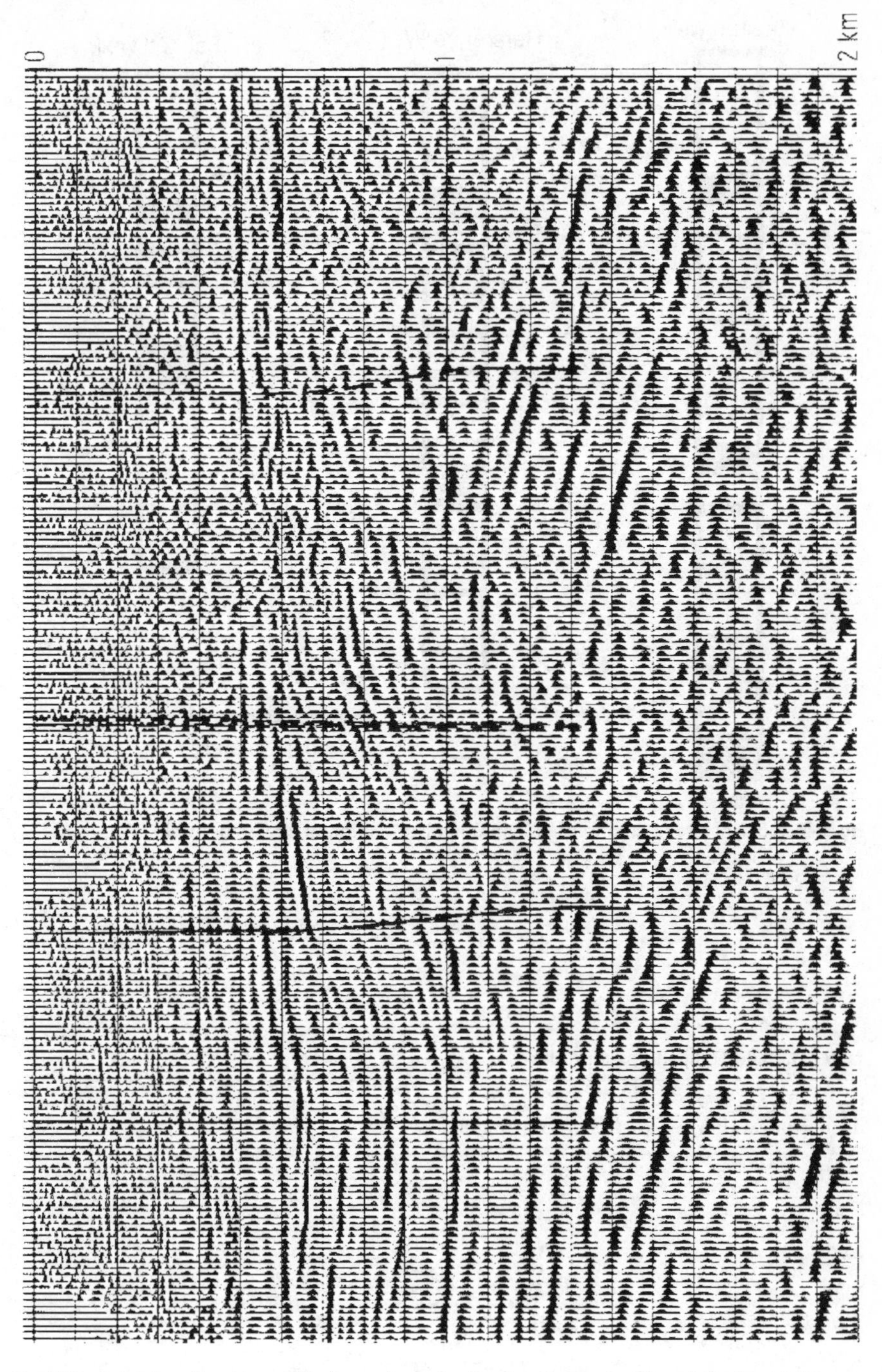

Fig. 5.39. Seismic reflection profile from the Ruhr district, Germany. (Drozdzewski 1983)

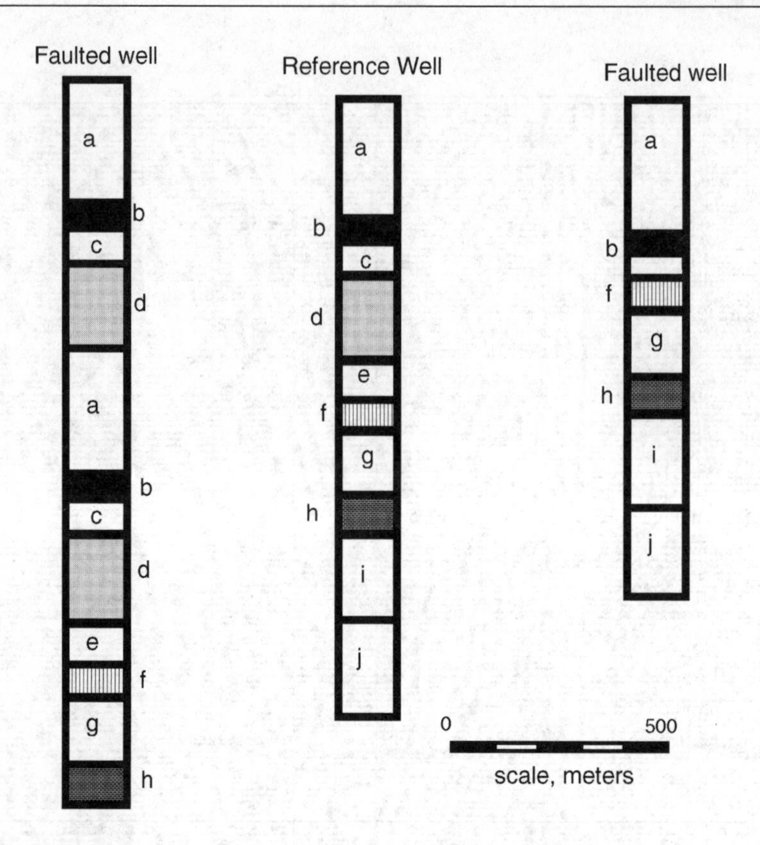

Fig. 5.40. Logs of two faulted wells and an unfaulted section in a reference well

Table 5.3. Bald Hill bedding attitudes

Distance from the northwest (km)	Attitude Dip, azimuth	T component	L component
0.54	60, 310		
0.70	90, 289		
0.90	55, 311		
1.10	40, 295		
1.30	14, 124		
2.38	12, 319		
2.68	26, 281		

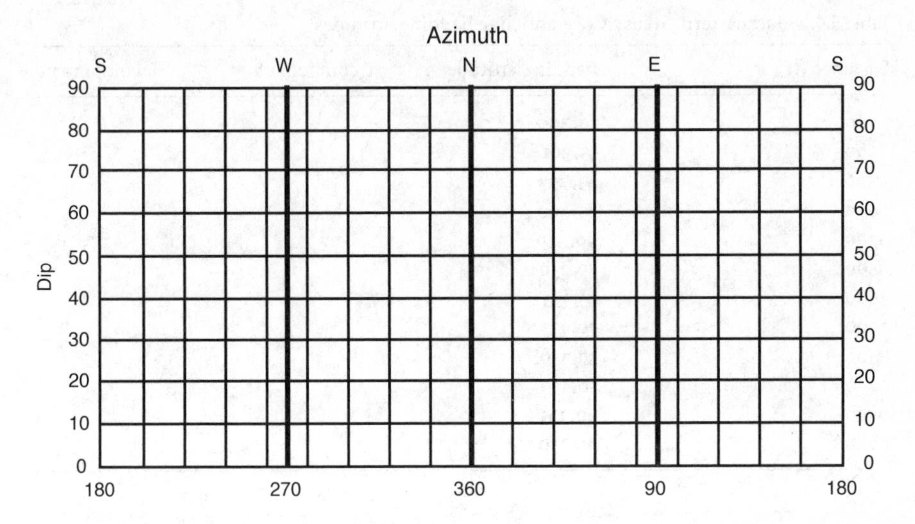

Fig. 5.41. Dip-azimuth graph

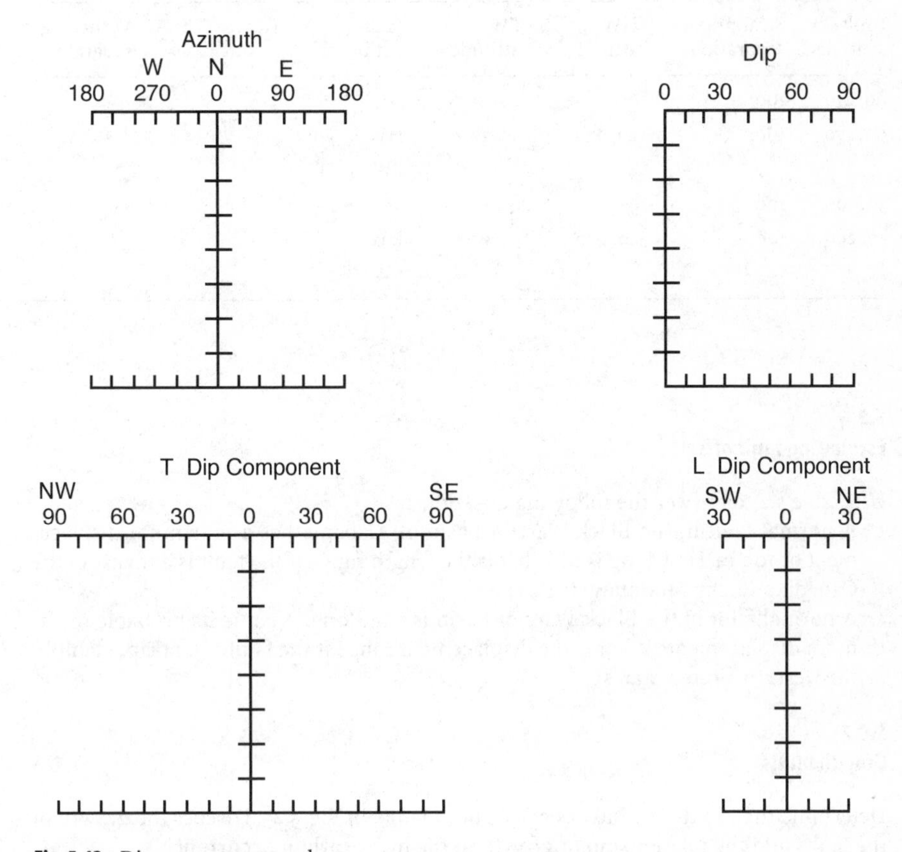

Fig. 5.42. Dip component graphs

Table 5.4. Southeastern Greasy Cove anticline bedding attitudes

Distance (ft)	bedding attitude	T component	L component
0	10, 300		
200	35, 302		
500	43, 293		
600	85, 300		
1200	90, 329		
1500	70, 130		
1600	45, 133		
2000	70, 133		
2100	30, 130		
2200	50, 130		
3500	60, 133		
4000	40, 150		

Table 5.5. Fault attitude and separation data

Fault attitude	Stratigraphic separation	HW attitude	FW attitude	fault cut in	ρ	Heave	Throw	Vertical separation
60, 270	100	0	0					
60, 270	100	20, 070	30, 200	HW				
				FW				
30, 200	100	0	0					
30, 200	100	20, 070	30, 070	HW				
				FW				

5.8.6 Estimating fault offset

Use Table 5.2 to answer the following questions:

1. A normal fault in the Black Warrior basin of Alabama has a maximum displacement of 100 m. How long is it? What is the length range if the fault is a thrust in the Canadian Rocky Mountains?
2. A normal fault in the Black Warrior basin is 5 kft long. What is its probable maximum displacement? What is the displacement range if the fault is a normal fault in the western United States?

5.8.7 Growth faults

Determine the expansion indices across both faults in Fig. 5.43. Discuss the growth of the faults and the relationship of growth to the hydrocarbon occurrence.

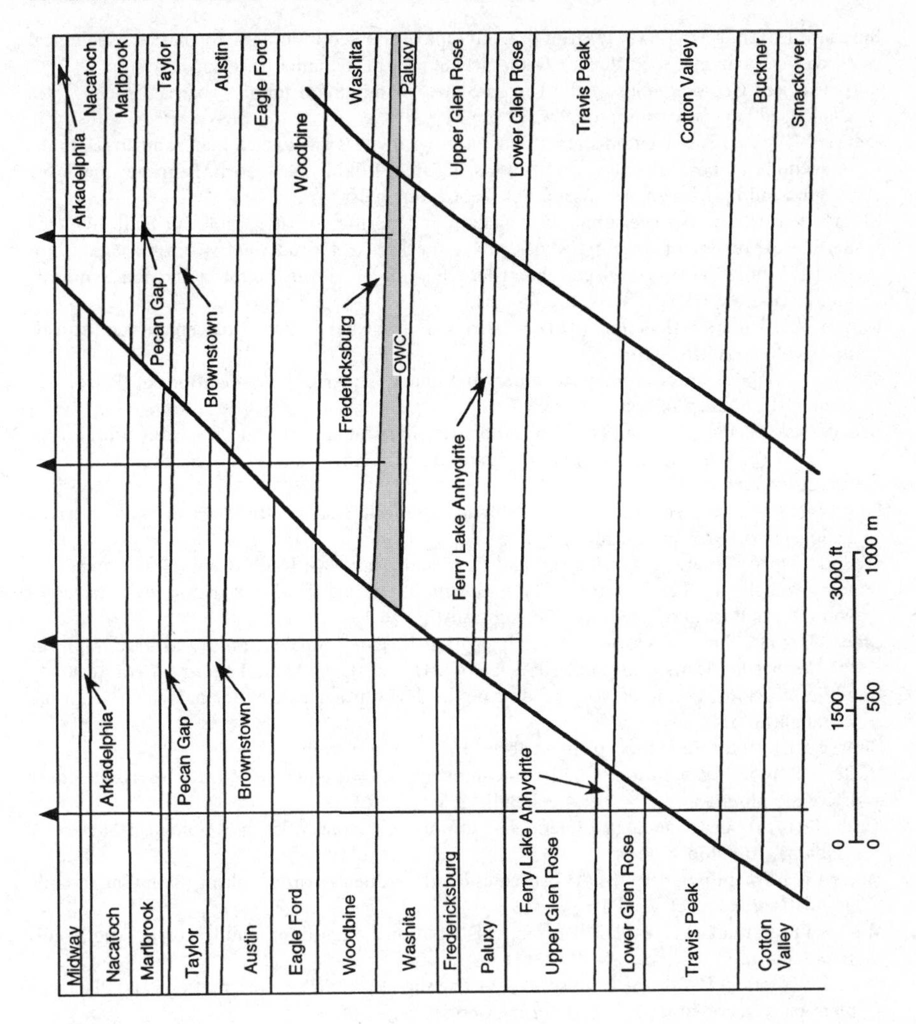

Fig. 5.43. Cross section of the Talco field, Texas. OWC oil–water contact. (After Galloway et al. 1983)

References

Barnett JAM, Mortimer J, Rippon JH, Walsh JJ, Watterson J (1987) Displacement geometry in the volume containing a single normal fault. Am. Assoc. Pet. Geol. Bull. 71: 925–937

Bates RL, Jackson JA (1987) Glossary of geology, 3 rd edn. American Geological Institute, Alexandria, Virginia, 788 pp

Becker A (1995) Conical drag folds as kinematic indicators for strike-slip fault motion. J. Struct. Geol. 17: 1497–1506

Bengtson CA (1981) Statistical curvature analysis techniques for structural interpretation of dipmeter data. Am. Assoc. Pet. Geol. Bull. 65: 312–332

Billings MP (1972) Structural geology. 3 rd edn. Prentice Hall, Englewood Cliffs, 606 pp

Boyer S, Elliott D (1982) Thrust systems. Am. Assoc. Pet. Geol. Bull. 66: 1196–1230

Burchard EF, Andrews TG (1947) Iron ore outcrops of the Red Mountain Formation. Geological Survey of Alabama, Spec. Rep. 19. Geological Survey of Alabama, Tuscaloosa, 375 pp
Butts C (1910) Geologic atlas of the United States, Birmingham folio, Alabama. United States Geological Survey, Washington, DC, 24 pp
Calvert WL (1974) Sub-Trenton structure of Ohio, with views on isopach maps and stratigraphic sections as basis for structural myths in Ohio, Illinois, New York, Pennsylvania, West Virginia, and Michigan. Am. Assoc. Pet. Geol. Bull. 58: 957–972
Christensen AF (1983) An example of a major syndepositional listric fault. In: Bally AW (ed) Seismic expression of structural styles. Am. Assoc. Pet. Geol., Stud. Geol. 15, 2: 2.3.1-36–2.3.1-40
Cowie PA, Scholz CH (1992) Displacement-length scaling relationship for faults: data synthesis and discussion: J. Struct. Geol. 14: 1149–1156
Dawers NH, Anders MH, Scholz CH (1993) Growth of normal faults: Displacement-length scaling. Geology 21: 1107–1110
Dennis JG (1967) International tectonic dictionary. American Association of Petroleum Geologists, Mem. 7, 196 pp
Drozdzewski G (1983) Tectonics of the Ruhr district, illustrated by reflection seismic profiles. In: Bally AW (ed) Seismic expression of structural styles. Am. Assoc. Pet. Geol., Stud. Geol. 15, 3: p.3.4.1-1–3.4.1-7
Elliott D (1976) The energy balance and deformation mechanisms in thrust sheets: Philos. Trans. R. Soc. Lond. Ser. A, v. 283: 289–312
Freund R (1970) Rotation of strike-slip faults in Sistan, southeast Iran. J. Geol. 78: 188–200
Galloway WE, Ewing TE, Garrett CM, Tyler N, Bebout DG (1983) Atlas of major Texas oil reservoirs. Texas Bureau of Economic Geology, Austin, 139 pp
Gibbs AD (1989) Structural styles in basin formation. In: Extensional tectonics and stratigraphy of the North Atlantic margins. Tankard AJ, Balkwill HR (eds) Am. Assoc. Pet. Geol. Mem. 46: 81–93
Hamblin WK (1965) Origin of "reverse" drag on the downthrown side of normal faults. Geol. Soc. Am. Bull. 76: 1145–1164
Hansen E (1971) Strain facies. Springer, Berlin Heidelberg, 207 pp
Hurley NF (1994) Recognition of faults, unconformities, and sequence boundaries using cumulative dip plots. Am. Assoc. Pet. Geol. Bull. 78: 1173–1185
Kidd JT (1979) Aerial geology of Jefferson County, Alabama, Atlas 15. Geological Survey of Alabama, Tuscaloosa, 89 pp
Marrett R, Allmendinger RW (1991) Estimates of strain due to brittle faulting: Sampling of fault populations. J. Struct. Geol. 13: 735–738
Muraoka H, Kamata H (1983) Displacement distribution along minor fault traces. J. Struct. Geol. 5: 483–495
Nicol A, Watterson J, Walsh JJ, Childs C (1996) The shapes, major axis orientations and displacement patterns of fault surfaces. J. Struct. Geol. 18: 235–248
Nicolas A, Poirier JP (1976) Crystalline plasticity and solid state flow in metamorphic rocks. John Wiley, London, 444 pp
Ramsay JG, Huber MI (1987) The techniques of modern structural geology, Vol 2: folds and fractures. Academic Press, London, 700 pp
Reches Z, Eidelman A (1995) Drag along faults. Tectonophysics 247: 145–156
Redmond JL (1972) Null combination in fault interpretation. Am. Assoc. Pet. Geol. Bull. 56: 150–166
Rippon JH (1985) Contoured patterns of the throw and hade of normal faults in the Coal Measures (Westphalian) of north-east Derbyshire. Proc. Yorkshire Geol. Soc. 45: 147–161
Schlische RW, Young SS, Ackermann RV, Gupta A (1996) Geometry and scaling relations of a population of very small rift-related normal faults. Geology 24: 683–686
Sibson RH (1977) Fault rocks and fault mechanisms. J. Geol. Soc. 133: 191–213
Tearpock DJ, Bischke RE (1991) Applied subsurface geological mapping. Prentice Hall, Englewood Cliffs, 648 pp
Thibaut M, Gratier JP, Léger M, Morvan JM (1996) An inverse method for determining three-dimensional fault geometry with thread criterion: application to strike-slip and thrust faults (Western Alps and California). J. Struct. Geol. 18: 1127–1138

Thorsen CE (1963) Age of growth faulting in southeast Louisiana. Trans.Gulf Coast Assoc. Geol. Soc. 13: 103–110

Watterson J (1986) Fault dimensions, displacements, and growth: Pure and Applied Geophysics, v. 124, 365–373

Williams WD, Dixon JS (1985) Seismic interpretation of the Wyoming overthrust belt. In: Gries RR, Dyer RC (eds) Seismic exploration of the Rocky Mountain region. Rocky Mountain Association of Geologists and the Denver Geophysical Society, Denver, pp 13–22.

Chapter 6
Mapping Faults and Faulted Surfaces

6.1
Introduction

This chapter covers how to correlate individual fault cuts into thorough-going faults, how to construct and validate fault maps, the integration of fault maps with horizon maps, mapping the displacement transfer zones between faults, mapping cross-cutting faults, and interpreting faults from isopach maps. The topics are covered in the sequence normally followed in order to define, map, and validate the interpretations of faults and faulted horizons. The key to the correct interpretation of faults and faulted horizons is to map the faults together with the displaced horizons. The fault throw and heave shown by the map must match the observed stratigraphic separations observed at the fault cuts.

6.2
Fault-Cut Correlation Criteria

Faults are commonly mapped on the basis of observations made at a number of separate locations, called fault cuts. If more than one fault is, or may be, present, a very significant problem is to establish which observations belong to the same fault (Fig. 6.1). This problem arises in surface mapping where the outcrop is discontinuous, in subsurface mapping based on wells, and in constructing maps from two-dimensional seismic data. Correlating faults in an area of multiple faults is one of the most difficult problems in structural interpretation. Typically a variety of interpretations seem to be possible. A number of different properties of the faults can be used to establish the correlation between the fault cuts. The criteria include the fault trend, the sense of throw across the fault, the smoothness of the fault surface, the amount of separation, and the growth history. The rules for correlation that are presented below may have exceptions and should be used in combination to obtain the best result.

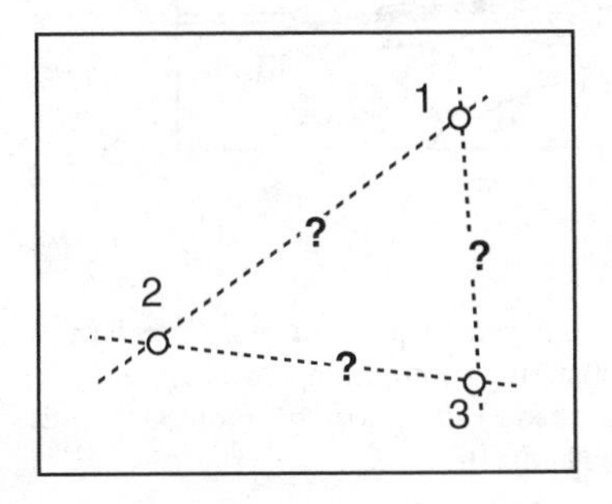

Fig. 6.1. Map showing three locations where faults have been observed. Which, if any, points are on the same fault? *Numbered circles* are observation points, *dashed lines* are some possible fault correlations

6.2.1 Trend and Sense of Throw

The consistency of the fault trend and the consistency of the sense of throw are usually the first two criteria applied in the correlation of fault cuts. The fault trend can sometimes be established at the observation points by direct measurement in outcrop, by physically tracing the fault, or by dip sequence analysis of outcrop traverses or dipmeter logs (Sect. 5.3). If, for example, the fault trend in the map area of Fig. 6.1 can be shown to be east – west, then observation points 2 and 3 are more likely to be on the same fault than are points 1 and 3 or 1 and 2 (Fig. 6.2a), and the fault through point 1 is also likely to trend east – west. A preliminary structure contour map on a horizon for which there is good control may give a clear indication of the fault trends (Fig. 6.2b). Faults that are directly related to the formation of the map-scale folds are commonly parallel to the strike of bedding, especially in regions where the contours are unusually closely spaced and may represent unrecognized faults.

The high and low areas on a structure contour map of a marker horizon should be explained by the proposed faults. The upthrown and downthrown sides of a fault will commonly be the same along strike. For example, the proposed fault linking observation points 2 and 3 in Fig. 6.2b is downthrown on the north at both locations. A constant sense of vertical separation is a common and reasonable pattern, but not the only possibility on a correctly interpreted fault.

The fault trend can be obtained by direct measurement in the field or from dipmeters in wells (Sect. 5.3). If the fault itself is not visible or measurable, the trend of a drag fold close to the fault should have the same trend as the fault.

Faults at a high angle to the map-scale fold trend may produce complex patterns of horizontal and vertical separations. Vertical displacement on a fault that strikes at a high angle to fold axes produces an apparent strike-slip displacement of the fold limbs for which the sense of slip reverses at the fold hinge line (Fig. 6.3a); the sense of throw is constant across the fault. Actual strike-slip displacement of the folds produces throw

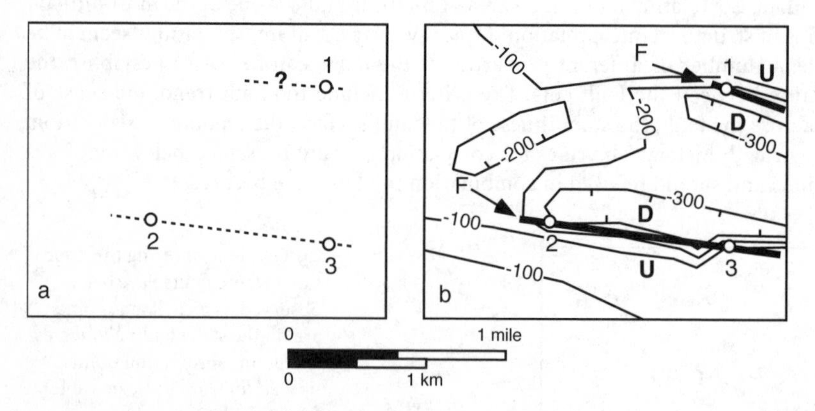

Fig. 6.2. Maps showing possible correlations between fault cuts at points 1–3. **a** Correlations *(dashed lines)* along an east – west trend. **b** Preliminary structure contour map on a marker horizon (after Fig. 5.5). Fault correlations (*heavy solid lines* marked F) along structure contour trends are confirmed by the consistent sense of throw across the faults

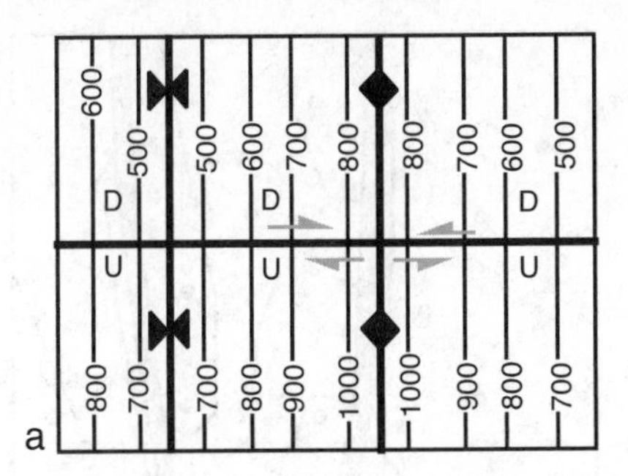

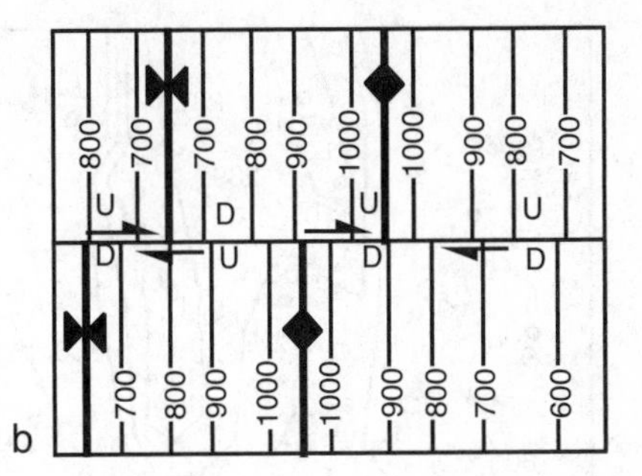

Fig. 6.3. Displacement along faults oblique to the fold trend. North – south lines are structure contours labeled with elevations. **a** Pure vertical displacement. The implied strike slip *(shaded arrows)* is incorrect. **b** Pure strike-slip displacement. The implied vertical slip *(U, D)* is incorrect

directions that change sense along the fault (Fig. 6.3b); the sense of strike separation is constant. Faults for which the sense of throw reverses along the trend of the fault are called scissors faults. Such faults may form by rotation around the points of zero separation or, as in Fig. 6.3b, may be caused by the strike-slip displacement of folds.

Determination of the true displacement on a fault requires matching pre-existing linear features across the fault. Fold hinge lines on identical marker horizons are potentially correlative marker lines. The offsets of the hinge lines in Fig. 6.3 give the true displacement for both the dip slip and strike slip examples. Offset hinge lines are good markers of displacement only if the folds developed prior to faulting. Tear faults present at the beginning of folding may divide a region into separate blocks in which the folds can develop independently. The folds in adjacent blocks may have never been connected and so the separation of their hinge lines is not a measure of the displacement. If more than one fold hinge line can be matched across the fault and more than one hinge line gives approximately the same displacement, the probability is high that the fault displaces pre-existing folds and the displacement determination is valid. The best features for determining true displacements are the offsets of pre-existing linear stratigraphic trends or the intersection lines produced by cross-cutting features, for example, a dike-bedding plane intersection line. See section 7.7.2 for further discussion of displacement determination.

An example from the Deerlick Creek coalbed methane field in the Black Warrior basin of Alabama will be used to illustrate the steps in a typical subsurface investigation involving faults. The Pennsylvanian strata of the basin consist of numerous coal cycles. The cycles are parasequences and their boundaries approximate time lines, making them ideal surfaces for structural mapping. Figure 6.4a is a preliminary structure contour map of the top of the Gwin coal cycle, contoured without faults. Two elongate synclines and an anticline are present. Only one well (solid circle, Fig. 6.4a) cuts a fault within the Gwin cycle. This fault has a stratigraphic separation at 200 ft, corresponding to two contour intervals on the map, yet a fault is not shown on the map (Fig. 6.4a).

Faults mis-mapped as folds will show up on structure contour maps as lines of closely spaced contours. The northwest-southeast trends of closely spaced contours on Fig. 6.4a might be faults that connect to the faulted well. The associated closed lows may also

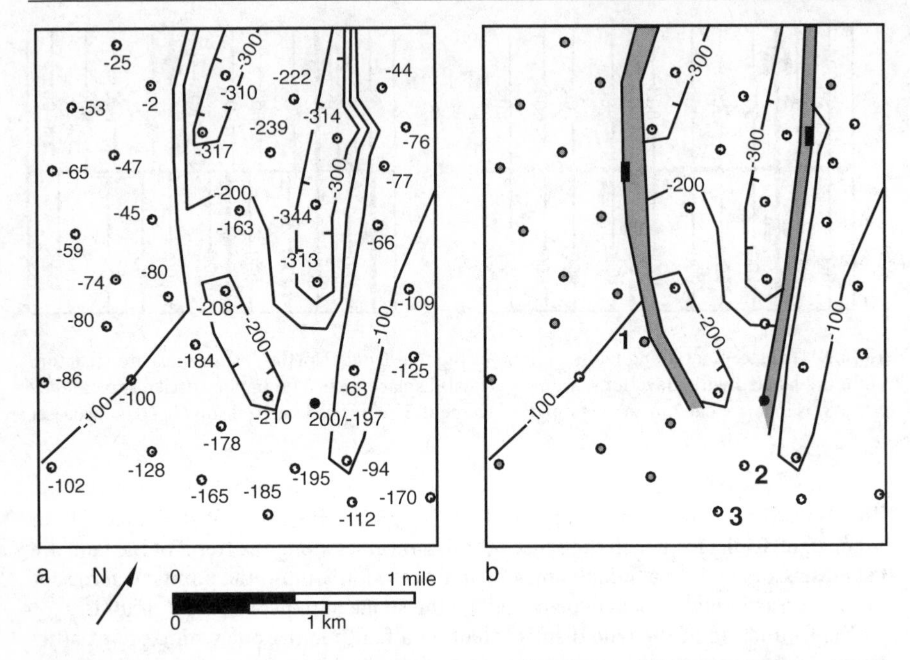

Fig. 6.4. Structure contours on the top of the Gwin coal cycle, Deerlick Creek coalbed methane field, Alabama. (Based on Wang 1994; Smith 1995). Elevations are in feet below sea level; the contour interval is 100 ft. The single well with a fault cut in the Gwin interval is a *solid dot.* Fault information is given as amount of missing section (in feet) / elevation of fault cut. **a** Top of the Gwin contoured without faults. **b** Top of the Gwin based on the wells in **a** but contoured with two hypothetical normal faults *(shaded).* Wells 1–3 contain fault cuts in units below the Gwin

be indicative of mis-mapping and could be indications of faults. If no other data are available, the lines of closely spaced contours can be replaced by faults, in regions where faulting is known to dominate over folding (Low 1951). A possible map of the Deerlick Creek field generated by this method (Fig. 6.4b) has two normal faults that form a graben. The best approach to improving the definition of the faults cutting a given horizon is to use the data available from additional horizons. All wells in the vicinity of the known fault cut should be examined for faults. Examination of the complete logs of wells near the well that shows the Gwin to be faulted reveals three additional wells with fault cuts, all in deeper units (wells 1–3, Fig. 6.4b). The four fault cuts in Fig. 6.4b could belong to one or more faults. Each of the fault cuts shows missing section, indicating normal faulting, in agreement with the preliminary interpretation of Fig. 6.4b. Whether or not the fault cuts fall on one or more than one fault will be indicated by the implied fault shape(s) and separation(s) determined by mapping the faults.

6.2.2 Shape

The inferred shape of a fault is an important criterion in correlating fault cuts and in validating fault interpretations. A valid fault surface is usually planar or smoothly curved and has an attitude that is reasonable for the local structural style. Contours on fault surfaces follow the same rules as contours on bed surfaces (Sect. 3.2.2) with the exception that the fault contours can end in the map area where displacement on the fault ends (Bishop 1960). The dip of the fault is obtained from a structure contour map with Eq. (2.16).

Smooth contours on the fault surface demonstrate that the correlations between the fault cuts are acceptable (Fig. 6.5a). Because active faults are typically planar or gently curved, it is reasonable to assume that both the strike and dip of the fault are approximately constant unless the data require otherwise. Surface undulations, if present, are most likely to be aligned in the slip direction. Unfaulted points or wells near an inferred fault provide additional constraints on the geometry because they must lie entirely within the hangingwall or footwall. The dips implied by the contours must be reasonable for the structural style. Very irregular contours on a fault surface suggest an incorrect correlation of the control points. On a fault surface, a data point surrounded by circular contours (Fig. 6.5b) probably belongs to another fault. Alternatively, the fault surface itself may be folded.

Return to the Deerlick Creek example (Fig. 6.4b) to test whether the four faulted wells at the southern end of the map area define an internally consistent fault surface or belong to more than one fault. To examine the possible correlations between fault cuts and fault planes, the fault plane attitude is determined from alternative sets of three adjacent points, for example, 1,2,4 and 4,2,3 (Fig. 6.6a). The fault attitudes so determined (Fig. 6.6b) dip south and southwest, both possible directions for the area. A fault plane through points 1,2,4 should cut the well located within the shaded triangle. The lack of a fault cut in this well eliminates this possible correlation of the fault cut in well 4 with the fault cuts in wells 1 and 2. A normal fault dipping to the southwest through points 4,2,3 should displace the top of the Gwin down in the dip direction of the fault. Elevations adjacent to the implied upthrown side range from -112 to -63 ft and on the downthrown side range from -185 to -210 ft (Fig. 6.6a). The sense of displacement is correct and the magnitude of the vertical displacement of 73 to 101 ft is close enough to the fault cuts of 130 to 150 ft so that points 4,2,3 can be considered to correlate into an acceptable fault. A normal fault through point 4 should dip and be

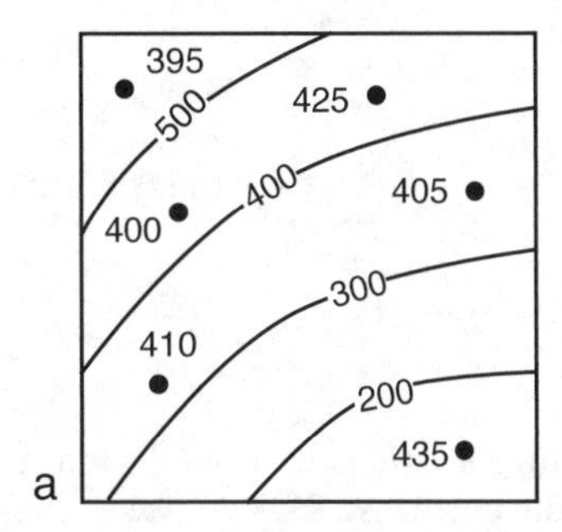

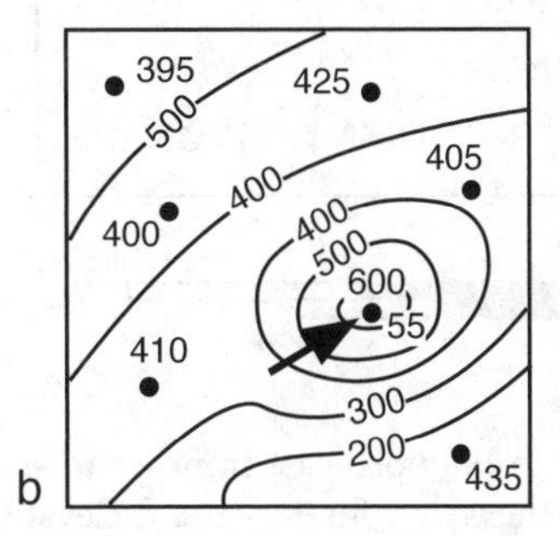

Fig. 6.5. Correlating fault cuts in wells. Dots are the positions of wells with fault cuts, *contours* are on the fault surface, and the stratigraphic separation is posted next to the well. **a** Good correlation: the fault surface is smooth and variation in the magnitude of the separation is small. **b** Poorly correlated fault cuts. One well *(arrow)* does not fit the same surface as the others and its stratigraphic separation is too small

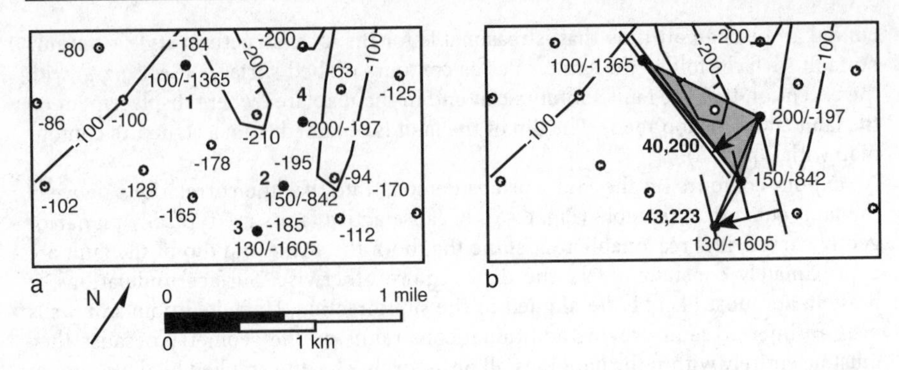

Fig. 6.6. Southern portion of the Deerlick Creek map area. Wells with fault cuts are indicated by *solid dots*. **a** In faulted wells 1–4, data are separation (in feet) / elevation of fault cut (feet below sea level). **b** Three-point dip determinations *(arrows)* on wells 1, 2, 4 and 4, 2, 3. *Bold numbers* are computed fault-plane attitudes. The *shaded triangle* contains an unfaulted well

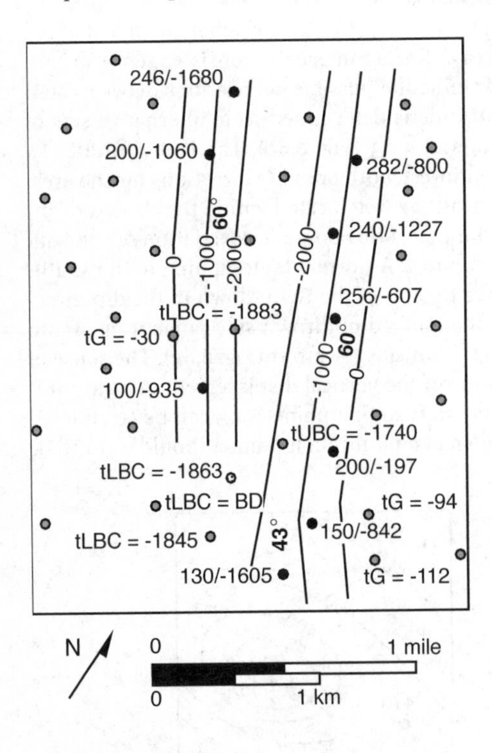

Fig. 6.7. Fault-plane maps for a portion of the Deerlick Creek field. In faulted wells the data are separation (in feet) / elevation of fault cut (feet below sea level). Critical data in unfaulted wells are the elevations of the highest unfaulted contact in the footwalls, the top of the Gwin (*tG*), and the lowest unfaulted contact in the hangingwalls, the top of the upper Black Creek (*tUBC*) and the top of the lower Black Creek (*tLBC*). BD Bad data. Dips on fault planes are in bold

downthrown to the northeast in order to separate the elevations of around –100 ft southwest of the well from the –200 ft elevations to the northeast.

Examination of all the wells in the Deerlick Creek map area reveals fault cuts in a total of nine wells (Fig. 6.7). The southernmost four points were discussed above. The new data make it clear that there are two faults, both trending northwest – southeast and together forming a full graben. This confirms the qualitative prediction of Fig. 6.4b. For both faults, the fault-cut data closely constrain the position of the –1000 ft fault-plane contour. The 3-point dip determined from the northeastern three points is 60°, an expected value for normal faults. The contour spacing reflects the 60° dip except in the southeast where the previous analysis indicated a 43° dip. The resulting contours agree with the adjacent unfaulted wells. The stratigraphic separations on both faults decrease reasonably smoothly to the southeast. The southwestern fault must end to the southeast because two unfaulted wells lie along its trend, preventing the contours from being continued straight or from being curved to the east to join the other fault.

6.2.3 Separation

The stratigraphic separation along a fault surface is normally a smoothly varying function, with a maximum near the center of the fault and decreasing to zero at the tip line (Sect. 5.6.2). This simple pattern can be perturbed if the fault grew by linking separate faults (Sect. 6.5), but the variation is nevertheless likely to remain smooth. The stratigraphic separation on the fault is the component usually known, and is used as a proxy for the displacement. The distribution of the separation can be displayed by posting the separations on a map of the fault surface (Fig. 6.8a). A point on a fault for which the location of the fault cut and the stratigraphic separation do not agree with the points around it (Fig. 6.8b) is likely to be miscorrelated with the fault.

Another method for displaying the variation of the stratigraphic separation along a fault is by means of a stratigraphic separation diagram (Fig. 6.9; Elliott and Johnson 1980; Woodward 1987). The vertical axis of the diagram is the scaled stratigraphic column and the horizontal axis is the distance along the fault. The curves on the diagram show the stratigraphic unit that the fault is in at any particular point. A thrust fault usually places older over younger units and so the footwall block should be the upper curve (Fig. 6.9). The opposite is true for a normal fault. For example, line A (Fig. 6.9a) represents a particular geographic point on the fault, one at which the footwall is at the base of unit 4 and the hangingwall is in unit 9. If the complete fault has been observed, the hangingwall and footwall curves will join at the fault tips. Both hangingwall and footwall curves will be smooth if they are not broken by oblique faults (Fig. 6.9a). An oblique structure will produce a rapid change in the stratigraphic level

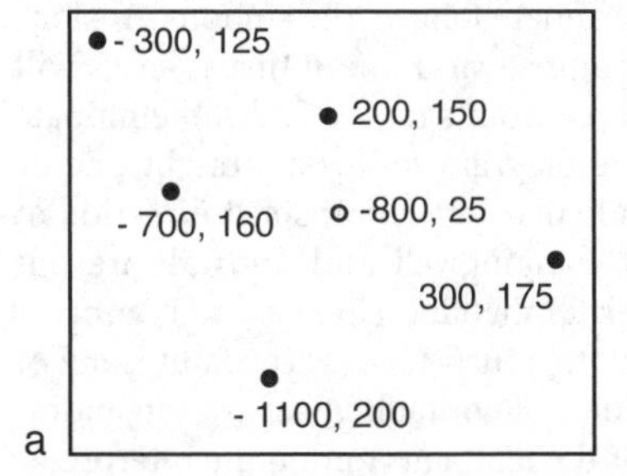

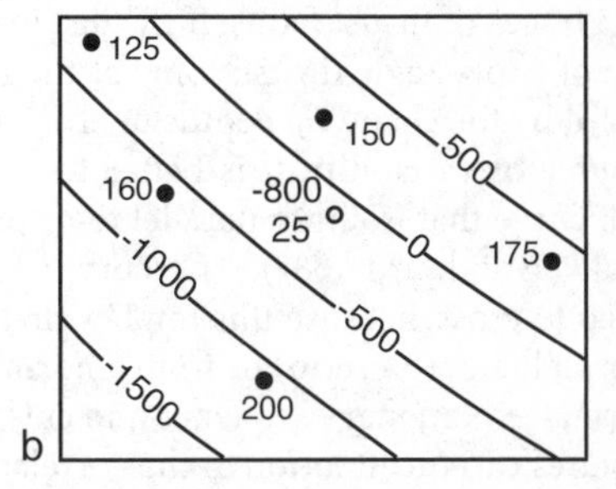

Fig. 6.8. Fault-surface maps. **a** Data: elevation of fault cut, amount of fault cut. **b** Structure contour map on the fault surface; amount of fault cut is posted The contour interval is 500 units, and negative contours are below sea level. The fault cut (*open circle*) does not correlate with the contoured fault

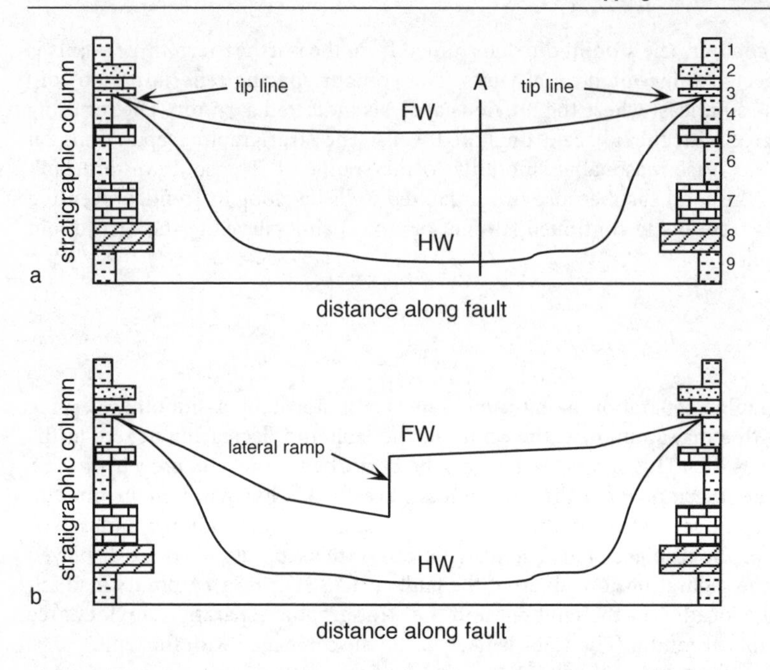

Fig. 6.9. Stratigraphic separation diagram representative of relationships seen along the strike direction of a thrust fault. *FW curve* Stratigraphic position of the thrust in the footwall; *HW curve* stratigraphic position of the fault in the hangingwall. **a** Single unbroken thrust sheet. **b** Thrust sheet broken by a lateral ramp

of the curve of the block that contains the feature (Fig. 6.9b). If the rapid change is in the footwall, it is likely to have been caused by a lateral ramp (Fig. 6.9b). If the oblique structure is in the hangingwall, it is likely to be a tear fault that subdivides only the hangingwall. If the feature is present to about the same degree in both the hangingwall and the footwall curves, it may have been caused by a later fault that cuts and displaces the fault in question.

Stratigraphic separation diagrams are most widely used for data derived from the outcrop traces of thrust faults. If the distance coordinate follows the sinuous erosional trace of a low-angle thrust, some of the stratigraphic variation in thrust levels will be caused by the changing depths of erosion and reentrants on the fault. To eliminate this source of variability, it is better to make the diagram follow a straight line or smooth curve that is either parallel to or perpendicular to the transport direction of the fault (Woodward 1987). The curves for the hangingwall and footwall are not expected to cross, because this implies that either (1) the fault changes from a thrust to a normal fault (Fig. 6.10) or from a normal fault to a thrust, or (2) the fault is out of the normal evolutionary sequence and cuts an older fold or fault. A separation anomaly requires careful attention to the correlation of the fault cuts and to the interpretation of the fault if the correlation is accepted.

The major pitfall in using the smooth variation of stratigraphic separation as a correlation criterion is that the stratigraphic separation is not equal to the displacement. The strike-slip displacement of pre-existing folds (Fig. 6.3b) will produce a very irreg-

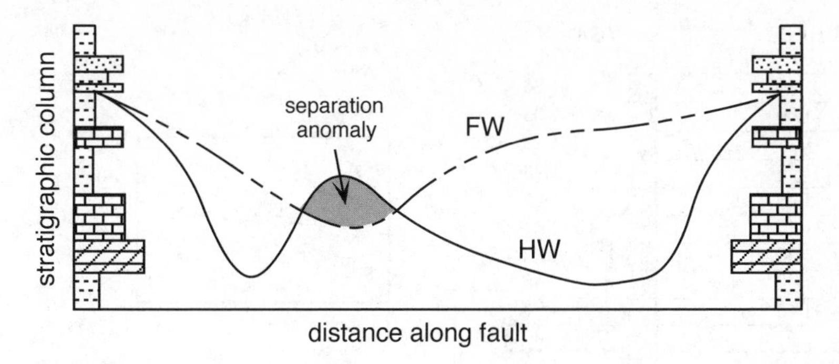

Fig. 6.10. Stratigraphic separation diagram showing a separation anomaly. *FW curve* Stratigraphic position of the thrust in the footwall; *HW curve* stratigraphic position of the fault in the hangingwall

ular distribution of stratigraphic separation although the displacement is constant. Wrench faults and faults with separations partly inverted by later displacements of opposite sense are likely to give separation anomalies even when correctly correlated.

6.2.4 Growth History

The stratigraphic evolution of a growth fault provides another criterion for correlating fault cuts. A single fault can be expected to have a similar stratigraphic growth history along its trend. Expansion index diagrams (Sect. 5.7.2) clearly illustrate the details of the growth history and are valuable in fault correlation (Fig. 6.11). In an example from Louisiana, the expansion index diagrams have the same form across the same fault at an interval about 2.5 miles apart (Fig. 6.11a, b). The values of E for the same units are not the same, but the forms of the expansion index curves are the same. Both crossings of the same fault show the maximum growth interval to be of the same age. Three different faults in the same area give three different expansion index diagrams over similar distances (Fig. 6.11c, d). The maximum growth interval becomes younger to the south. Thus, similar growth – history relationships can indicate that the same fault has been crossed and different relationships can indicate that different faults have been crossed.

The form of the expansion index plot can be very helpful in developing and testing fault interpretations. The appropriate form of the expansion index plot is determined from the local structural style. For example, a reasonable assumption in the Gulf of Mexico basin-fill sequence is that the sense of displacement on the normal faults does not reverse and, consequently, that the expansion index must always be greater than one. A value of less than one means that the downthrown side received less sediment than the upthrown side which, in turn, implies that the downthrown side was high (upthrown) during the interval having an $E < 1.0$. In this situation, an $E < 1.0$ indicates a miscorrelation, either of the units or of the fault cuts. In Fig. 6.12, if the top three units are correlated, E is less than one in unit 2. A better interpretation is that the thin unit 2 on the hangingwall is part of a thicker interval that correlates with unit 2 on the footwall (interpretation 2). Another possibility is that the abnormally thin unit in well

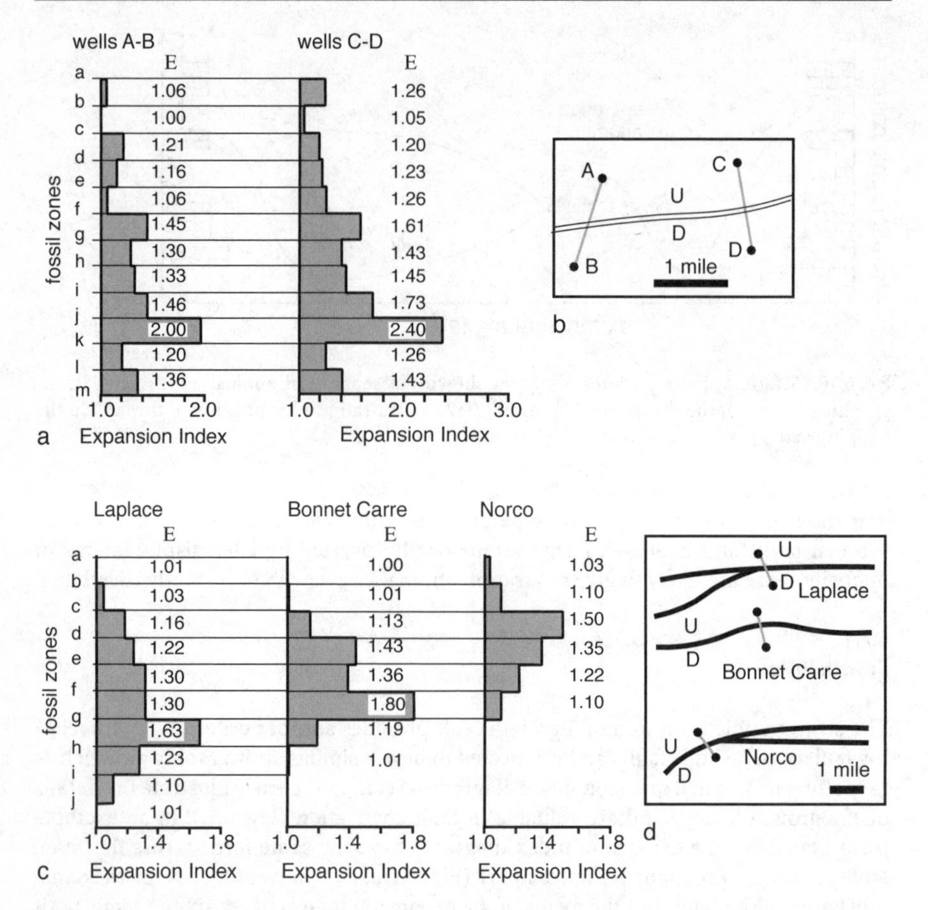

Fig. 6.11. Expansion index (*E*) diagrams across Louisiana growth faults. **a** Expansion indices from two crossings of the same fault. **b** Locations of wells used to determine the expansion indices in **a**. **c** Expansion indices across three adjacent faults. **d** Locations of the wells used to determine the expansion indices in **c** (after Thorsen 1963)

B (Fig. 6.13a) on the downthrown side of a fault has been truncated by another fault. Inserting fault f_2 at the proper location (Fig. 6.13b) results in the correct form of the expansion-index diagram. The expansion index increases steadily downward for the reinterpreted fault blocks.

The growth history should agree with the stratigraphic separation on the fault. Within the growth interval on a normal fault, the separation is expected to increase down the dip. The throw across the fault can be plotted against depth (Fig. 6.13). A misinterpreted growth stratigraphy (Fig. 6.13a) shows the separation to be decreasing downward and then increasing again. The correct interpretation (Fig. 6.13b) shows the continuing increase of vertical separation downward.

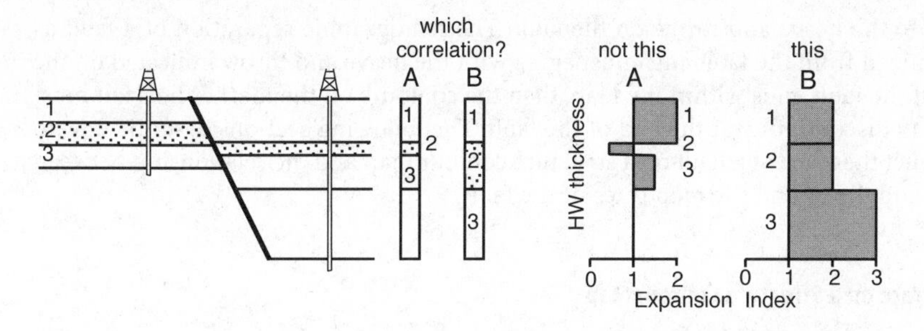

Fig. 6.12. Use of the expansion index to test unit correlations. An expansion index of less than one implies reverse fault movement or a miscorrelation across a normal fault

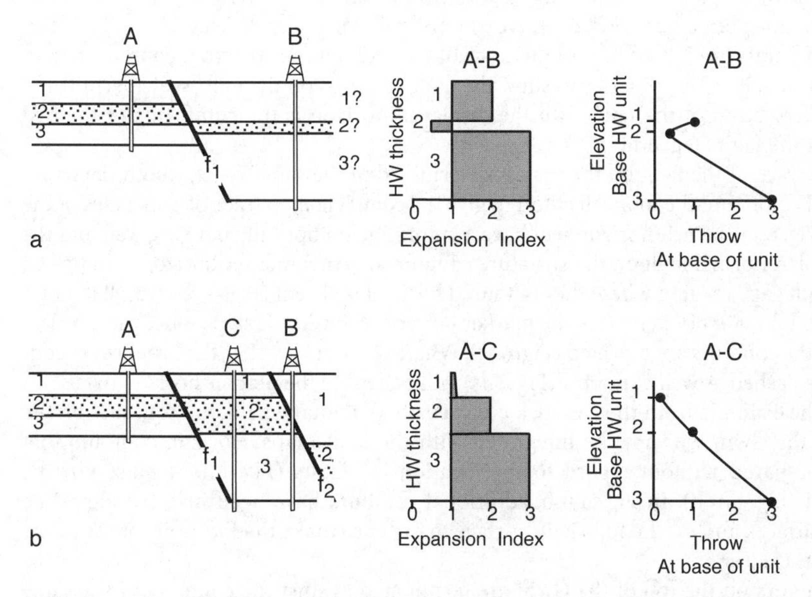

Fig. 6.13. Presence of a fault interpreted from anomalies in the expansion index and vertical separation. **a** A misinterpretation of the hangingwall of fault f_1 that gives an expansion index of less than one for unit 2 and a fault throw that decreases and then increases with depth. **b** The anomalous thinness of unit 2 in well *B* is explained by normal fault f_2. *C* is a well that would penetrate an unfaulted section in the hangingwall of the fault f_1

6.3 Faulted Surfaces

Faults are discontinuities in the map of a marker horizon and produce either gaps, overlaps, or vertical offsets of the surface. The only accurate way to determine the trace of a fault on a structure contour map of a marker horizon is to intersect the structure contours on the fault surface with the contours on the horizon surface. The offset shown by the structure contour map of a marker horizon is directly proportional to the heave and throw on the fault. The stratigraphic separation of a fault as determined from the fault cuts must agree with the heave and throw indicated on the map. If the fault ends within the map, then the contours on the marker horizon must show no discontinuity at the end of the fault. The following sections describe how to construct the trace of a fault on a structure contour map and the relationship between heave and throw and the map trace of the fault.

6.3.1 Fault Trace on a Structure Contour Map

Determination of the trace of a fault on the structure contour map of a marker horizon requires mapping the fault, mapping the marker horizon in the hangingwall and footwall, and intersecting the maps (Fig. 6.14). The preliminary horizon and fault contours in the region of intersection are drawn so that they overlap (Fig. 6.14a). The intersection points occur where the contours cross at the same elevation. The intersection points are joined to find the trace of the fault on the structure contour map of the horizon. It is best practice to show the fault contours on the map, scale permitting. The intersection of the fault with the displaced horizon is the cutoff line of the bed against the fault (Fig. 6.14a).

The trace of the fault on the map is determined in the same way for both the hangingwall cutoff and the footwall cutoff to give the complete map trace of both sides of the fault (Fig. 6.14b). Each horizon will have a cutoff line on both the hanging wall and the footwall. A normal fault on the structure contour map of a marker horizon is indicated by a fault gap, a region where the contoured horizon is absent (Fig. 6.14). A well at point 1 (Fig. 6.14b) will not penetrate the marker horizon. A reverse fault produces an overlap, where the contours are repeated (Fig. 6.15). Where contours overlap, the lower set of contours is dashed. A well at point 1 (Fig. 6.15) will penetrate the marker horizon twice.

In the example from the Deerlick creek coalbed methane field, the contours on the top of the Gwin will now be integrated with the fault maps. Horizon contours that were prepared without regard to the locations of faults (Fig. 6.16a) must now be revised (Fig. 6.16b). In Fig. 6.16b, additional contours have been inserted along the fault surfaces and on the top Gwin surface in order to make the locations of the intersections clearer.

Contours on the top of the Gwin are terminated against the contours of the same elevation on the fault surface (figs. 6.16b, 6.17a). The lines of intersection of the fault surfaces and the horizon surface are the traces of the fault on the map (dotted lines in Fig. 6.14a). The fault contours can be removed for clarity (Fig. 6.17b), although leaving the contours in the fault gap or overlap region (figs. 6.14b, 6.15) proves the internal consistency of the interpretation. The resulting map is qualitatively similar to that of the preliminary interpretation (Fig. 6.4b) but is much more accurate. On the new map (Fig. 6.17b) the fault gaps are smaller indicating smaller displacements, and the southwestern fault dies out to the southeast.

6.3.2 Measures of Separation

Throw and heave are the components of fault separation directly visible on a structure contour map. Throw is the vertical component of the dip separation; heave is the hor-

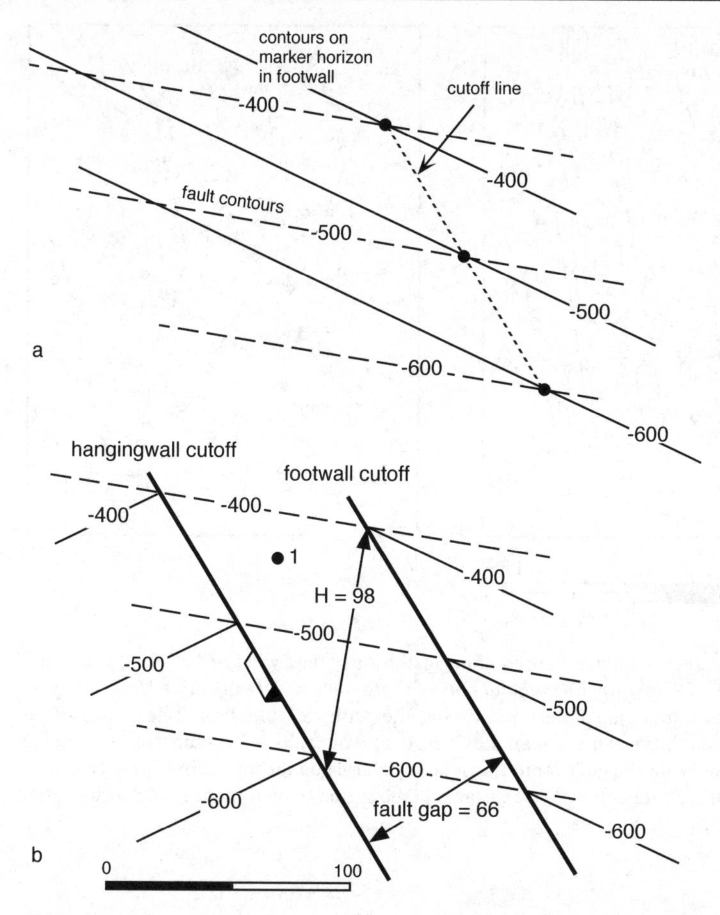

Fig. 6.14. Construction of the trace of a normal fault on an offset marker horizon. Structure contours on the fault are *dashed* and contours on the marker horizon are *solid lines*. **a** Intersection line (*dotted*) between the footwall marker and the fault. Intersection points are *solid dots*. **b** Hangingwall intersected with the fault plane to complete both sides of the fault trace. Throw on the fault is 210 and heave is 98 units. The fault gap is 66 units

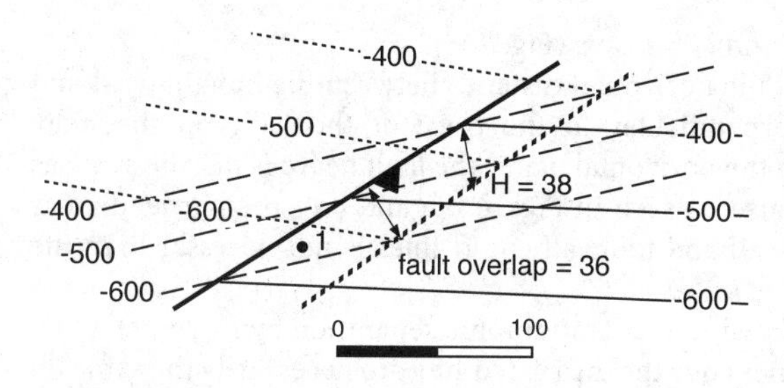

Fig. 6.15. Intersection of structure contours on a reverse fault (*dashed*) with contours on an offset bed (hangingwall, *solid lines*; footwall, *dotted lines*) determines the fault trace (*heavy lines*: hangingwall *solid*, footwall *dashed*) on a structure contour map. The throw is 125 units, heave 38 units, and overlap 36 units

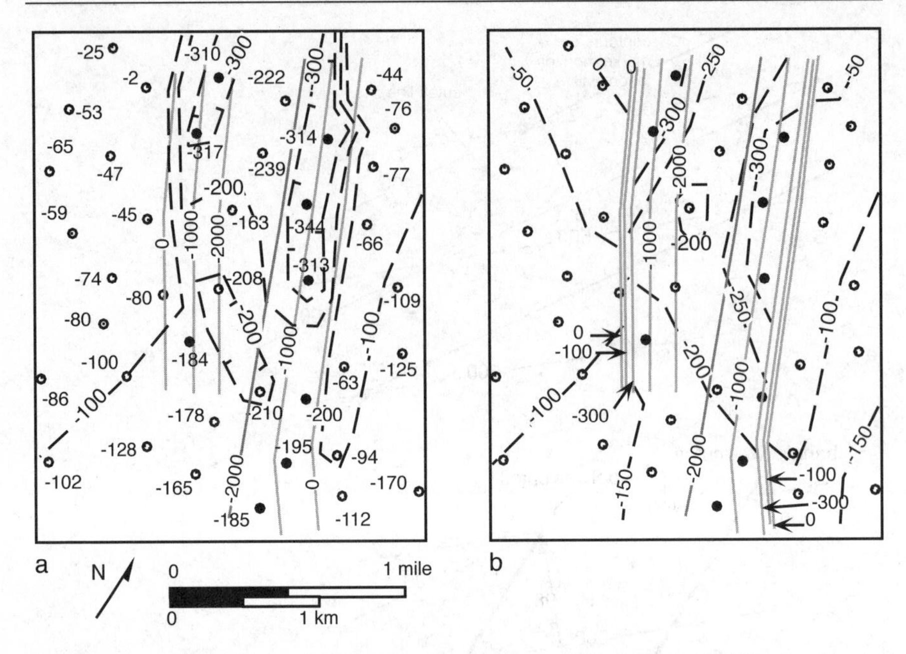

Fig. 6.16. Deerlick Creek field, structure contours on the top of the Gwin (*dashed lines*) and structure contours of the faults (*gray lines*). *Solid dots* indicate wells with fault cuts. **a** Unfaulted preliminary structure contour map of the top Gwin together with the faults. Posted elevations are on the top Gwin. Contour interval is 100 ft on the Gwin, 1000 ft on faults. **b** Structure contours on the top Gwin intersected with the fault contours. Contour interval on the top Gwin is 50 ft. Contours at –100 and –300 ft are added to the fault surfaces in the region of their intersection with the top of the Gwin to improve resolution

izontal component of the dip separation (Fig. 6.18), both components being defined in the dip direction of the fault (Billings 1972). On a structure contour map the heave is the map distance between the hangingwall and footwall cutoffs, measured in the direction of the fault dip (figs. 6.14b, 6.15). The throw is the change in vertical elevation between the hangingwall and footwall cutoffs in the direction of the fault dip. The dip of the fault can be determined from the heave and throw as

$$\phi = \arctan (T / H), \tag{6.1}$$

where ϕ = fault dip, T = throw, and H = heave (Fig. 6.18).

The fault gap (or overlap) is the horizontal distance between the hangingwall and footwall cutoffs, measured perpendicular to the trace of the fault on the map. Although both are measured in the horizontal plane, the fault heave is not the same as the fault gap (or overlap) because, as seen in Figs. 6.14b and 6.15, the perpendicular distance between the hangingwall and footwall cutoff lines is not necessarily in the direction of the fault dip.

The throw and heave are related to the stratigraphic separation by Eqs. (5.5)–(5.8). Neither the stratigraphic thickness nor the dip of bedding are necessarily the same on both sides of a fault, which is important if the throw and heave are calculated from

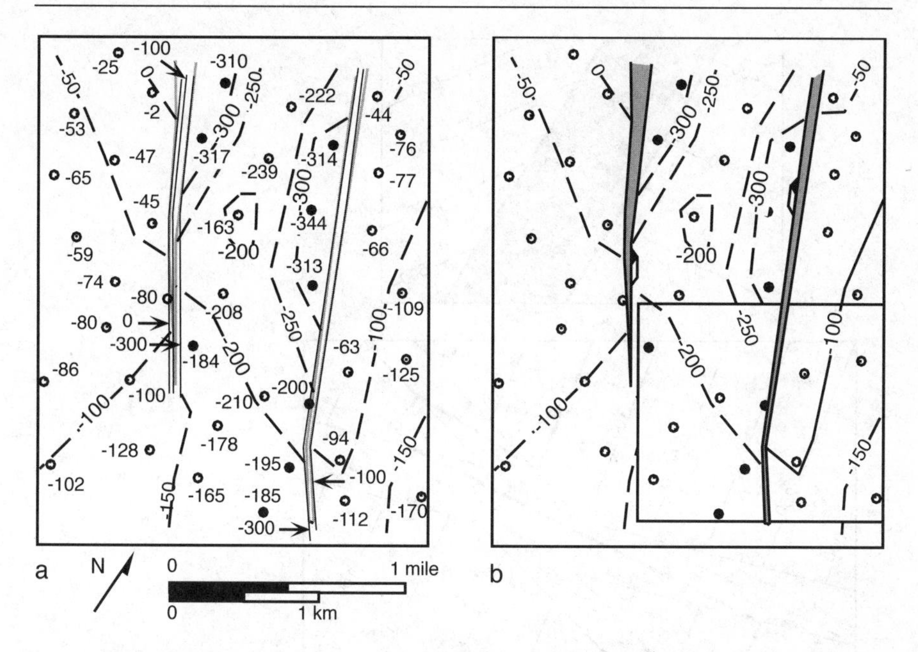

Fig. 6.17. Deerlick Creek field, structure contours on the top of the Gwin (*dashed lines*). *Solid dots* indicate wells with fault cuts. **a** Structure contours on the faults are *thin solid lines* and bed cutoff lines are dotted. Posted elevations are on the top Gwin. **b** Structure contours on the top of the Gwin showing fault gaps (*shaded*) constructed from the bed cutoff lines. The area in the *box* is enlarged in Fig. 6.24

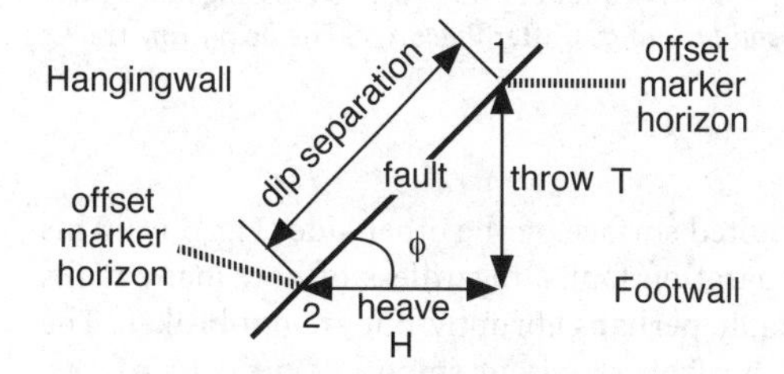

Fig. 6.18. Vertical cross section in the dip direction of a fault showing throw and heave as components of the dip separation of an offset marker horizon

Eqs. (5.5) and (5.6). The appropriate values belong to the side of the fault opposite to that of the position of the marker used to determine the stratigraphic separation. For example, in Fig. 6.18, if the fault cut is at point 1 and the stratigraphic separation is determined with respect to the footwall marker, the dip required to determine the throw and heave (the location of point 2) is that of the hangingwall.

6.3.3 Map Validation

A valid map of a faulted surface can be contoured as a single surface that includes the fault. The structure contours must be continuous across the faults from the faulted

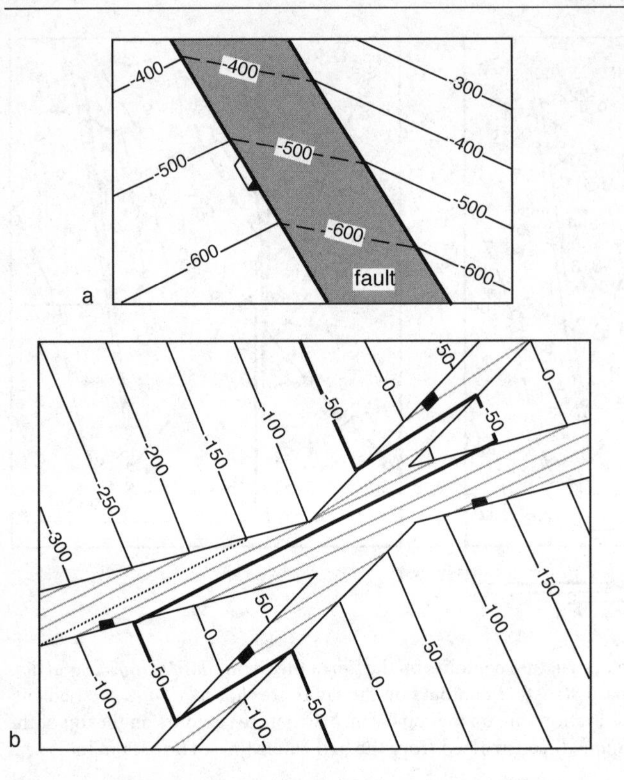

Fig. 6.19. Structure contours on faults and faulted surfaces. **a** A single fault (*shaded with dashed contours*) displacing a marker horizon (*solid contours*), after Fig. 6.14b **b** Two crossing faults (*dotted contours*) displacing a marker horizon (*solid contours*), after Fig. 6.37c. The *heavy line* tracks one elevation across the map

surface on one side of the fault to the faulted surface on the other side (Fig. 6.19). This applies to both normal and reverse separation faults, regardless of how many faults are present. The contours bend at the fault, perhaps abruptly, but are not broken. The contours on the faults and the faulted horizon represent smoothly curved surfaces. Therefore the fault contours are approximately parallel to one another and the marker horizon contours are approximately parallel to one another but may have different trends on each side of the fault.

Structure contours on a fault surface are easily constructed by connecting the points of equal elevation where the marker horizon intersects the fault surface on opposite sides of the fault (Fig. 6.20). This is called the implied fault surface (Tearpock and Bischke 1991). Contouring the implied fault surface is a very simple and effective test of the interpretation. Unless the map has been constructed by intersecting the fault surface map with the horizon map, it is unlikely to be correct. Incorrect interpretations of the fault-bedding relationships will show unreasonable or impossible contour patterns for the fault surface. The implied fault surface in Fig. 6.20 has a spiral shape, possible perhaps, but not very likely. In more complex situations, crossing contours may occur on the implied fault, an even less likely result.

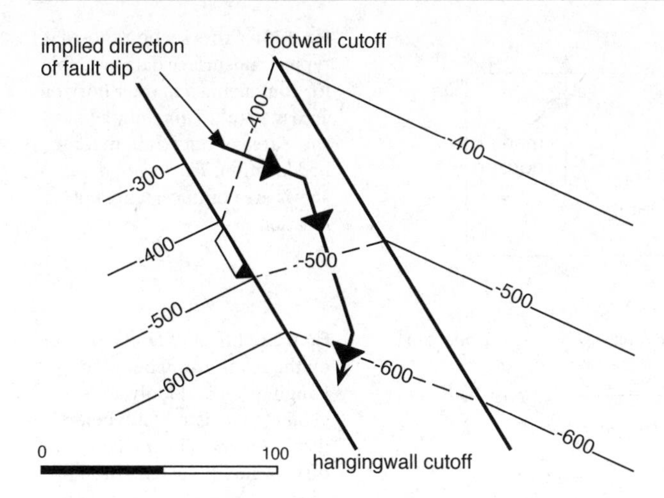

Fig. 6.20. Implied structure contours on a fault. The implied fault contours (*dashed*) are found by connecting equal elevations of the hangingwall and footwall bed cutoffs. A spiral fault shape is implied on this fault

6.4 Contouring Across Faults

Contouring a marker across a fault is one of the major problems in structural mapping. In a typical situation, a marker horizon is to be contoured on both sides of a fault but neither the position of the fault nor the geometry of the marker bed is known in the vicinity of the fault (Fig. 6.21). To construct a map it may be necessary to assume (1) the fault dip, (2) the fault displacement, (3) the hangingwall geometry, (4) the footwall geometry, and (5) the relationship between the hangingwall and footwall geometry. The gap or overlap shown on the map must be consistent with the stratigraphic separation on the fault. If the fault has been cut, then the stratigraphic separation should be known, although not necessarily for the horizon being mapped, and must be included in the interpretation. On an existing structure contour map, the trace of a fault and the locations of the hangingwall and footwall cutoffs can be validated or refined with the stratigraphic separation data from the fault cuts.

Where no information is available about the dip of the fault and the fault separation is unknown, the structure contours on the fault surface can be generated from the fault geometry most reasonable for the local structural style, for example, a 60° dip for an unrotated normal fault. The marker beds are extrapolated from both the hangingwall and footwall until they intersect the fault. The structure contours on the hypothetical fault and the extrapolated marker horizon are intersected to produce the fault trace on the map. The resulting map will be internally consistent but is, of course, hypothetical.

Where the amount of the fault separation is known, it provides an important constraint on the interpretation. The maps of faulted horizons can be improved by including the heave and throw information obtained from the stratigraphic separation. The

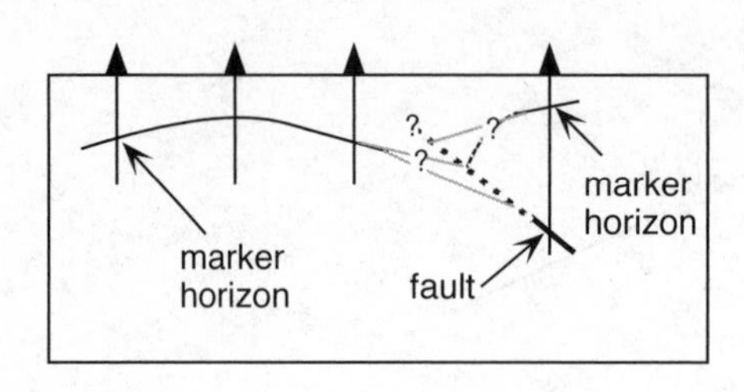

Fig. 6.21. Cross section showing typical subsurface data available for contouring a marker horizon across a fault. *Thin dotted lines* show area of uncertain marker bed location. *Thick dotted line* shows area of uncertain fault location

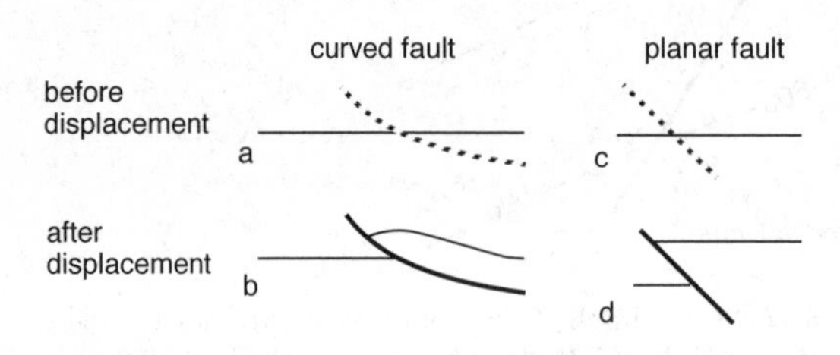

Fig. 6.22. Effect of fault curvature on the relationship between hangingwall and footwall bed geometry. **a** Listric fault before displacement. **b** Listric fault after displacement: the hangingwall shape is changed **c** Planar fault before displacement. **d** Planar fault after displacement: the hangingwall shape is unchanged

heave and throw provide control on the location of the offset marker across the fault. Developing the best interpretation is commonly an iterative process. The map may be used to provide the dips and then the heave and throw used to improve the map in the vicinity of the fault. Changes in the map may give different values for the dips, requiring another revision of the heave and throw calculations, and so on.

Where the shape change, or the lack of shape change, across a fault is known, this information can be incorporated into the interpretation. If the fault is curved or if the structures on opposite sides of the fault developed independently using the fault as a displacement discontinuity, then the marker surfaces may be completely unrelated across the fault (Fig. 6.22a,b) and should be contoured independently by the method of projected fault cutoffs. This is the most general method, and is the first of the two given below. This method is appropriate where the shape of the marker surface may have been changed by the faulting. If a planar fault displaces a pre-existing structure, then the shape of the marker horizon will be unaffected by the fault (Fig. 6.22c,d) and should be contoured so that the hangingwall and footwall bed geometries are the same across the fault. This is the method of restored vertical separation (restored tops), and is the second method given below. The method of restored tops is appropriate where the shape of the marker horizon is not changed by the faulting.

6.4.1. Projected Fault Cutoffs

The stratigraphic separation measured on a fault can be used to control the position of the contours across the fault (Sebring 1958), whether or not the shapes of the marker surfaces are related across the fault. The geometry of the marker surface must be known or inferred on one side of the fault, and the marker cutoff point is projected to the other side. In Fig. 6.23, a hangingwall dip from well 4 is used to project the mark-

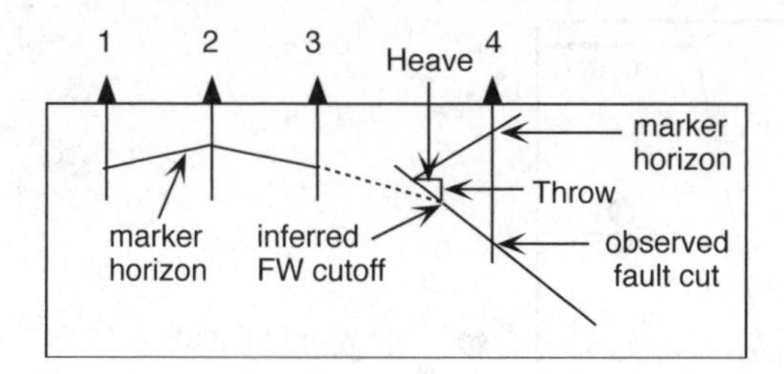

Fig. 6.23. Cross section in the direction of fault dip, illustrating use of projected fault cutoff to control the position of the marker horizon. The shape of the marker horizon is significantly different from that obtained by the method of restored vertical separation (Fig. 6.25b)

er horizon from the well to the fault. The geometry of the fault must also be known so that the location of the cutoff point can be found. From the known or inferred cutoff point, the heave and throw are calculated from Eqs. (5.5) and (5.6) using the cross-fault marker dip determined in the vicinity of well 3. The footwall cutoff point is then projected from the location of the hangingwall cutoff point and the projected point is included in the contour map of the footwall. If the dip near the fault is not known for either the hangingwall or the footwall, then it is necessary to assume the dips. Reasonable dips may be known from elsewhere in the area. If not, for a first approximation it is reasonable to assume zero dips and use Eqs. (5.7) and (5.8) to calculate the heave and throw.

The strengths of this method are that it can be used regardless of the true slip on the fault because it depends only on the geometry in the direction of fault dip, and that it works where the marker horizon shapes are different across the fault. When fault blocks are contoured independently, use of this method will ensure that the fault separation between the blocks is correct. The weakness in the method is that the location of the fault cutoff and the cross-fault dip must be known or inferred before the cutoff can be projected across the fault.

The procedure is illustrated by the Deerlick Creek example (Fig. 6.24). The data required to find the heave and throw in hangingwall well number 1 from Eqs. (5.5) and (5.6) are the stratigraphic separation (200 ft), the cross-fault bedding dip (the footwall dip: about 1°), the fault dip (60°), and the angle between the fault dip and the bed dip directions (0°). The calculated heave is 117 ft, in agreement with the width of the fault gap in the fault dip direction within the accuracy of the map. The calculated throw is 202 ft, significantly larger than the ~ 160 ft that is shown by the original map (Fig. 6.17b). The throw of 202 ft indicates that the top of the Gwin must reach an elevation a little over −50 ft in the footwall adjacent to well 1. A −50 ft contour is thus placed on the footwall close to the fault in Fig. 6.24. This contour is consistent with the elevations in nearby wells in the footwall and with the inferred dip of bedding of about 1°. The −50 ft contour could have been placed on the original interpretation as it is consistent with the data available; it was left off because there seemed to be no data requiring it to be there. The throw determined from the stratigraphic separation of the fault in well 1 is the required evidence for the −50 ft contour. For well 2, the computed heave is 88 ft and throw is 152 ft. Best agreement with the map is obtained by extending the −50 structure contour on the footwall of the fault until it is also across from this well. The dip of the fault in the vicinity of well 3 is 43°, significantly less than the 60° dip in the vicinity of the top of the Gwin and so this well is not used to provide direct control for the top of the Gwin. Because fault dip and displacement may change along strike and up and

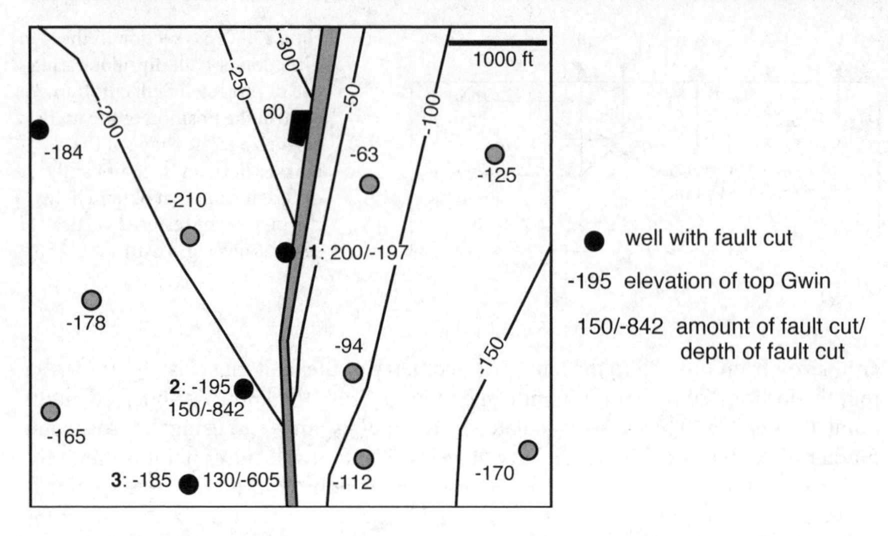

Fig. 6.24. Structure contour map of the southeast corner of the Deerlick Creek field (revised from Fig. 6.17b). Contours are on the top of the Gwin, with an interval of 50 ft. Elevations noted for wells are on the top of the Gwin except for well *1*, where the top Gwin has been faulted out. The –50 ft contour has been added to agree with the fault throw in wells *1* and *2*.

down the dip, conversion of stratigraphic separation into heave and throw should only be done for the horizons located near the fault cut. Utilization of data located at a long distance from the fault cut is best accomplished with cross sections (Chap. 7).

6.4.2 Restored Vertical Separation

Contouring based on restored vertical separation (restored tops) is appropriate where the geometry of the hangingwall and footwall beds is unchanged by faulting except for the dip separation on the fault (Jones et al. 1986; Tearpock and Bischke 1991). This method is used to find the shape of a surface before faulting and its configuration after a displacement amount equal to the vertical separation of the fault (Fig. 6.25). Begin the interpretation by finding the vertical separation from the stratigraphic separation at a fault cut (well 4, Fig. 6.25a). To calculate the vertical separation (Eq. 5.3) it is necessary to know or assume the dip of bedding at the fault cut. For dips below 25°, the stratigraphic separation is similar in magnitude to the vertical separation and can be used as a substitute. The stratigraphic separation is always less than the vertical separation, but below a marker dip of 25° the difference is under about 10%. Next, remove the separation by adding it to all marker elevations on the downthrown side of the fault (wells 1-3, Fig. 6.25a). Then contour the marker horizon as if no fault were present to obtain the restored top (Fig. 6.25a). Then, subtract the vertical separation from all contour elevations on the downthrown side of the fault (Fig. 6.25b). Re-contour the data while maintaining the shapes of the structure contours from the restored-top map. Break the map at the fault by intersecting the contoured fault surface with the marker horizon on both the upthrown and downthrown sides of the

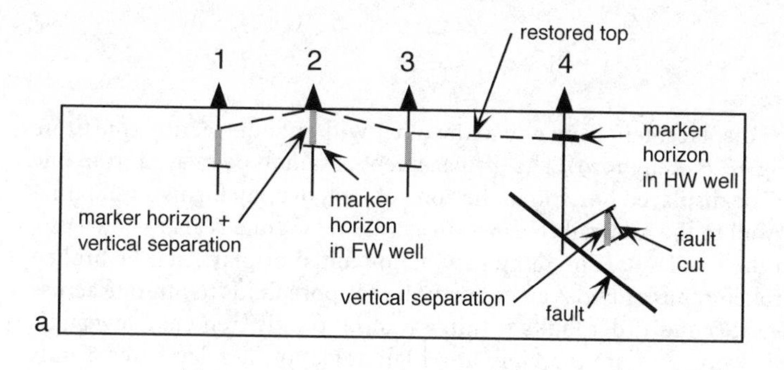

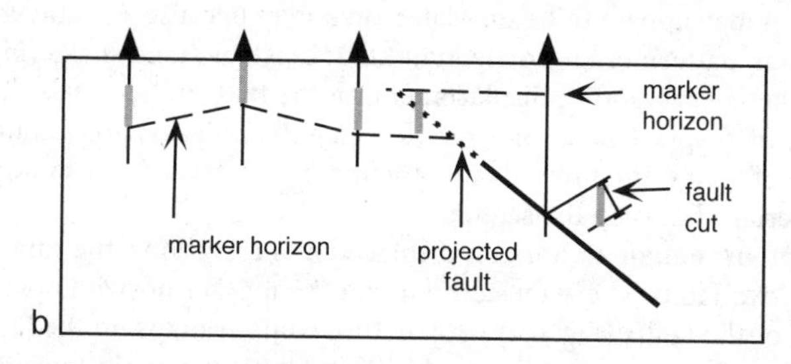

Fig. 6.25. Cross sections in the fault dip direction, showing the method of restored vertical separation to control position of the marker horizon. The dip shown at the fault cutoff is used to determine the vertical separation. **a** Fault separation is added to all marker elevations on the downthrown side of the fault and elevations are contoured to give the restored top. **b** Vertical separation is subtracted from the downthrown side and the marker horizon re-contoured to give the inferred geometry

fault. This method ensures that the vertical separation on the fault is included on the final map and that the marker geometry is related across the fault.

The strengths of this method are that it uses the maximum amount of elevation data (from both sides of the fault) and guarantees that the marker horizon shape is related across the fault. The weakness in the method is that in many structural styles the hangingwall and footwall geometries should not have the same shape (Fig. 6.22a,b). If there is any strike-slip component to the fault displacement, it must be removed before the points are contoured in order to correctly relate the hangingwall and footwall geometry.

6.5 Displacement Transfer

As a fault dies out it may transfer its displacement to another fault. The two faults may link across a relay zone without ever intersecting or may intersect and join along a branch line. Here the basic patterns of fault-to-fault displacement transfer are examined. All types of faults show the same forms of displacement transfer. The examples given here are normal faults because they are easy to visualize, but the same principles apply to reverse and strike-slip faults.

6.5.1
Relay Overlap

Faults that transfer displacement from one to the other without intersecting constitute a relay pattern (Fig. 6.26; Kelly 1979). The displacement is said to be relayed from one fault to the other. The displaced horizon in the zone of fault overlap forms a ramp that joins the hangingwall to the footwall (Fig. 6.26, 6.27a). A relay zone is also known as a soft link between the two faults. The ramp may be unfaulted or may itself be broken by faults. An unbroken ramp provides an opportunity for pore fluids to migrate across the main fault zone. Second-order faults within a ramp typically trend at a low angle to the strike of the ramp and are therefore at a high angle to the relay faults. Faults exhibiting a relay pattern may appear to be unrelated on a map because a single fault will have the displacement pattern of an isolated fault that dies out along strike (Figs. 5.31, 5.32a). The sympathetic variation of displacement on the two faults reveals their relationship to be that of displacement transfer (for example, Fig. 5.34c). Detailed examination of large fault zones often reveals that the main fault consists of multiple segments with relay overlaps between the segments.

On the structure contour map of an horizon displaced in a relay zone, the ramp is the region between the two faults where the contours on the marker horizon are at a high angle to the trend of the faults (Fig. 6.27a). Structure contour maps on the faults show them to overlap in the relay zone (Fig. 6.27b). Both faults end at tip lines that define the ends of the relay zone. The extent of the faults may be large, both updip and downdip, and the tip lines may be straight or curved.

6.5.2
Branching Fault

The intersection of two faults (Fig. 6.28) occurs along a branch line (Boyer and Elliott 1982). Some of the displacement is transferred from the main fault to the branch fault at the intersection. Faults that intersect are said to be hard linked. Displacement is conserved at the branch line (Ocamb 1961). As can be seen from Fig. 6.28, at the branch line, the throw of the largest fault will be equal to the sum of the throws on the two smaller faults. All the faults may change throw independently away from the branch line.

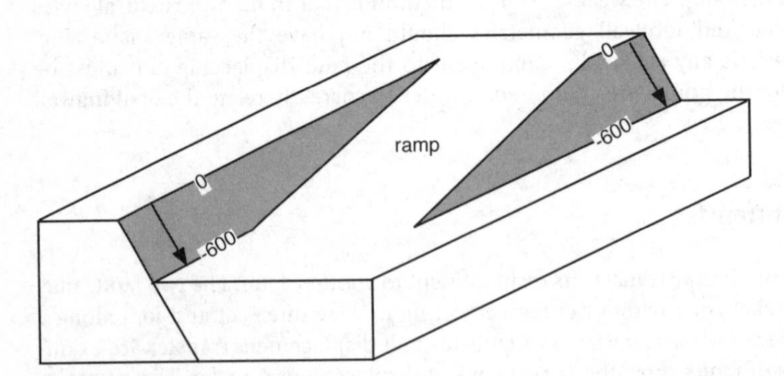

Fig. 6.26. Displacement transfer at a relay overlap. *Arrows* indicate amount of dip separation on the relay faults

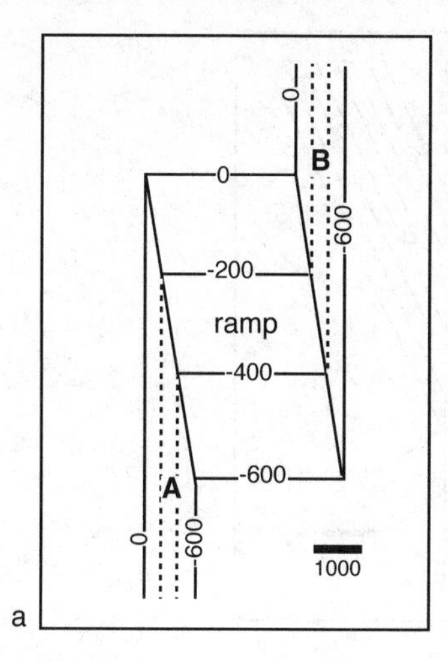

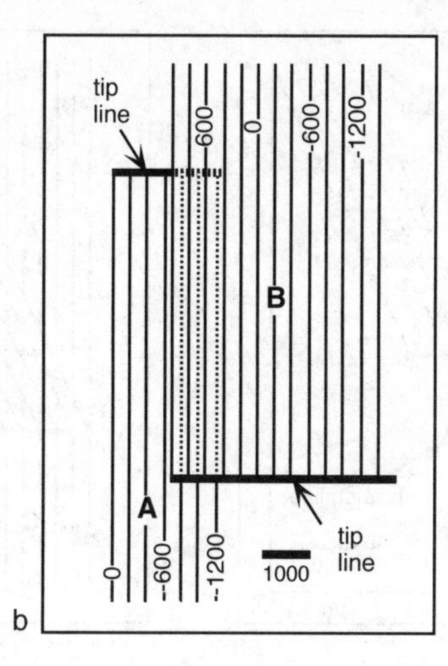

Fig. 6.27. Structure contour maps based on the relay overlap geometry in Fig. 6.26. Both maps are at the same scale. **a** Displaced stratigraphic horizon. *Solid contours* are on the offset marker, *dashed contours* are on faults *A* and *B*. **b** Portions of the two fault planes; contours on fault *A* are *dashed* where they lie below fault *B*

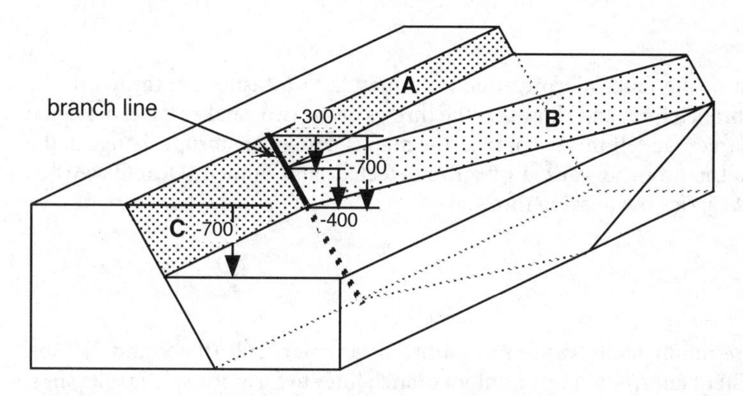

Fig. 6.28. Displacement transfer at a branch line on a normal fault. Fault surfaces are shaded. The total throw is constant at the branch line. Faults *A* and *C* are a single plane; fault *B* is the branch

The structure contour map of a marker horizon displaced by a branching fault (Fig. 6.29a) shows two intersecting faults and an abrupt change in throw at the intersection. The throw of fault A plus that of fault B is equal to that of C at the branch line. In this example, the fault plane of A is continuous with the plane of fault C. Contours on the faults (Fig. 6.29b) show a straight branch line and a region of fault overlap. One fault does not necessarily continue straight at the branch line as shown in Figs. 6.28 and 6.29. All three faults at the branch line could have different strikes.

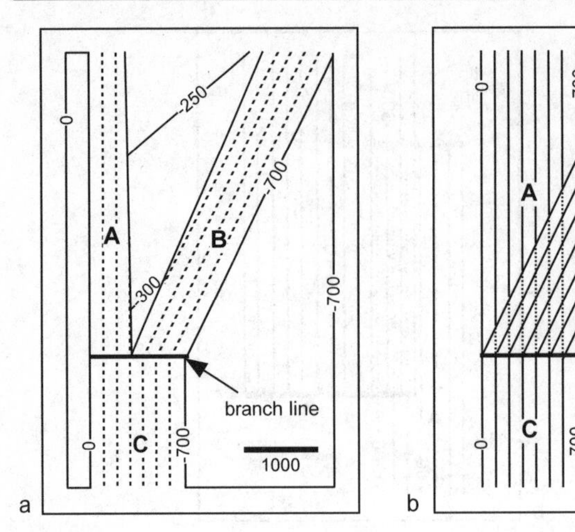

Fig. 6.29. Structure contour maps of a branching normal fault, based on the geometry of the structure in Fig. 6.28. Both maps are at the same scale. **a** Structure contours on the displaced marker horizon (*solid lines*) and on the faults (*dashed lines*). **b** Structure contours on portions of the three fault planes; contours on fault *A* are *dashed* where the plane lies below fault *B*. Fault planes *A* and *C* are the same

A good test of the fault interpretation on a map is to measure the throws in the vicinity of all branch lines and show that the throws of the two smaller faults are equal to that of the larger fault. If this is not true, the map is wrong. An abrupt change in the throw of a fault is probably caused by an undetected fault branch (Ocamb 1961) or relay fault that carries the missing throw.

6.5.3 Splay Fault

Splay faults are minor faults at the extremities of a major fault (Bates and Jackson 1987). The master fault splits at one or more branch lines to form the splay faults (figs. 6.30, 6.31). The splays then die out along strike at tip lines. Splay faults distribute the total offset over a wider area, which makes it possible for more of the displacement to be accommodated by folding rather than faulting. This helps a large master fault die out into a fold. The master fault may continue straight in the region of fault splays (Fig. 6.30), or all the splays may have different orientations from that of the master fault.

Splay faults are ultimately bounded in all directions by branch lines and tip lines. The leading edge of a structure is its termination in the transport direction and the trailing edge is in the opposite direction (after Elliott and Johnson 1980). For a thrust fault, the leading edge of a splay is a tip line and the trailing edge is a branch line (Fig. 6.32a). The leading hangingwall and footwall cutoff lines of a unit carried on a splay join and end at the tip line of the leading fault (Fig. 6.32b). The trailing bed cutoff line continues along the trailing thrust beyond the tip of the splay.

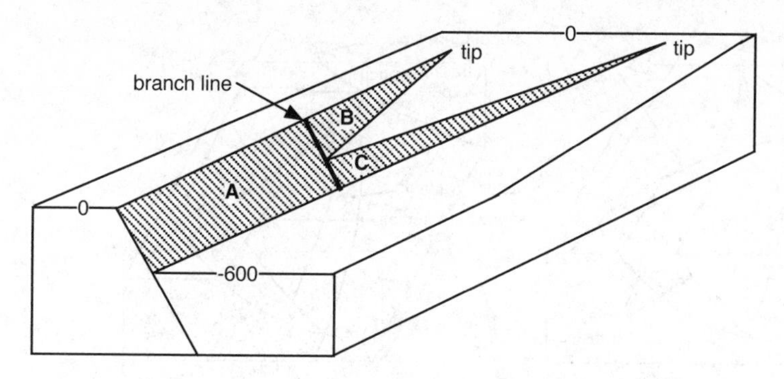

Fig. 6.30. Splay faults (*B* and *C*) at the end of a master normal fault (*A*)

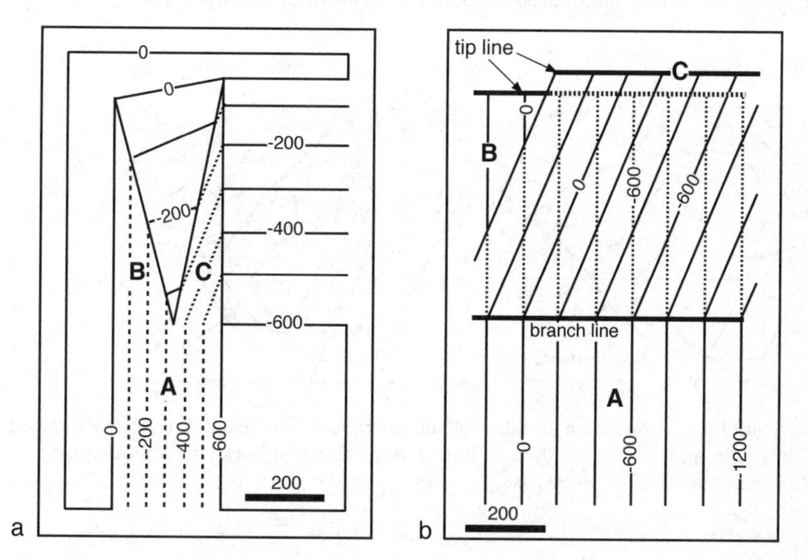

Fig. 6.31. Structure contour maps of a splaying normal fault based on geometry of the structure in Fig. 6.30. The master fault is *A*, its continuation into the region of splay faulting is *B*, and the other splay fault is *C*. Both maps are at the same scale. **a** Offset marker horizon. *Solid contours* are on the marker horizon and *dashed contours* are on the faults. **b** Portions of the faults. Contours are *dotted* where they lie below another fault.

6.5.4 Fault Horse

A fault horse (Fig. 6.33) is a body of rock completely surrounded by fault surfaces (Dennis, 1967; Boyer and Elliott 1982). The geometry can be thought of as a splay fault that rejoins the main fault. A horse is completely bounded by a branch-line loop along which the two fault surfaces join (Fig. 6.33a). A stratigraphic unit within a horse will have leading and trailing cutoff lines at the two bounding fault surfaces (Fig. 6.33b). Multiple fault horses form a duplex (Boyer and Elliott 1982).

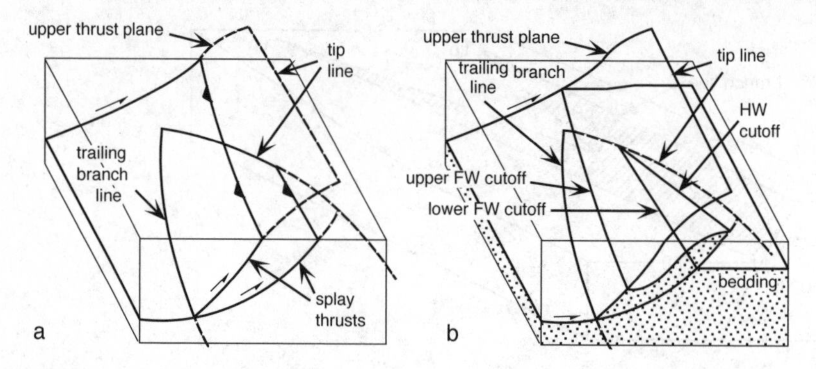

Fig. 6.32. Splay faults at the leading edge of a thrust fault. **a** Fault surfaces in three dimensions. **b** Bedding cutoff lines of a unit transported on the lower splay. (After Diegel 1986)

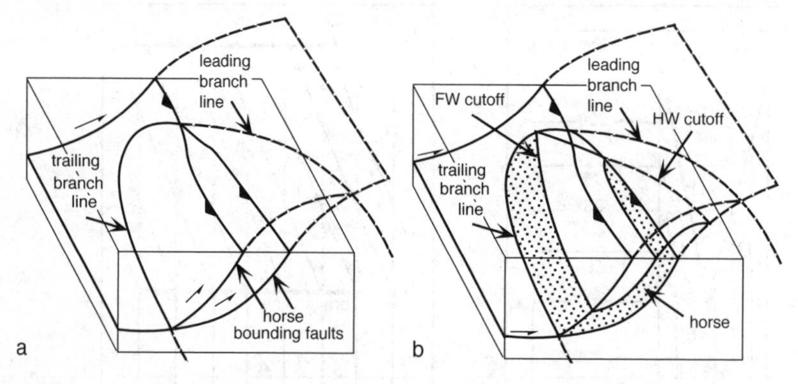

Fig. 6.33. Fault horse bounded on all sides by fault surfaces that end in all directions at a closed branch-line loop. **a** Main fault and branch lines. **b** Bedding cutoff lines for a stratigraphic unit within the fault horse. (After Diegel 1986)

6.6 Crossing Faults

Two intersecting faults may both continue past the line of intersection, a relationship here termed crossing faults. Where one fault offsets the other, the line of intersection between the two faults is a cutoff line of one fault by the other, not a branch line. Crossing faults may be sequential or contemporaneous. Sequential faulting means that an older fault is cut and displaced along a younger fault. Contemporaneous faulting means that the faults form at approximately the same time and as part of the same movement picture.

6.6.1 Sequential Faults

Crossing dip-slip normal faults are illustrated in Fig. 6.34. Both faults have constant (but different) displacement, but the shaded marker horizon is displaced to four dif-

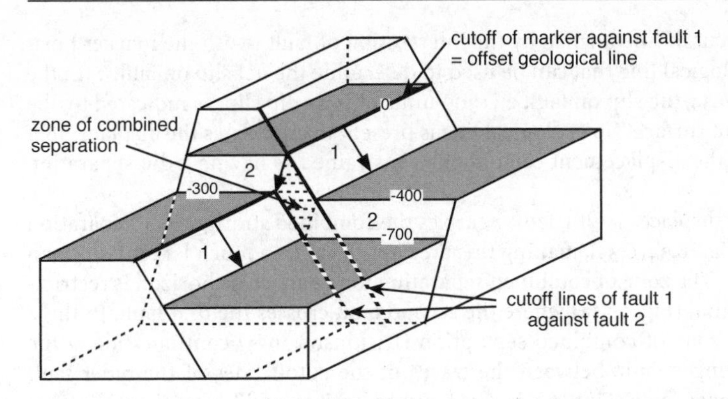

Fig. 6.34. A young normal fault (*2*) cuts and displaces an old normal fault (*1*). An offset marker horizon is *shaded*. Displacement on each fault is dip slip of a constant amount. The displacement of the offset geological line is the slip on fault 2

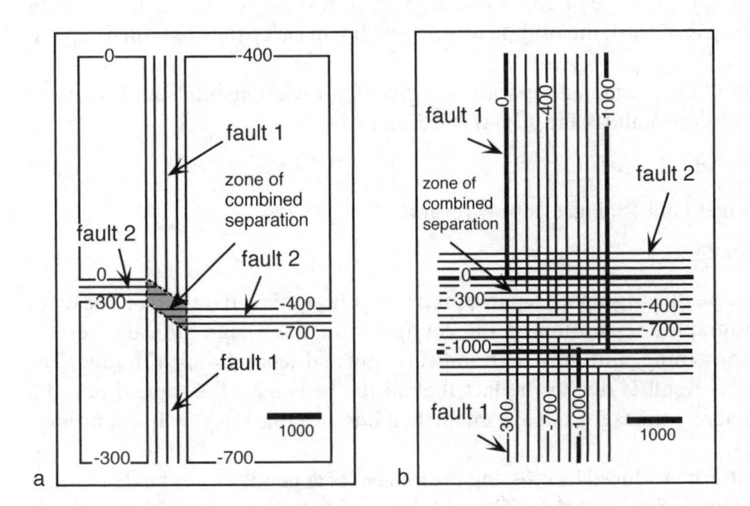

Fig. 6.35. Structure contour maps based on the geometry of the structure in Fig. 6.34. Both maps are at the same scale. Fault *2* is through-going and displaces fault *1* down dip 300 units to the south. **a** The marker horizon. **b** Portions of both faults in the vicinity of the offset marker horizon

ferent elevations, reflecting the four different combinations of displacement from the original elevation (no displacement, fault 1 alone, fault 2 alone, fault 1 + 2). At first glance the structure contour map (Fig. 6.35a) appears to imply that fault 1 is younger and is a strike-slip fault because its trend is straight whereas the trend of fault 2 appears to be offset. Comparison with Fig. 6.34 shows that fault 2 is the through-going plane. It is the different elevations of the marker horizon along fault 2 that make this fault appear to bend at the intersection. The fault-plane map (Fig. 6.35b) shows fault 2 to be unbroken and fault 1 to be displaced. In the area near the intersection of the two faults, both faults would be penetrated by a vertical well (Fig. 6.35b).

The footwall cutoff line formed by the intersection of fault 1 with the marker horizon forms a geological line that can be used to determine the net slip on fault 2. In the example in Fig. 6.34, the slip on fault 2 is 300 units of pure dip slip, as indicated by the arrow on the fault surface. No geological line is present that predates the displacement of fault 1; hence the displacement can only be constrained as having a dip separation of 400 units.

Where fault 2 displaces fault 1, fault 2 carries the combined stratigraphic separation of both faults (Figs. 6.34, 6.35), making this area appear to be a much larger fault than either fault 1 or 2. The zone of combined separation for a particular horizon is restricted to a small region (Fig. 6.35a) where the second fault crosses the first fault. In three dimensions, the zone of combined separation (Dickinson 1954) continues along the plane of the younger fault, between the traces of the cutoff lines of the older fault against the younger fault (Figs. 6.35b, 6.36). Other horizons will have their zones of combined separation along the same trend, but offset from one another. Given low bedding dips, a vertical well drilled into the zone of combined separation for two normal faults will cut one fault that carries the combined separation (Fig. 6.36a). A vertical well drilled into the zone of combined separation for two reverse faults will have three fault cuts (Fig. 6.36b), the middle one of which will carry the combined separation.

The amount of the combined separation is given by a relationship from Dickinson (1954). If the younger fault (B, Fig. 6.36a) is normal,

$$t = -b + (\pm a), \tag{6.2}$$

and if the younger fault (B, Fig. 6.36b) is reverse,

$$t = +b - (\pm a), \tag{6.3}$$

where t = combined stratigraphic separation, a = stratigraphic separation on the older fault, b = stratigraphic separation on the younger fault, the + sign indicates reverse separation = thickening, and the – sign indicates normal separation = thinning. The simplicity of this result is due to the fact that all the beds have the same dip. If dip changes occur across the faults but are small, then Eqs. (6.2) and (6.3) will still provide good estimates.

The map pattern produced by crossing normal faults depends on the angle of intersection of the faults, the dip of the faults and the marker horizon, and the magnitude

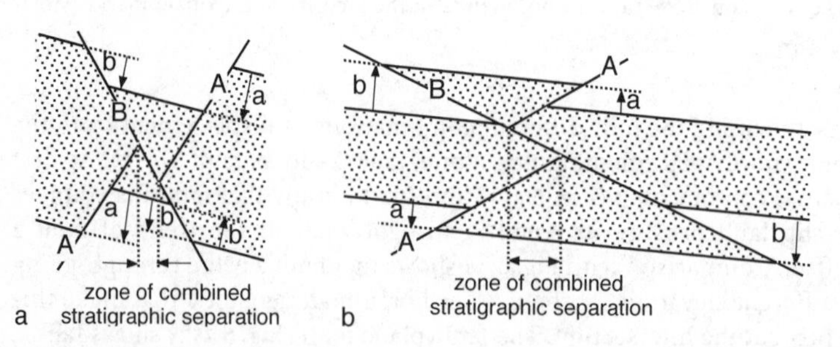

Fig. 6.36. Zone of combined stratigraphic separation on fault *B*. Fault *A* is displaced by fault *B*. **a** Two normal faults. **b** Two reverse faults. For explanation of **a** and **b**, see text (Sect. 6.6.1)

and sense of slip of the faults. The trace of the fault on the marker horizon depends on the dip of the marker as well as on the attitude of the fault surface. The following examples illustrate some of the possibilities. In the three parts of Fig. 6.37, the orientation of the older fault (A) is changed while the orientation of the younger fault (B) and the attitude of the marker horizon remain constant. The structure changes from a geometry that could be described as tilted steps (figs. 6.37a,c) to a horst that changes into a graben along a northwest – southeast trend (Fig. 6.37b). Note that the direction of the fault-bedding intersection is not parallel to the fault strike in any of the examples because the strike of bedding is not parallel to the strike of the faults.

The width of the zone of combined stratigraphic separation is reduced to a line for certain combinations of the displacement directions (Fig. 6.38a,b). An apparent strike separation on the older fault will be caused by slip on the younger fault that is in the strike direction of the younger fault. In Fig. 6.38c the later, through-going fault (B)

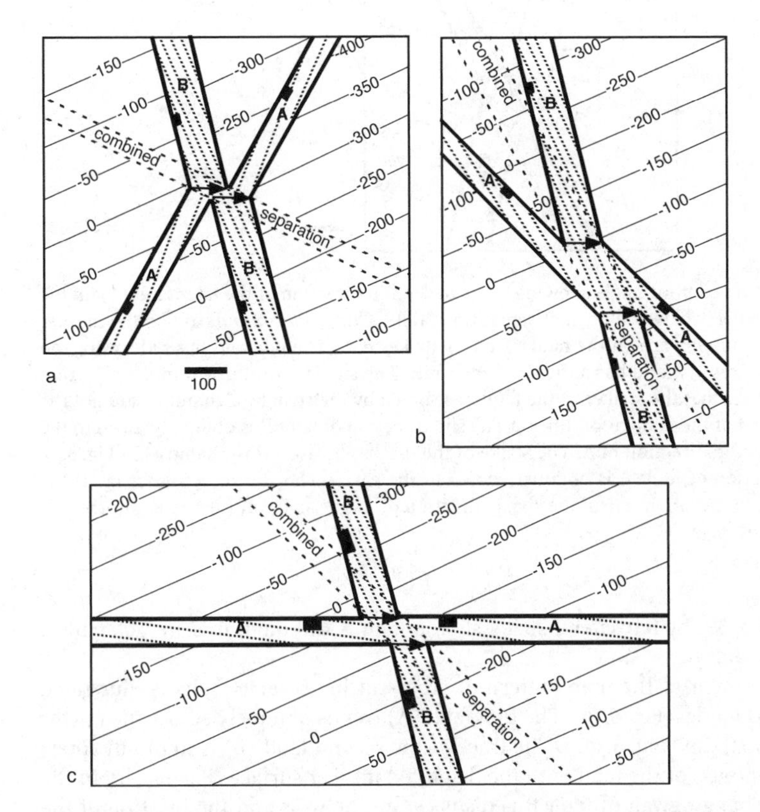

Fig. 6.37. Structure contour maps showing the effect of the angle of fault intersection on the geometry of sequential dip-slip normal faults. Fault *A* is older and has a pure dip-slip of 100 units; fault *B* is younger, strikes 335°, and has a slip of 200 units down to the east. *Arrows* give the slip direction of fault B. Structure contours on the displaced marker horizon are *thin solid lines*; contour interval is 50 units. *Dotted contours* are on the fault planes. Note that the strikes of the faults, as shown by the structure contours, are not the same as the trend of the fault cutoff lines of the marker horizon. **a** Fault *A* strikes 020°. **b** Fault *A* strikes 326°. **c** Fault *A* strikes 276°

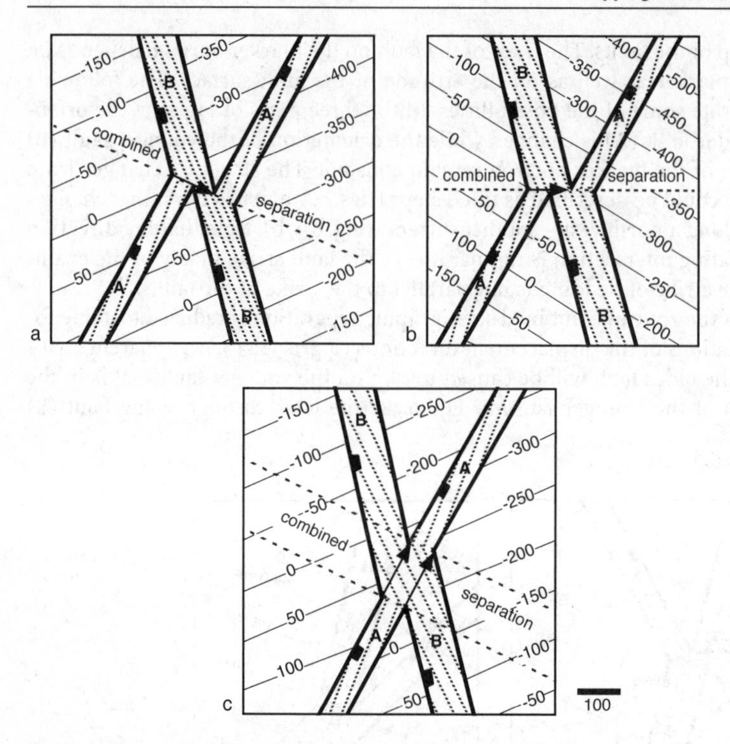

Fig. 6.38. Structure contour maps showing effect of the slip direction of the later fault (*B*) on the intersection geometry between normal-separation faults. *Thin solid contours* are on the marker horizon, *dotted contours* are on the faults. Fault *A* strikes 020°, is older, and has a throw of 100 units. Fault *B* strikes 335° and has a throw of 200 units. The slip is downthrown in the direction of the *arrows*. Note that the strikes of the faults, as shown by the structure contours, are not the same as the trend of the fault cutoff lines. **a** The slip direction of fault B is obliquely down to the southeast, in the dip direction of A. The strike of the marker horizon is the same as in Fig. 6.37. **b** The slip direction of fault B is obliquely down to the east. **c** The slip direction of fault B is obliquely down to the northeast, in the strike direction of A. The strike of the marker horizon is the same as in Fig. 6.37

appears to be offset by left-lateral strike-slip on fault A, although it is, in fact, fault B that displaces fault A.

The development of the map pattern of cross-cutting reverse faults is illustrated with a forward model (Fig. 6.39). The first fault (A) trends obliquely across the northwesterly regional dip (Fig. 6.39a). The trace of the second fault (B) is found by intersecting the contours of the fault with the displaced marker surface (Fig. 6.39b). In the final step, the hangingwall of fault B is displaced to the west and the location of the hangingwall cutoff and the contours are drawn as an overlay on the footwall (Fig. 6.39c). Three fault cuts are present in the vertical zone of combined stratigraphic separation where the older fault is repeated by the younger fault (Dickinson 1954).

A cross section (Fig. 6.40) shows the evolution of two reverse faults having the same dip direction, like those in the previous map (Fig. 6.39c). In the zone of combined stratigraphic separation between the dashed lines, a vertical well would cut

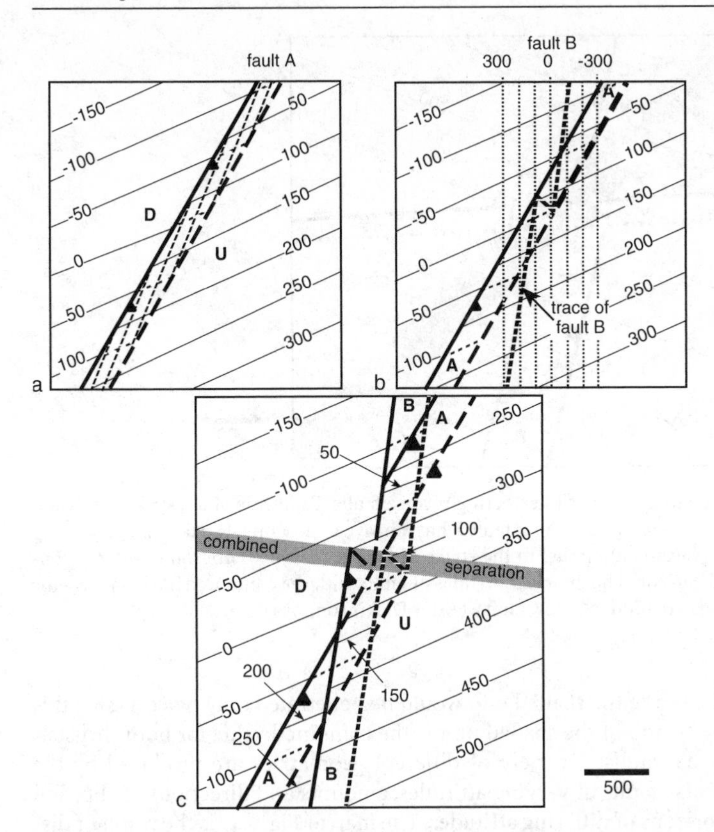

Fig. 6.39. Structure contour maps showing evolution of intersecting reverse faults having approximately the same dip direction and sense of displacement. Contours on the fault planes are *dotted*. Contours on hidden surfaces are *dashed*. *Thin solid lines* are contours on the marker horizon. *D* Downthrown block, *U* upthrown block. **a** First displacement. The northeast-striking thrust (*A*) has a displacement of 100 units up to the northwest. The hangingwall cutoff is a *thick solid line* and the footwall cutoff is a *thick dashed line*. **b** Structure contours on incipient fault *B* are added. The trace of thrust *B* on the marker surface is *dotted*. **c** Second displacement, enlarged to show detail. Thrust *B* displacement is 200 units up to the west. *Dotted* bed contours are in the footwall of fault *A*; contours indicated by *arrows* are in the footwall of fault *B*

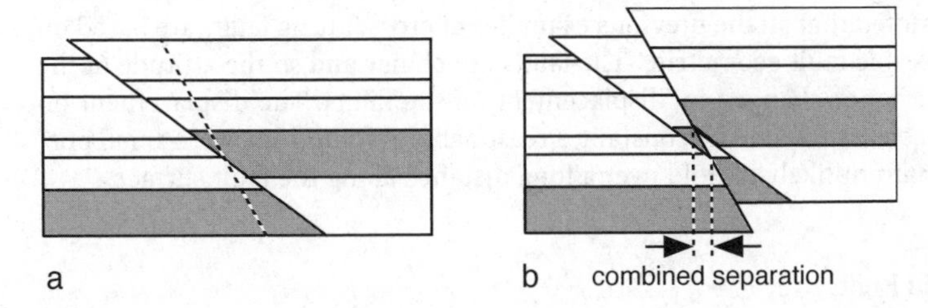

Fig. 6.40. Evolution of crossing reverse faults having the same dip direction. The cross sections approximate the evolution along an east – west cross section through the center of Fig. 6.39c. **a** First fault displacement with the *dotted* trace of the second fault. **b** Displacement on the second fault. *Dotted lines* bound the zone of combined stratigraphic separation for the *shaded* horizon

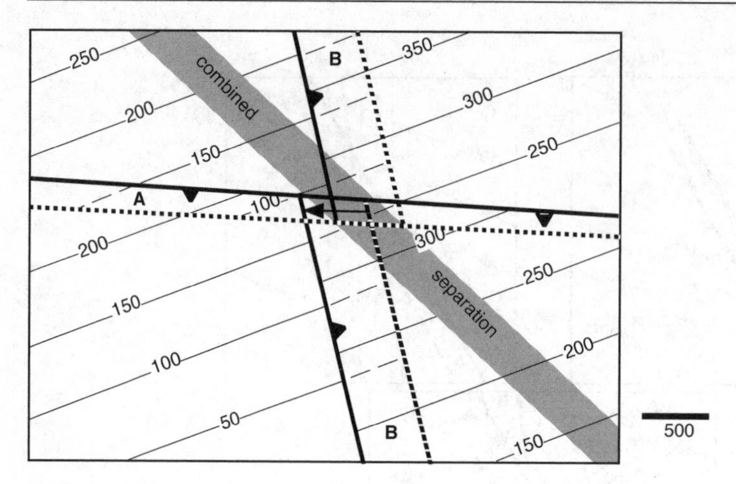

Fig. 6.41. Structure contour map of intersecting reverse faults. Fault *A* is older, strikes 273°, and has a heave of 100 units, up on the south; fault *B* has a heave of 200 units, strikes 347° and is up on the east with displacement parallel to the strike of fault A (*arrow*). *Thin solid lines* are contours on the marker horizon. The intersection lines of faults with the contoured horizon are *wide lines*. *Dashed contours* are hidden below faults. (After Dickinson 1954)

three faults, but the top of the shaded unit would be penetrated only twice. Inside this zone the throw on the top of the shaded unit is the combined value for both thrusts.

Just as for normal faults, a variety of different geometries are produced by the intersection of reverse faults of varying attitudes, amounts and directions of slip, and that cut marker horizons of differing attitudes. The map of Fig. 6.41 is the result of displacement of a southeast dipping bed by two orthogonal reverse faults, fault A being the older. The map of Fig. 6.42 is of two parallel thrusts having opposite dips. The evolution of an east – west cross section of a structure approximately like that of Fig. 6.42 is shown in Fig. 6.43. A vertical well in the zone of combined throw would have three fault cuts and penetrate the top of the shaded unit twice. Even though the two faults thicken the section, a unit within the zone of combined separation can be reduced in thickness by the second fault. In the region labeled T (Fig. 6.43b) the shaded unit is thinned by the crosscutting reverse faults and the middle fault cut could be mistaken for a normal fault downthrown to the left.

It should be noted that all the previous examples of crosscutting faults are based on the simplest possible fault geometries. The faults are planar and so the attitude of the marker horizon is not changed by displacement on the faults. The displacement on each fault has been assumed to be constant, a reasonable assumption over a small portion of a fault, but not likely to hold over a long distance along the fault surface.

6.6.2 Contemporaneous Faults

Intersecting faults may be of the same age, that is, contemporaneous. [The term contemporaneous faulting has also been used for faults in which the displacement is contemporaneous with deposition (Hardin and Hardin 1961) but the term growth fault is

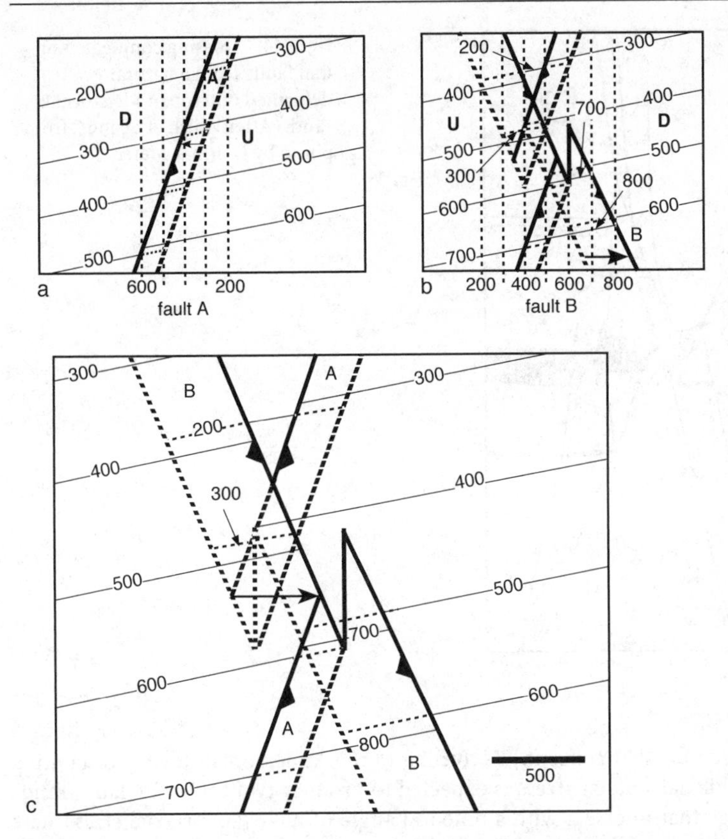

Fig. 6.42. Evolution of the structure contour map of reverse faults having equal and opposite dips. North – south *dashed contours* are on faults *A* and *B*. *Thin solid lines* are contours on the marker horizon. *Wide lines* are fault traces. Contours are *dotted* where hidden. **a** Fault *A* strikes north – south and is displaced 100 units up due west. **b** Fault *B* strikes north – south and is displaced 200 units up due east. **c** Final map, enlarged to show detail. *Arrow* shows the displacement direction of fault *B*

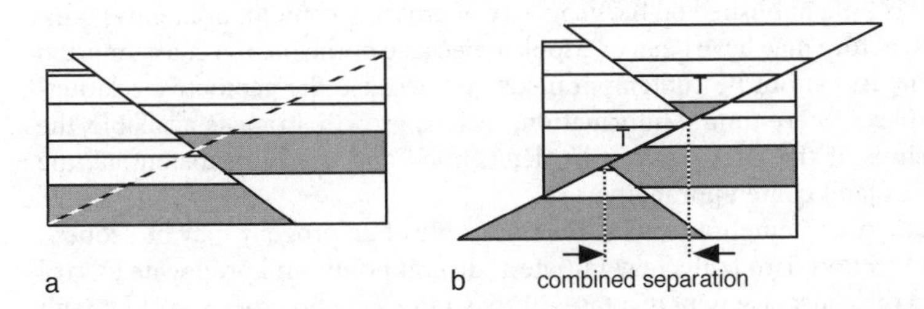

Fig. 6.43. Evolution of crossing reverse faults having opposed dips, similar to those in Fig. 6.42c. **a** Displacement on the first fault, with location of the second fault *dashed*. **b** Displacement on the second fault. *Dotted lines* bound the zone of combined stratigraphic separation for the *shaded* horizon. *T* indicates regions of local structural thinning

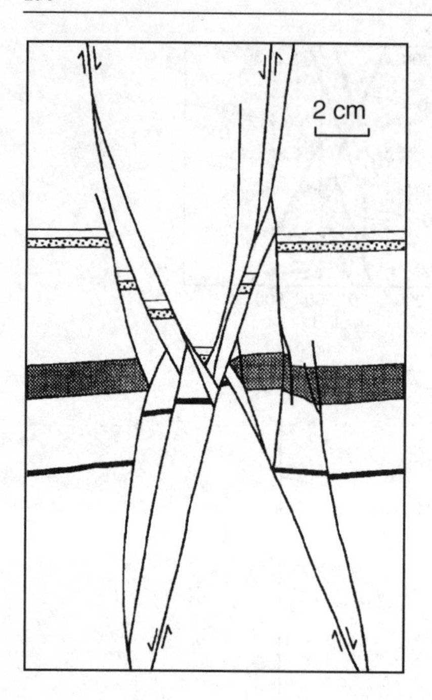

Fig. 6.44. Crossing conjugate normal faults from a naturally deformed outcrop of Pleistocene sand. (After Walsh et al. 1996, from a photo by Dietmar Meier)

now widely used for that concept.] According to the Andersonian theory of faulting (Sect. 1.6.4) a biaxial state of stress is expected to produce two conjugate fault trends of the same age that intersect with a dihedral angle of 40–65°. A triaxial stress state may cause three or four fault trends (Oertel faults) to form simultaneously.

The fault pattern in the zone of intersection of contemporaneous faults tends to be more complex than the patterns for sequential faults discussed in the previous section. Multiple small faults may occur in the zone of intersection (Fig. 6.44). In the experimental studies of normal faults by Horsfield (1980), two crossing conjugate faults formed initially and with continuing extension the initial horst and graben were segmented by smaller conjugate faults. The net displacement across fault zones consisting of contemporaneous crossing faults is likely to be small, or even zero, as in Fig. 6.44.

As yet there is little published on the geometry of crossing contemporaneous faults. This style of structure may have been overlooked because of the incorrect assumption that crossing faults cannot be contemporaneous or because the geometric relationships are complex. Where timing information, such as growth strata, is available, the time relationships of the fault sets can be determined and used to substantiate the interpretation of fault contemporaneity.

Faults which are contemporaneous at the time scale of an orogeny may be sequential at any one location. Two faults may initiate at distant points and propagate toward one another as their displacement increases. If they intersect and cross, the older fault will be displaced by the younger, even though the age difference may be small.

6.7 Faults on Isopach Maps

Faults cause characteristic thickness variations on isopach maps (Hintze 1971). If unrecognized, these variations could be misinterpreted as being stratigraphic in origin. Recognized as fault related, they provide a tool for fault interpretation. A normal fault thins the stratigraphic unit (Fig. 6.45) and a reverse fault thickens the unit. A single fault produces two parallel bands of thickness change (figs. 6.45, 6.46), one associated with the cutoff of the top of the unit and one with the cutoff of the base of the unit. The affected width on the isopach map is appreciably wider than the fault gap on the structure contour map of a single horizon. The affected zone extends from the fault cutoff of the upper surface of the affected bed on one side of the fault to the fault cutoff on the lower surface of the affected bed on the other side of the fault (Fig. 6.45). The zone of thinning (normal fault) or thickening (reverse fault) is centered on the trace of the fault through the middle of the unit. Such elongate zones of thinning or thickening can indicate the presence of faults having displacements that are too small to completely separate the map unit and that might otherwise go unrecognized.

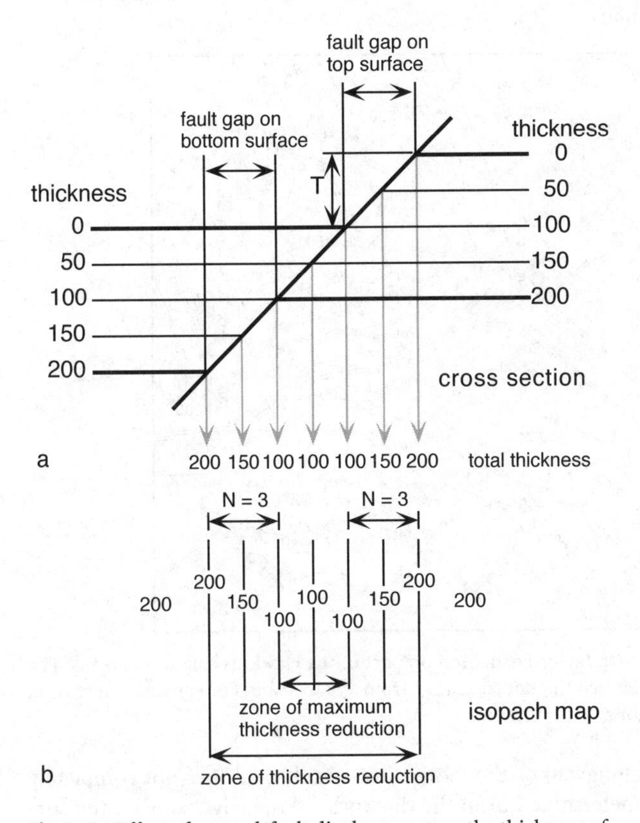

Fig. 6.45. Effect of normal-fault displacement on the thickness of a unit. **a** Cross section. *T* throw on the fault. **b** Isopach map. *N* number of contours across which the thickness change occurs. (After Hintze 1971)

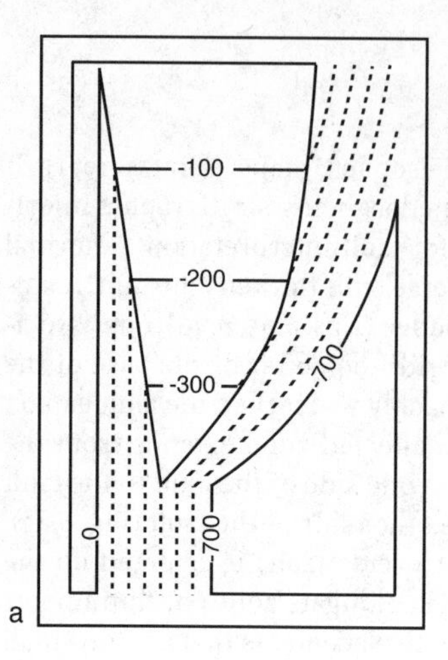

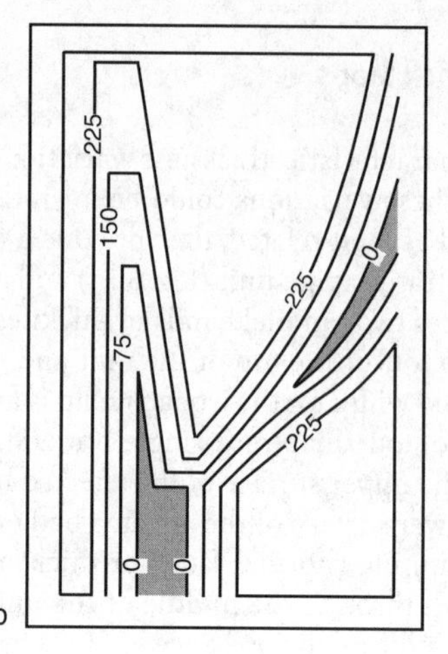

Fig. 6.46. Maps of a branching normal fault. **a** Structure contours on the fault (*dashed contours*) and on the upper surface of a unit (*solid contours*). **b** Isopach map of the faulted unit having an unfaulted thickness of 225 units

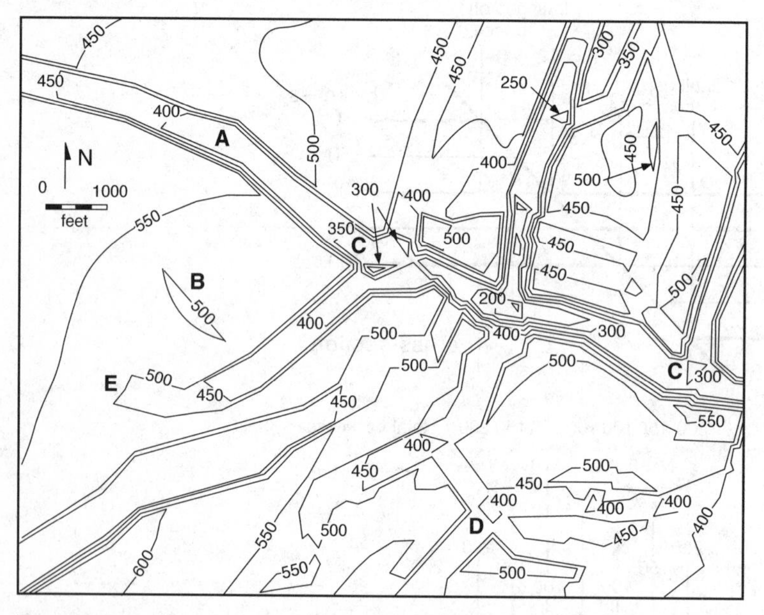

Fig. 6.47. Isopach map of lower Taylor Formation on part of the Hawkins salt dome, Texas. The contour interval is 50 ft. Unfaulted thicknesses range from 470 to 584 ft. For explanation of *A–E*, see text (Sect. 6.7). (After Hintze 1971)

The throw on the hangingwall or the footwall of a fault that does not completely separate the unit can be determined from the thickness of the edge zone on the corresponding side of the fault. From Hintze (1971), the relationship is

$$T = (N-1)\, I, \tag{6.4}$$

where T = fault throw, N = the number of contours in the edge zone from the most reduced (or increased) thickness to the regional thickness, and I = the contour interval. In Fig. 6.45b, the throw is three contour lines minus one, times 50, equal to 100, the amount of throw in Fig. 6.45a. Fault intersections are marked on a map by localized triangular (Fig. 6.46) to rectangular areas of thinning or thickening.

Maps on a portion of a salt dome provide an example of the effect of normal faults on an isopach map. Elongate zones of isopach thinning follow the trends of the normal faults (Fig. 6.47). At point A on the isopach map (Fig. 6.47) the fault throw on the upper surface is 100 ft from the isopach contours on the north side of the fault (2 contours times the 50-ft contour interval), the amount recorded on the structure contour map (Fig. 6.48). On the south side of the fault, the throw on the lower surface of the unit is 150 ft (3 contours times 50 ft), implying a growth fault. The D-shaped isopach pattern at point B in Fig. 6.47 represents a fault that dies out within the isopach interval. Fault branching is indicated by the triangular thin areas at the points labeled C. The rectangular isopach thin at point D is caused by crossing faults. Where a fault dies out as at E, the width of the zone of thinning is quite wide relative to the amount of the thinning.

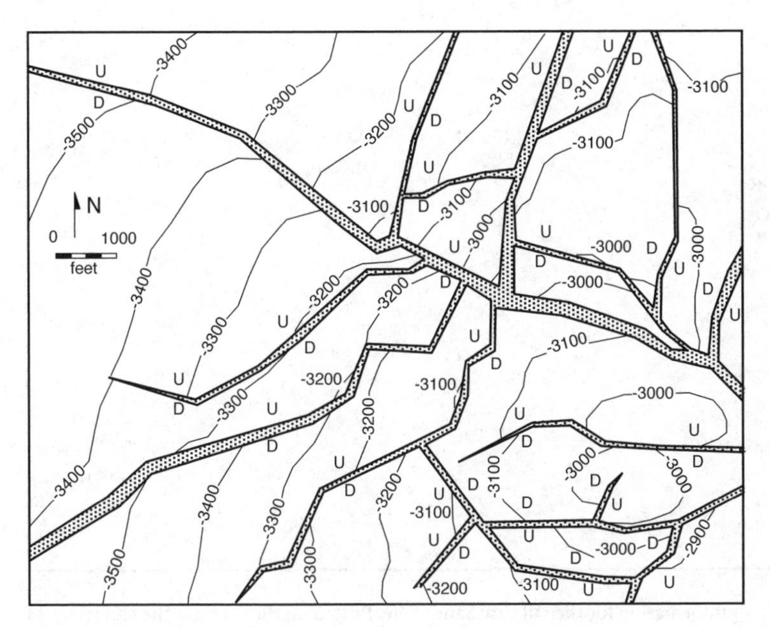

Fig. 6.48. Structure contour map of the top of the lower Taylor Formation on part of the Hawkins salt dome, Texas. The contour interval is 100 ft. The faults (*shaded*) are normal and dip about 45°. *D* Downthrown block, *U* upthrown block. (After Hintze 1971)

6.8 Problems

6.8.1 Normal Fault

Determine the structure of the Oil City Sandstone in Fig. 6.49.

1. Is a single fault present? What kind? What is the evidence?
2. What is the attitude of the fault? of bedding in the hangingwall? of bedding in the footwall?
3. What is the heave and throw on the fault?
4. Does the map agree with the attitude information?
5. Explain the reason for the hydrocarbon trap.

6.8.2 Reverse Fault

Map the fault (f) and the shale marker horizon (s) using the data in Fig. 6.50.

1. Is a single fault present? What kind? What is the evidence?
2. What is the attitude of the fault? of bedding in the hangingwall? of bedding in the footwall?
3. What is the heave and throw on the fault?

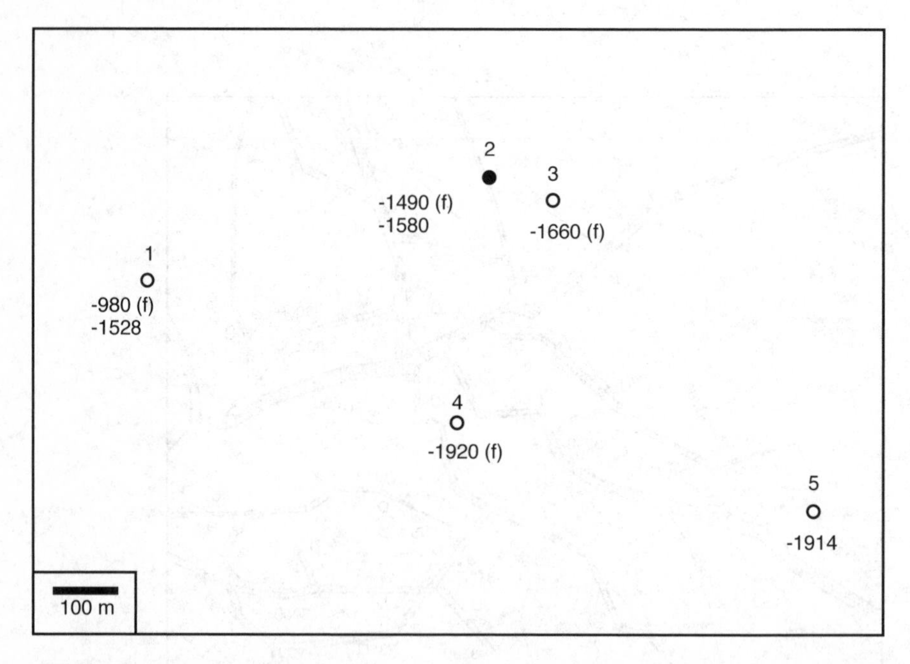

Fig. 6.49. Map of information for the Oil City Sandstone. Posted on the map are the elevations of fault cuts (*f*) and the top of the sandstone. Well *2* is an oil well. Dipmeters in well *1* indicate a bedding attitude of 10, 315 and a fault strike of 045; in well 5 the bedding attitude is 8, 315. Elevations are in meters

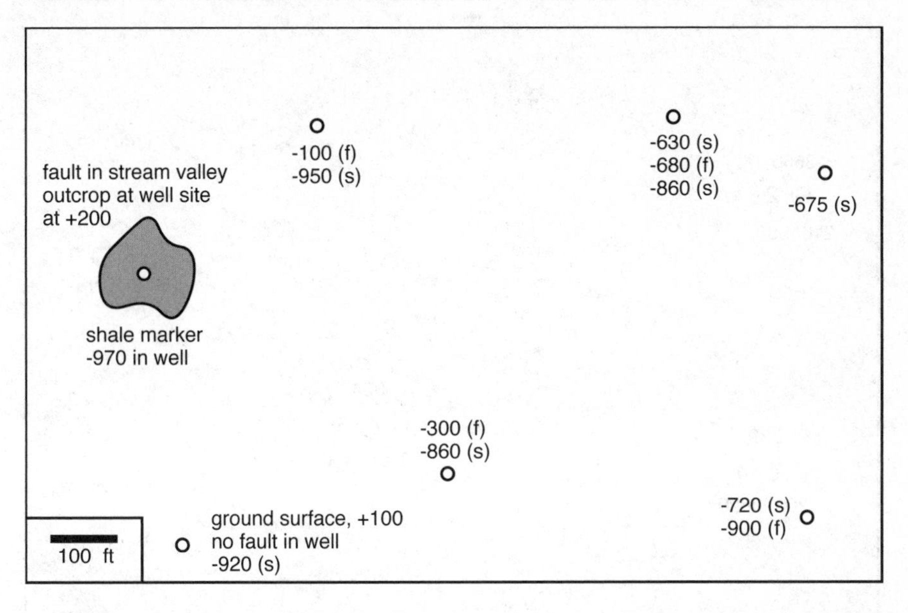

Fig. 6.50. A shale marker and fault cuts in a groundwater basin. Posted on the map are the elevations of fault cuts (*f*) and of the shale tops (*s*) in each well where present. Elevations are in feet, negative below sea level

4. If a heavy liquid is spilled in the stream valley in the shaded area, could the fault provide a barrier to the fluid movement in porous units above the shale marker? Explain the reason behind your answer.
5. Where would a fault trap for fluids lighter than water be located?

6.8.3 Reservoir Structure

Use the subsurface information in Fig. 6.51 to make a structure contour map of the top of the Hamner Sandstone reservoir.

1. What type of fault or faults are present? Are there any faults in the reservoir?
2. What is the dip of the fault or faults?
3. What is the calculated heave and throw at each fault cut?
4. Do the fault map and fault separation data agree with the map?

6.8.4 Method of Projected Fault Cutoffs

Using the data in problem 6.8.3, construct a map of the top of the Hamner Sandstone using the method of projected fault cutoffs. For this method, the structure on one side of the fault must be known or assumed to serve as the reference. Which side will you use as the reference side and why?

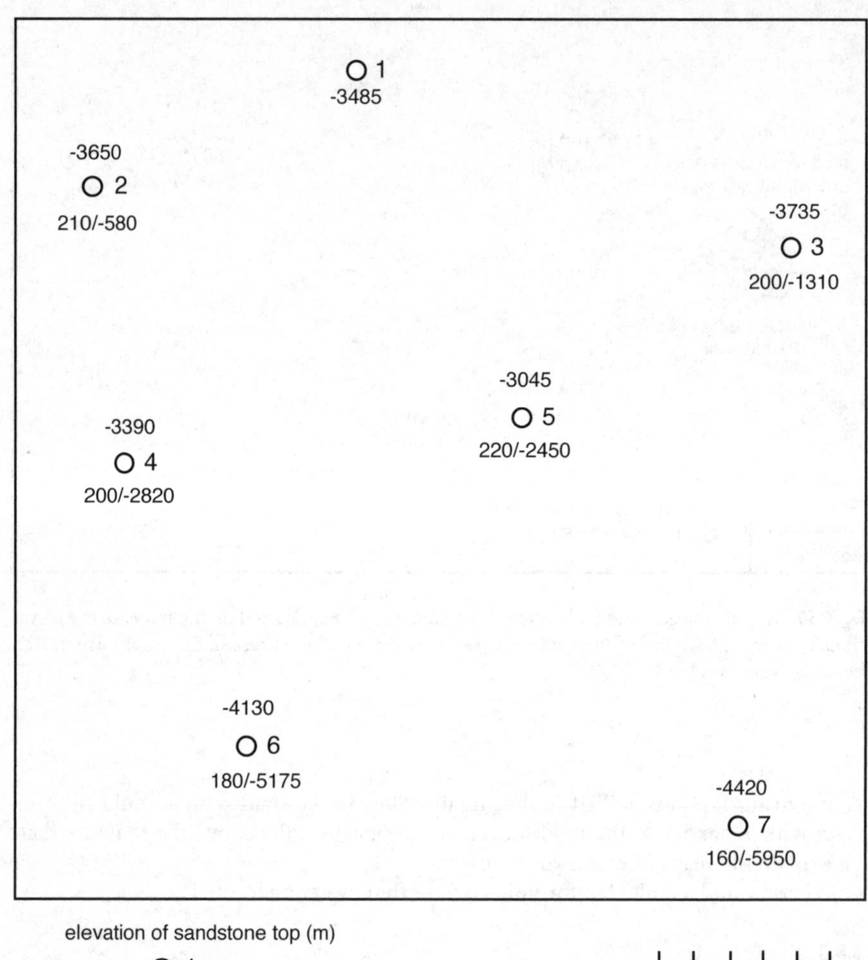

Fig. 6.51. Well data on the Hamner Sandstone reservoir. Wells (*numbered*) penetrate the Hamner Sandstone (*single number* is the elevation of the top contact). Fault cuts are indicated by a *pair of numbers* (amount/elevation). Elevations are in meters, negative below sea level

6.8.5 Method of Restored Tops

Using the data in problem 6.8.3, construct a map of the top of the Hamner Sandstone using the method of restored tops. For this method, the value of the vertical separation must be known at the marker horizon. What value will you use for the vertical separation and why?

1. Is the map produced by this method different from that given by the method of projected fault cutoffs (problem 6.8.4)?
2. Which method is better for this example? Why?

6.8.6
Thrust-Faulted Fold

Based on the map in Fig. 3.28, answer the following questions:
1. What is the 3-point dip of the fault at the surface?
2. Construct a structure contour map of the fault from its surface dip.
3. Intersect the previously-constructed structure contour map of the top Fairholme with the map of the fault.
4. Does the projected structure contour map agree with the drilled depths to the top of the Fairholme?
5. The wells to the Fairholme were drilled to find a hydrocarbon trap but were not successful. What is a structural reason for drilling the wells and what is a structural reason that they were unsuccessful?

6.8.7
Correlating Fault Cuts from Fault Attitude

Based on the map of Fig. 6.52, answer the following questions:
1. Contour a fault that includes all four wells that cut faults. Is the fault surface obtained both possible and reasonable?
2. What is the attitude of the fault plane given by the three points 1,4,2?
3. What is the attitude of the fault plane given by the three points 2,3,4?

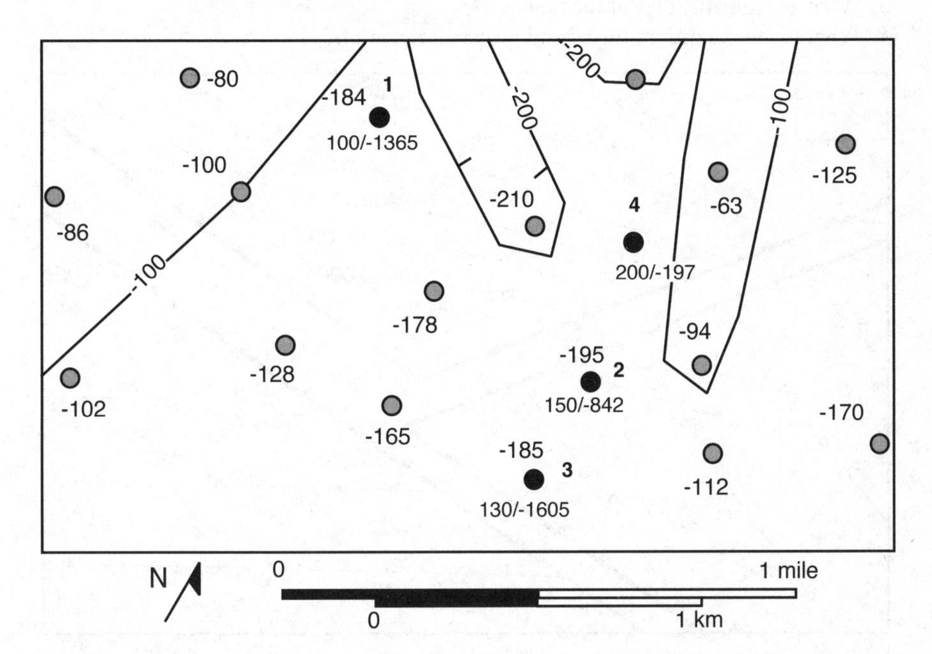

Fig. 6.52. Map of the southern portion of the Deerlick Creek coalbed methane field. *Circles* are wells and the *small number* near the well is the elevation of the top of the Gwin coal cycle (in feet). *Solid circles* are wells with fault cuts, given as stratigraphic separation / depth to fault cut. Structure contours shown on the top of the Gwin are tentative. Elevations below sea level are negative

4. What is the attitude of a fault plane through wells 1, 2, and 3? Is this a possible fault plane? Why or why not?

6.8.8 Map validation

The structure contour map of the top of a faulted limestone (Fig. 6.53) fits the well information and explains the hydrocarbon trap. Is the interpretation valid?

1. Based on the map as presented, what kind of faults are present? In which direction do they dip?
2. Where are the structural closures on the map? Could any of them be new hydrocarbon traps? Why or why not?
3. Draw implied structure contour maps on the faults. Is the original map correct?
4. Construct an improved map that honors all the well data.

6.8.9 Relay Zone

Map the faults and the top of the Northriver Sandstone on the map of Fig. 6.54.

1. Where is the relay zone?
2. What is the attitude of the sandstone away from the faults?
3. What is the attitude of the sandstone between the faults?
4. What are the attitudes of the faults?
5. What is the maximum throw and heave on the faults?

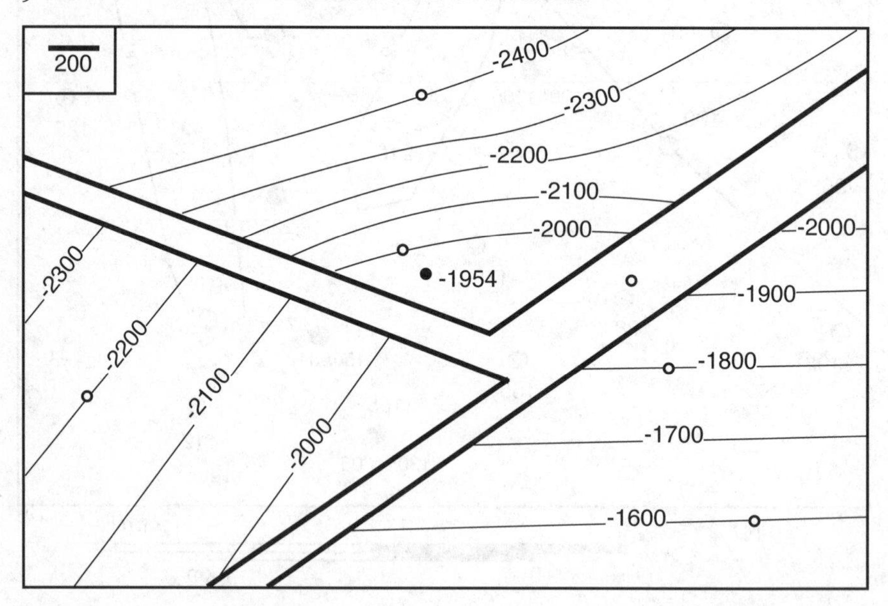

Fig. 6.53. Hypothetical structure contour map of the top of the Appling Bend Limestone gas reservoir. Contours are depths below sea level. The *solid circle* is a gas well, *open circles* are dry holes. The limestone is missing in the well in the fault gap. Choose the measurement units to be in either feet or meters. Elevations below sea level are negative

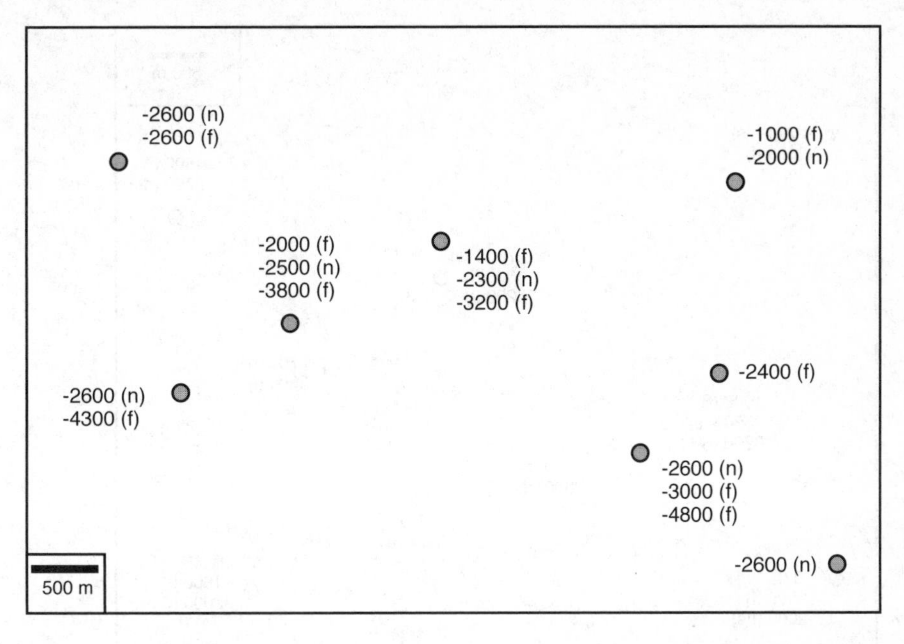

Fig. 6.54. Top of the Northriver Sandstone (*n*) and fault-cut elevations (*f*) in wells. Elevations are in meters, negative below sea level

6.8.10 Branching Fault

Map the faults and the top of the Reef Limestone on the map of Fig. 6.55.
1. Where is the branch line?
2. What is the attitude of the limestone away from the faults?
3. What is the attitude of the limestone between the faults?
4. What are the attitudes of the faults?

6.8.11 Splay Faults

The water-well map of Fig. 6.56 shows a distinctive clay seam to be absent in some wells due to faulting. Map the faults and the top of the clay seam.
1. Where is the branch line?
2. What are the attitudes of the faults?
3. What is the maximum throw and heave on the clay seam?
4. If the clay seam is a barrier to ground water flow from the surface, where is this barrier absent?
5. Is a spill of toxic heavy liquid in the southwest corner of the map area likely to sink below the clay seam? Why or why not?
6. In which direction will a spill of heavy liquid in the southeast corner of the map migrate?

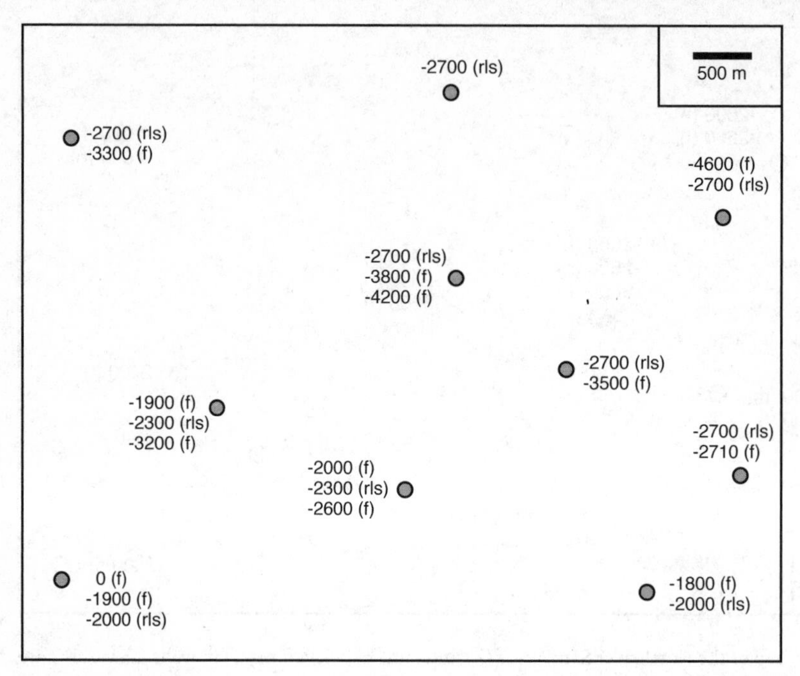

Fig. 6.55. Map giving the top of the Reef Limestone (*rls*) and faults (*f*) in wells. Elevations are in meters, negative below sea level

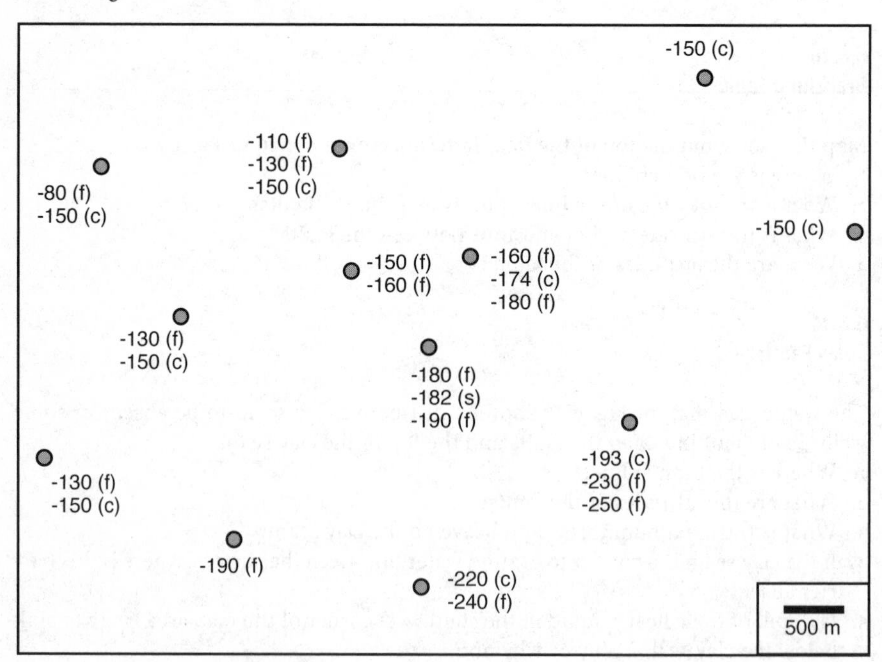

Fig. 6.56. Map of the top of a clay seam in water wells drilled into an alluvial aquifer. Elevations below sea level are negative

6.8.12
Sequential Faults 1

Two different fault trends occur in the area of Fig. 6.57. Map the faults and the A sand.
1. What is the reason for the hydrocarbon trap in the A sand?
2. What are the attitudes of the faults?
3. What is the throw and heave on each fault?
4. Which fault is older?
5. If the hydrocarbons migrated before the formation of the younger fault, would the trapping potential of the structures be the same?

6.8.13
Sequential Faults 2

Contour the Northport Dolomite in the map of Fig. 6.58, being careful to explain the fault cuts and the oil trap(s).
1. Is there one oil field or two?
2. What are the attitudes of the faults?
3. What is the throw and heave on each fault?
4. Which fault is older?

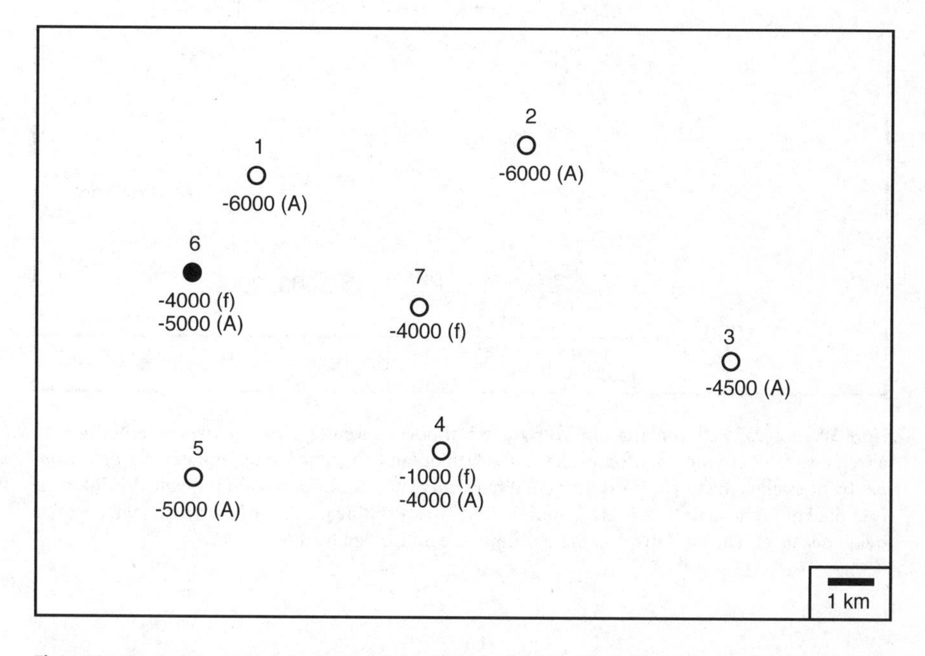

Fig. 6.57. Map of the top of the A sand (*A*) and faults (*f*) in wells drilled for oil. The *solid circle* is an oil well, *open circles* are dry holes. Everywhere away from the faults clear bedding dips are recorded on the dipmeter; they are about 27, 334. Close to the fault in well *4* the bedding dip is at azimuth 062. In well 6 the bedding dip close to the fault is at azimuth 189. In well 7 the dips of bedding are in all directions near the fault. Elevations are in meters, negative below sea level

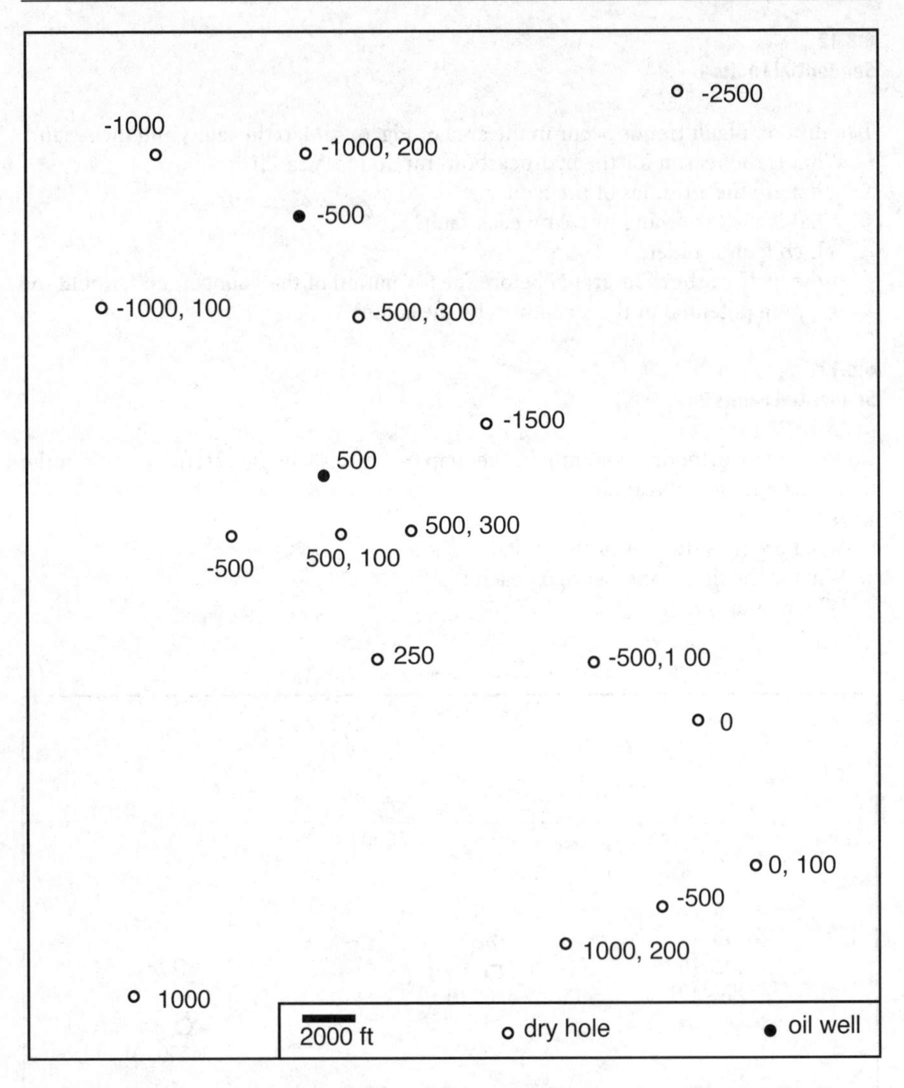

Fig. 6.58. Map of well information from the Northport Dolomite. *Two numbers* together next to a well give fault cut information: 500, 100 = depth of fault cut, amount of fault cut. A *single number* by the well is the top of the dolomite. Where only fault-cut information is given, the dolomite is faulted out. The fault trends are generally northwest–southeast. Elevations are in feet, negative below sea level. The fault displacements should be multiplied by ten

5. If the hydrocarbons migrated before the formation of the younger fault, would the trapping potential of the structures be the same?
6. Are there any additional hydrocarbon prospects?

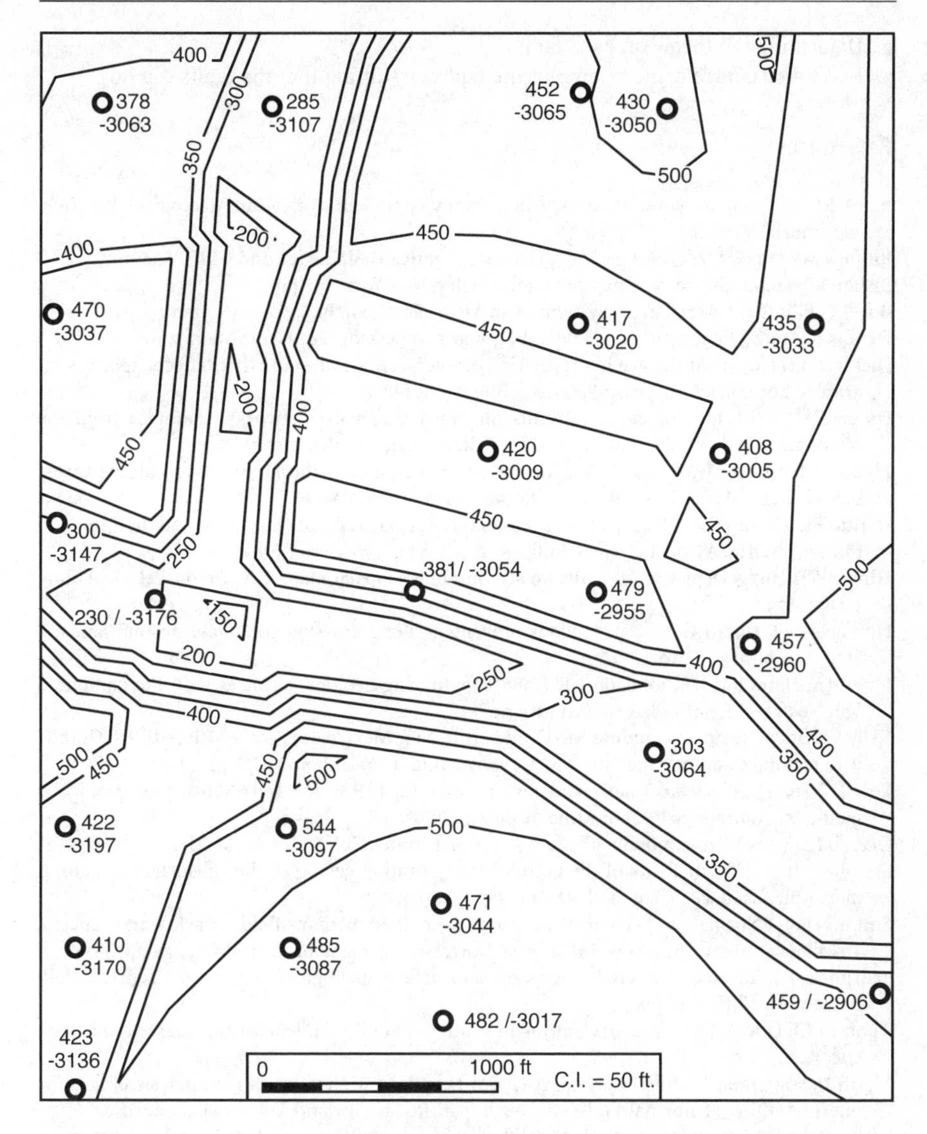

Fig. 6.59. Isopach map of the lower Taylor Formation above the Hawkins salt dome, Texas. Posted next to the wells (*circles*) are the thickness (upper or left-hand *number*) and the elevation of the formation top, in feet below sea level. (After Hintze 1971)

6.8.14
Faults on an isopach map

Figure 6.59 in an isopach map of the faulted lower Taylor Formation.

1. On an overlay, locate the faults that explain the thickness changes. Indicate the upthrown and downthrown sides of each fault.

2. Determine the throw on each fault.
3. Make a structure contour map of the faults, assuming that the faults dip 60°.

References

Bates RL, Jackson JA (1987) Glossary of geology, 3 rd edn. American Geological Institute, Alexandria, Virginia, 788 pp

Billings MP (1972) Structural geology, 3 rd edn. Prentice Hall, Englewood Cliffs, 606 pp

Bishop MS (1960) Subsurface mapping. John Wiley, New York, 198 pp

Boyer S, Elliott D (1982) Thrust systems. Am. Assoc. Pet. Geol. Bull. 66: 1196–1230

Dennis JG (1967) International tectonic dictionary. Am. Assoc. Pet. Geol., Mem. 7, 196 pp

Dickinson G (1954) Subsurface interpretation of intersecting faults and their effects upon stratigraphic horizons. Am. Assoc. Pet. Geol. Bull. 38: 854–877

Diegel FA (1986) Topological constraints on imbricate thrust networks, examples from the Mountain City window, Tennessee, USA. J. Struct. Geol. 8: 269–279

Elliott D, Johnson MRW (1980) Structural evolution in the northern part of the Moine thrust belt, N. W. Scotland. Trans. Roy. Soc. Edinb. Earth Sci. 71: 69–96

Hardin FR, Hardin GC Jr (1961) Contemporaneous normal faults of Gulf Coast and their relation to flexures. Am. Assoc. Pet. Geol. Bull. 48: 238–248

Hintze WH (1971) Depiction of faults on stratigraphic isopach maps. Am. Assoc. Pet. Geol. Bull. 55: 871–879

Horsfield WT (1980) Contemporaneous movement along crossing conjugate normal faults. J. Struct. Geol. 2: 305–310

Jones TA, Hamilton DE, Johnson CR (1986) Contouring geologic surfaces with the computer. Van Nostrand Reinhold, New York, 314 pp

Kelly VC (1979) Tectonics, middle Rio Grande Rift, New Mexico. In: Riecker RE (ed) Rio Grande Rift: tectonics and magmatism. Am. Geophys. Union, Washington, DC, pp 57–70

Low JW (1951) Subsurface maps and illustrations. In: LeRoy LW (ed) Subsurface geological methods. Colorado School of Mines, Golden, Colorado, pp 894–968

Ocamb RD (1961) Growth faults of south Louisiana. Trans. Gulf Coast Assoc. Geol. Soc. 11: 55–65

Sebring L Jr (1958) Chief tool of the petroleum exploration geologist: the subsurface structural map. Am. Assoc. Pet. Geol. Bull. 42: 561–587

Smith J (1995) Normal faults in Holt and Peterson coalbed methane fields, Black Warrior basin, Tuscaloosa County, Alabama. MS Thesis, University of Alabama, Tuscaloosa, 76 pp

Tearpock DJ, Bischke RE (1991) Applied subsurface geological mapping. Prentice Hall, Englewood Cliffs, 648 pp

Thorsen CE (1963) Age of growth faulting in southeast Louisiana. Trans. Gulf Coast Assoc. Geol. Soc. 13: 103–110

Walsh JJ, Watterson J, Childs C, Nicol A (1996) Ductile strain effects in the analysis of seismic interpretations of normal faults. In: Buchanan PG, Nieuwland DA (eds) Modern developments in structural interpretation, validation and modelling. Geol. Soc. Spec. Publ. 99: 27–40

Wang S (1994) Three-dimensional geometry of normal faults in the southeastern Deerlick creek coalbed methane field, Black Warrior basin, Alabama: MS Thesis, Univ. Alabama, Tuscaloosa, 77 pp

Woodward NB (1987) Stratigraphic separation diagrams and thrust belt structural analysis. 38 th Field Conf. 1987, Jackson Hole, Wyoming. Wyo. Geol. Assoc. Guidebook, pp 69–77

Chapter 7 Cross Sections

7.1 Introduction

A cross section shows the relationships between different horizons and allows the information from multiple map horizons to be incorporated into the interpretation. This makes a cross section a powerful tool for checking and improving the structural interpretation presented on a map. Poorly controlled map horizons may be improved with information from other horizons. Errors in mapping on one horizon can be recognized as incompatibilities with the structure of other horizons. Many of the mapping pitfalls noted in the previous chapters and their correct interpretations are immediately obvious on a cross section. A valid cross section must be compatible with the structure adjacent to the section. Accurate interpretation in three dimensions is demonstrated by means of maps and cross sections that are internally consistent and consistent with each other.

The chapter is organized in the sequence that is normally followed in the construction of a cross section from surface or subsurface geological maps. The initial choices that must be made are (1) the direction of the line of section, (2) the location of the section within the structure, (3) the dip of the section plane, and (4) the vertical exaggeration. The best choice for the most effective use of the cross section in structural interpretation is usually a line of section that is straight, perpendicular to the general structural trend in the area, in the central portion of the structure, and vertical or perpendicular to plunge. Ordinarily the section should have no vertical exaggeration. Other choices may be dictated by the nature of the data or by the type of result desired, but any other choice may introduce geometric complexities that can make it more difficult to compile the data, interpret the result, and validate the interpretation.

After the section plane and the vertical exaggeration are chosen, the data are transferred from the line of section on the map onto the cross section. This is a simple geometric step for data that are on the line of section but is a critical interpretive step if data are to be projected to the line of section from elsewhere on the map. Three techniques are presented for projecting data from the map to the line of section, with structure contours, plunge projection, and projection within dip domains. Once the data are on the line of section, the section is constructed by the drawing technique most appropriate for the structural style using planar dip domains, circular arcs, or cubic curves. After the section is constructed, its orientation and tilt can be changed by simple horizontal and vertical exaggerations as long as the structure is cylindrical.

The last topic in the chapter is the construction and use of fault cutoff maps. Fault cutoff maps are a type of cross section that shows the traces of the beds where they are cut by the fault surface. Superimposing hangingwall and footwall cutoff maps provides a method to determine fault displacement accurately and to determine fluid migration pathways in the vicinity of the fault.

7.2 Choosing the Line of Section

Cross sections constructed for the purpose of structural interpretation are usually oriented perpendicular to the fold axis, perpendicular to a major fault, or parallel to these trends. The structural trend to use in controlling the direction of the cross section is the axis of the largest fold in the map area or the strike of the major fault in the area. Good reasons may exist for other choices of the basic design parameters. For example, the cross section may be required in a specific location and direction for the construction of a road cut or a mine layout. If other choices of the parameters, such as the direction of the section line or the amount of vertical exaggeration, are required, it is recommended that a section normal to strike be constructed and validated first. A grid of cross sections is needed for a complete three-dimensional structural interpretation.

The reason that a structure section should be straight and perpendicular to the major structural trend is that it gives the most representative view of the geometry. The simplest example of this is a cross section through a circular cylinder (Fig. 7.1). If the entire cylinder is visible, then it would readily be described as being a right circular cylinder. The cross section that best illustrates this description is Fig. 7.1a, normal to the axis of the cylinder, referred to as the normal section. Any other planar cross section oblique to the axis is an ellipse (Fig. 7.1b). An elliptical cross section is also correct but does not convey the appropriate impression of the three dimensional shape of the cylinder. A section that is not straight (Fig. 7.1c) also fails to convey accurately the three-dimensional geometry of the cylinder, although, again, the section is accurate. Section c in Fig. 7.1 could be improved for structural interpretation by removing the segment parallel to the axis, producing a section like Fig. 7.1a.

The best cross sections are constructed using bed-thickness and fold-curvature relationships that are appropriate for the structural style. In order to use these geometric relationships, or rules, to construct and validate cross sections, it is necessary to choose the cross section to which the rules apply. Such a rule, in the case of the circular cylinder in Fig. 7.1, is that the beds are portions of circular arcs having the same center of curvature. In this simple and easily applicable form, the rule applies only to section a. More complex rules could be developed for the other cross sections, but it is quicker and less confusing to select the plane of the cross section that fits the simplest rule than to change the rule to fit an arbitrary cross section orientation.

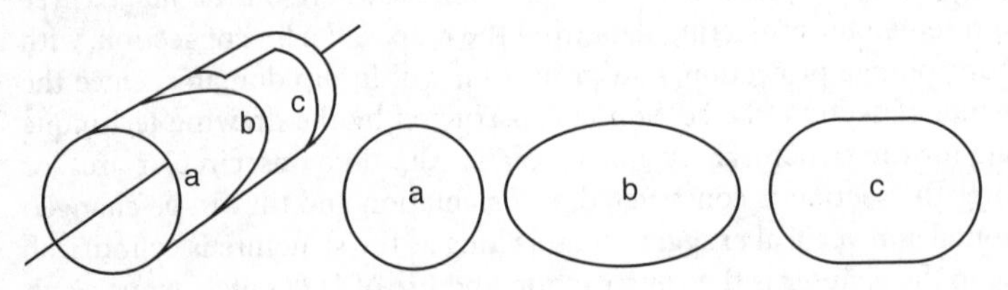

Fig. 7.1. Cross sections through a circular cylinder. **a** Normal section. **b** Oblique section. **c** Offset section

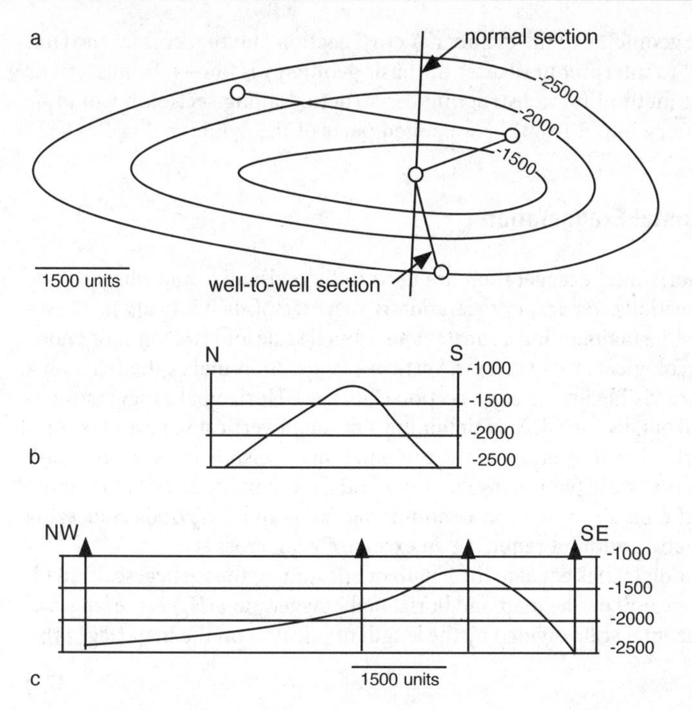

Fig. 7.2. Map and cross sections of an anticline. **a** Structure contour map. **b** Normal cross section through the crest of the structure. **c** Well-to-well cross section through the crest

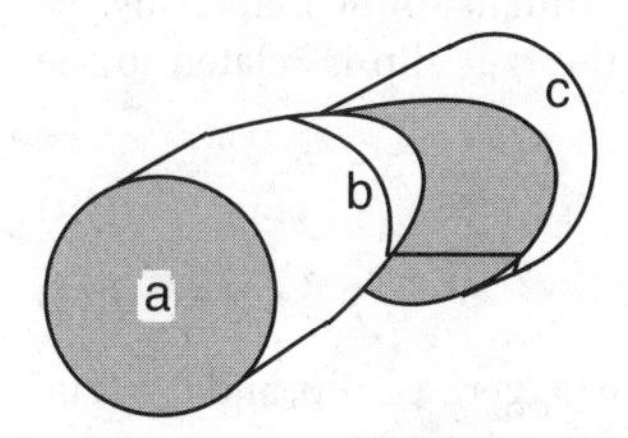

Fig. 7.3. Cross section lines across a cylinder offset along an oblique fault

The effect of the line of section on the inferred geometry of an elongate dome is shown in Fig. 7.2. The correct geometry of the structure (Fig. 7.2b) is shown by the normal section, a straight-line cross section perpendicular to the axial trace of the structure. A line of section that is not straight, such as one that runs through an irregular trend of wells or a seismic line that follows an irregular road, produces a false image of the structure. The zig-zag section across the map (Fig. 7.2c) incorrectly shows the anticline to have a long, gentle northwest limb instead of the correct, nearly symmetrical structure. This would be a serious problem if the cross-section geometry is used to infer the deep structure using the section drawing techniques described in Sect. 7.5.

The first line of section across a structure chosen for interpretation should avoid local structures, like tear faults, that are oblique to the main structural trend. Oblique structures introduce complexities into the main structure that are more easily interpreted after the geometry of the rest of the structure has been determined. Returning to the cylinder, now shown offset along a tear fault (Fig. 7.3), cross sections at a and c will reveal the basic geometry of the cylinder. A cross section at b that crosses the fault will be very difficult to interpret until after the basic geometry is known from sections a or c. The simplest method for constructing the structure along section b would be to project the geometry into it from the unfaulted parts of the cylinder.

7.3 Vertical and Horizontal Exaggeration

Both vertical and horizontal exaggeration are used to help visualize and interpret the structure on cross sections. Vertical exaggeration is a change of the vertical scale (usually an expansion) while maintaining a constant horizontal scale and is a common mode of presentation of geological cross sections. Vertical exaggeration makes the relief on a subtle structure more visible on the cross section (Fig. 7.4a). Horizontal exaggeration is a change of the horizontal scale while maintaining a constant vertical scale and is common, along with vertical exaggeration, in the presentation of seismic lines (Stone 1991). Reducing the horizontal scale (squeezing) makes a wide, low amplitude structure more visible and makes the break in horizon continuity at faults more obvious. Squeezing exaggerates the structure without requiring an excessively tall cross section.

Vertical exaggeration (V_e) is equal to the length of one unit on the vertical scale divided by the length of one unit on the map, and horizontal exaggeration (H_e) is the length of one unit on the horizontal scale divided by the length of one unit on the map (Fig. 7.5):

$$V_e = v_v / v; \tag{7.1}$$

$$H_e = h_h / h, \tag{7.2}$$

where v_v = exaggerated vertical dimension, v = vertical dimension at map scale, h_h = exaggerated horizontal dimension, and h = horizontal dimension at map scale. As derived at the end of the chapter (Eqs. D7.3 and D7.4), the true dip is related to the exaggerated dip by

$$\tan \delta_v = V_e \tan \delta; \tag{7.3}$$

$$\tan \delta_h = \tan \delta / H_e, \tag{7.4}$$

where δ_v = vertically exaggerated dip, δ_h = horizontally exaggerated dip, and δ = true dip. Equation (7.3) is plotted in Fig. 7.6.

In its effect on the dip, a vertical exaggeration is equivalent to the reciprocal of a horizontal exaggeration (from Eq. D7.6):

$$V_e = 1 / H_e. \tag{7.5}$$

The effect of exaggeration on the thickness of a unit is given by (from Eq. D7.8 and D7.10):

$$t_v / t = V_e (\cos \delta_v / \cos \delta); \tag{7.6}$$

$$t_h / t = \cos \delta_h / \cos \delta. \tag{7.7}$$

The symbols are the same as in Eqs. (7.1)–(7.4). Horizontal exaggeration has no effect on the thickness of a horizontal bed, whereas vertical exaggeration changes the thickness of a horizontal bed by an amount equal to the exaggeration. Horizontal exaggeration changes the thickness of a vertical bed by the full amount of the exaggeration,

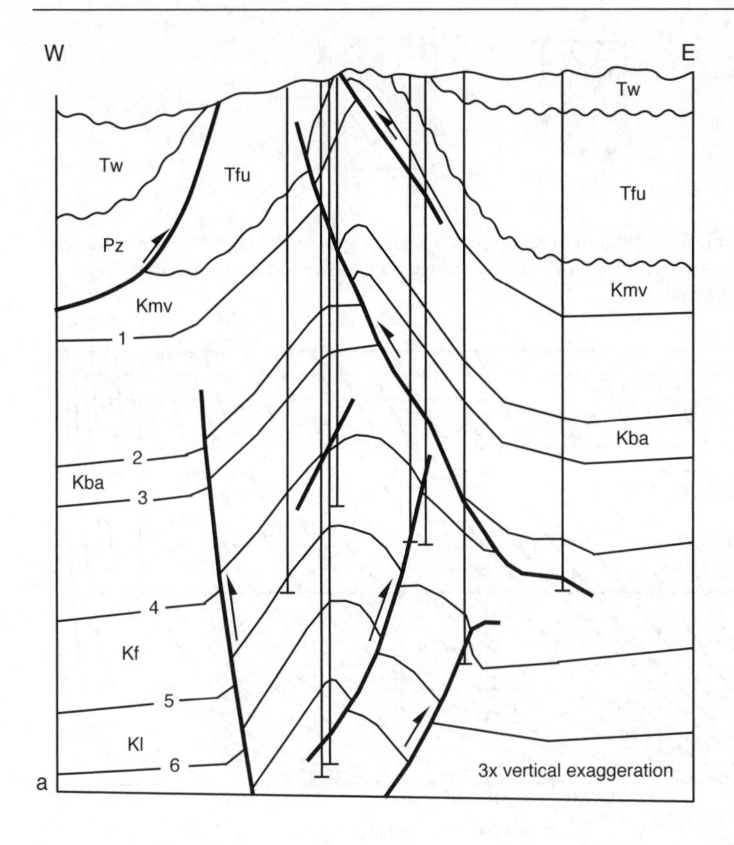

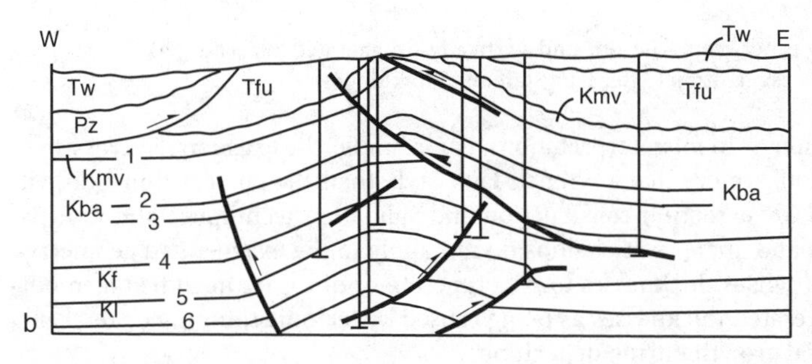

Fig. 7.4. Cross sections across Tip Top field, Wyoming thrust belt **a** 3:1 vertical exaggeration. **b** Unexaggerated cross section. (Modified from Groshong and Epard 1994, after Webel 1987)

whereas vertical exaggeration has no effect on the thickness of a vertical bed. For beds dipping between 0 and 90°, both horizontal and vertical exaggeration cause the apparent thickness to increase.

Exaggeration creates several problems in the interpretation of a cross section. The first is that a large vertical exaggeration or horizontal squeeze may so distort the structure that the structural style becomes unrecognizable. This will lead to difficul-

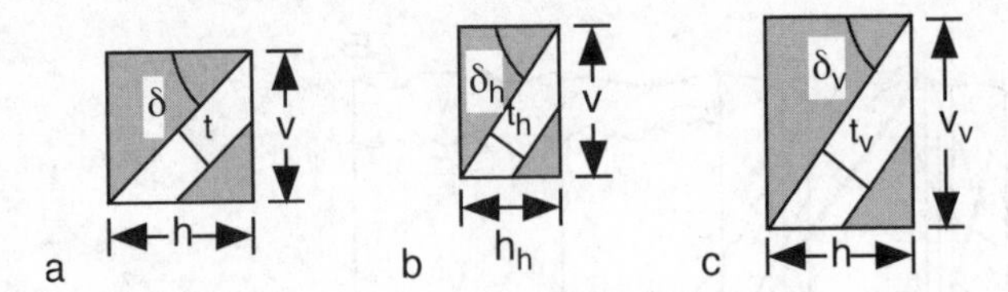

Fig. 7.5. Vertical and horizontal exaggeration. A bed of original thickness t is shown in *white*. **a** Unexaggerated cross section. **b** Horizontally exaggerated (squeezed) cross section. **c** Vertically exaggerated cross section

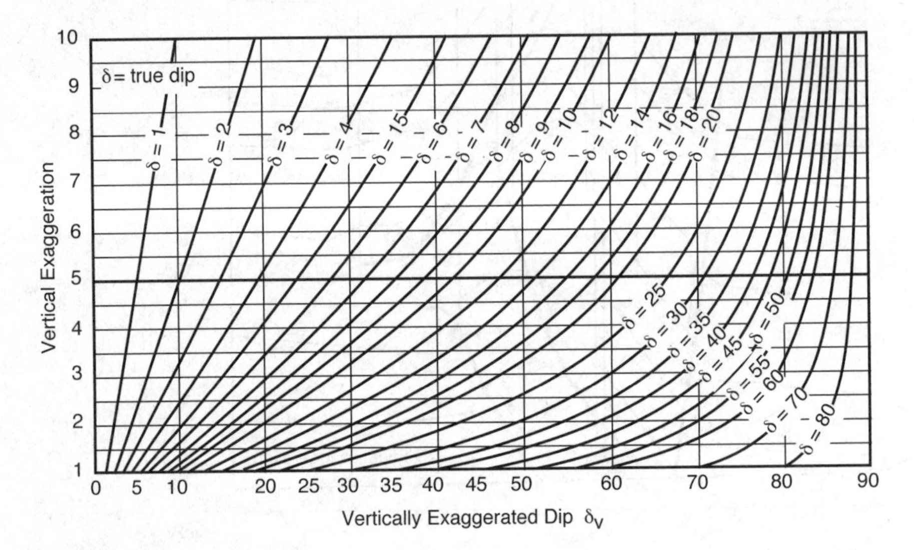

Fig. 7.6. Relationship between true dip and vertically exaggerated dip (Eq. 7.3) for various amounts of vertical exaggeration (After Langstaff and Morrill 1981)

ties in interpretation or to misinterpretations. For example, the exaggerated cross section in Fig. 7.4a looks more like a wrench-fault style than the correct thin-skinned contraction style. Cross section construction and validation techniques and models for the dip angles and angle relationships do not apply to the exaggerated geometry. Exaggeration also causes thicknesses to be a function of dip. Care must be taken not to interpret exaggerated thicknesses as being caused by tectonic thinning or thickening or by structural growth during deposition.

Whether or not the initial cross section is exaggerated, the next steps in the interpretation should be done on an unexaggerated profile. The profile can be easily corrected when the true horizontal and vertical scales are known. The correction factor is the inverse of the horizontal or vertical exaggeration. Create an unexaggerated profile by multiplying the correction factor times the scale of the exaggerated axis. If the cross section is in digital form, this is a simple operation using a computer drafting program.

One of the constraints that can be applied to a cross section is that of constant bed thickness. Departures from constant thickness may be the result of unrecognized fault cuts or incorrect dips. On the other hand, true thickness variations may indicate the locations of structural thickness changes or may represent important original stratigraphic variations. The critical significance of thickness variations to the interpretation requires that a cross section constructed for structural analysis shows the true unit thicknesses. The section should be unexaggerated in its dimensions and as nearly perpendicular to bedding as possible. Many structural modeling and validation techniques require straight cross sections having true dips and true thicknesses.

Seismic time sections are commonly displayed with both horizontal and vertical exaggerations (Stone 1991). Horizontal exaggeration may be applied to obtain a legible horizontal trace spacing. The amount of horizontal exaggeration is most conveniently determined by comparing the distance between shot or vibration points marked on the profile with the scale between the corresponding points on the location map. The vertical scale on a time section is in two-way-travel time, and the determination of the vertical exaggeration requires depth conversion as well as scaling. A few simple techniques can provide the necessary scaling information without geophysical depth migration. If the depth to a particular horizon is known independently, as from a well, then the vertical exaggeration at that well can be determined directly from the definition (Eq. 7.1). If the true dip is known for a unit or a fault, then the vertical exaggeration can be found by solving Eq. (7.3), given the exaggerated dip from the profile.

If there is a unit on a seismic time section that can be expected to have constant depositional thickness and minimal structural thickness changes, then any observed thickness change in the unit is caused by the exaggeration (Fig. 7.7a). The vertical exaggeration of the time section can be removed by restoring the bed thickness to constant (Stone 1991). A package of reflectors should be chosen that retains its reflector character and proportional spacing regardless of dip. The reflectors should be parallel to one another and not terminate up or down dip. The thickness changes of such a package are more likely to be caused by exaggeration than by deposition. A simple procedure for removing the vertical exaggeration is to change the vertical scale until the unit maintains constant thickness regardless of dip (Fig. 7.7b). This provides a quick depth migration that applies to the depth interval over which the unit occurs. Because seismic velocity varies with depth, the profile might remain exaggerated at other depths. Normally seismic velocity increases with depth and so the vertical exaggeration decreases with depth. This method does not take into account horizontal velocity variations. A further caution is that the thickness variations seen in Fig. 7.7a are just like those that are caused by deformation. The inferred vertical exaggeration should always be cross checked by other methods whenever possible. If the profile is also horizontally exaggerated, then this method will give the correct exaggeration

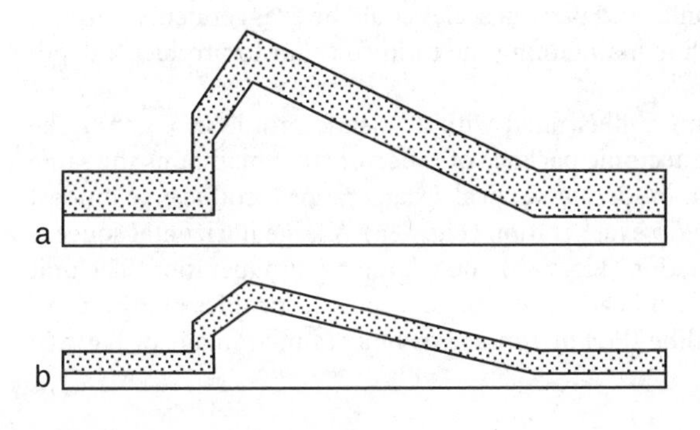

Fig. 7.7. Effect of a 2:1 vertical exaggeration on thickness. **a** Profile vertically exaggerated 2:1. Bed thickness increases as the dip decreases. **b** Unexaggerated profile. Bed thickness is constant

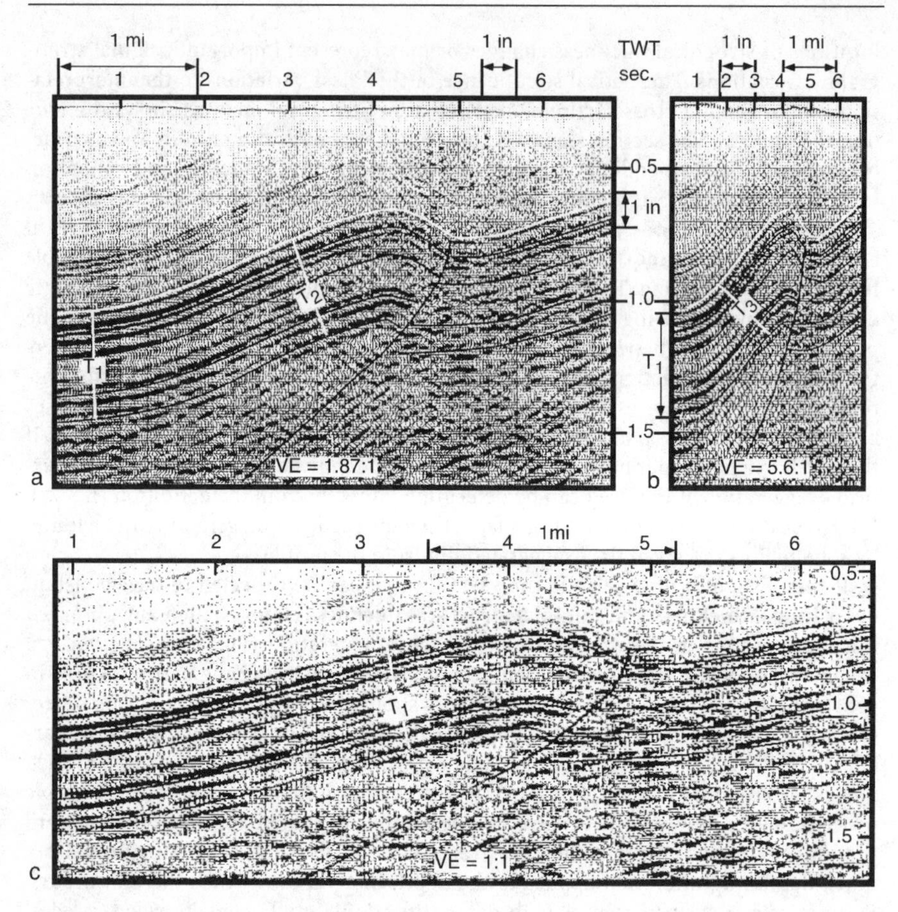

Fig. 7.8. Time migrated seismic profile from central Wyoming. *TWT* Two-way travel time; T_i interval thickness. **a** Original profile having a vertical scale of 7.5 in. per s. and a horizontal scale of 12 traces per in. Vertical exaggeration (*ve*) is 1.87:1. **b** The vertical scale is the same as in a, the horizontal scale is reduced by two-thirds. Vertical exaggeration is 5.6:1. **c** Unexaggerated version produced by expanding the horizontal scale. (after Stone 1991)

ratio of 1:1, but both the horizontal and vertical scales could be exaggerated. Eliminate the horizontal exaggeration while maintaining the ratio constant to produce a depth section.

The removal of exaggeration is illustrated with a seismic profile in Fig. 7.8. The thickness is measured on a pre-tectonic package of reflectors that maintains the same character regardless of dip. A moderate vertical exaggeration produces a modest thickness variation but a large dip exaggeration (Fig. 7.8a). A large horizontal squeeze produces a large thickness variation (Fig. 7.8b) and a large dip exaggeration. The unit thickness is constant in Fig. 7.8c, indicating no exaggeration. The profiles in Figs. 7.8b,c were produced by digitally scaling the horizontal axis of a scanned image of Fig. 7.8a (Stone 1991).

7.4 Transferring Data from Map to Cross Section

Begin by drawing the line of section on the map (A–A′, Fig. 7.9a). The line of section in Fig. 7.9a has been selected to be perpendicular to the crest of the anticline. Clearly show the end points of the section because they will serve as the reference points for all future measurements. The section will be compiled on a graph where the vertical axis represents elevations and the horizontal axis is the distance along the profile (Fig. 7.9b). For an unexaggerated profile, the vertical scale should be the same as the map scale and the lines spaced accordingly. To construct a vertically exaggerated profile, let the vertical scale be some multiple of the map scale.

Cross sections are usually drawn either to be vertical or to be inclined such that the plane of the section is perpendicular to the direction of plunge of the structure of interest. For many purposes, it is most convenient to construct a vertical profile. For a cylindrical structure, a vertical profile is easily transformed into a normal section (Sect. 7.6). The techniques of section construction will begin with vertical profiles. Construction of an initially tilted profile is considered in Section 7.4.2.2.

The next step in constructing a cross section is to transfer the data from the map to the profile. To be transferred are the elevations of topographic and structural surfaces, geologic contacts, and the attitudes of beds and faults. These data may be locat-

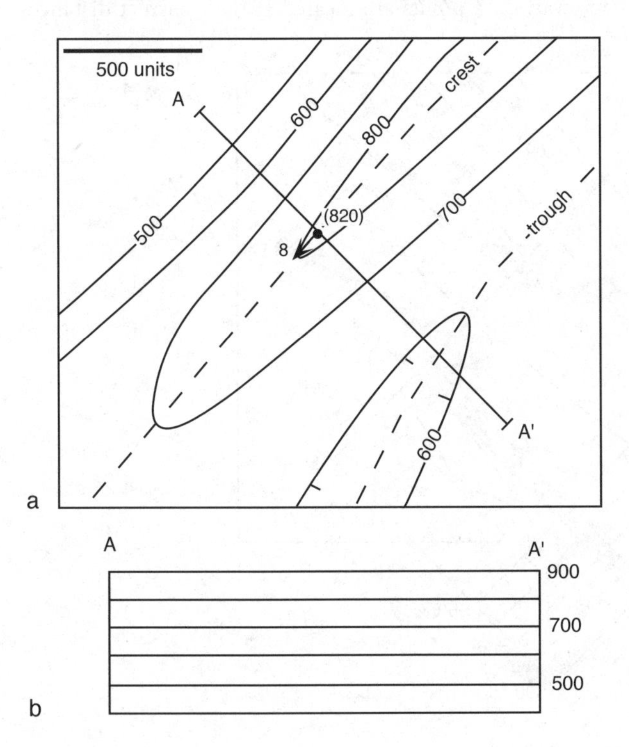

Fig. 7.9. Initial stage of cross section construction from a map. The line of section is *A–A′*. **a** Structure contour map and a dip measurement (at elevation 820). **b** Graph for cross section construction. For no vertical exaggeration both the horizontal and vertical scales are the same as the map scale

ed on the line of section or may be projected onto the line. Transferring data from the section line onto the cross section is a strictly mechanical process, described in the next section. Projecting data to the line of section, if required, is an interpretive process that will be discussed in Section 7.4.2.

7.4.1
Data Located on the Section Line

The data to be transferred from the map to the cross section are the topographic or structure contour profile, attitudes of bedding and faults, and contact locations. The first step is the projection of the data from the line of section to the cross section.

One convenient projection method is to align the cross-section graph parallel to the line of section, tape the map and section together so that they cannot slip, and project data points at right angles onto the cross section with a straight edge and a right triangle (Fig. 7.10). Any type of map information can be transferred to the cross section by this method. In computer drafting it is usually more accurate to draw straight lines vertically or horizontally; therefore the map should be rotated so that the projection direction is either horizontal or vertical (Adams et al. 1996). The example (Fig. 7.10) is a structure contour map, but the topographic profile from an outcrop map is constructed the same way. Data points are located at their correct distances

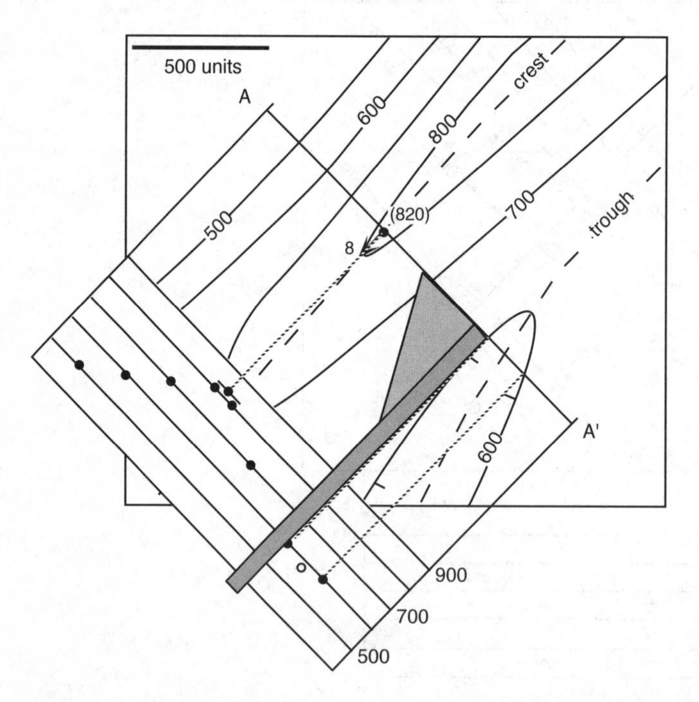

Fig. 7.10. Direct projection of data from structure contour map to cross section. *Dotted lines* are right-angle projection lines. *Circles* show projected points: *filled circles* are projected from known elevations; the *open circle* is an interpolated elevation. The drafting tools are *shaded*

from the ends of the section and at their proper elevations. The probable locations of turning points of the structure contour map or of the topography are also marked as points (Fig. 7.10, between the two adjacent 600 contours). The location of a turning point is constrained to be between the next higher and lower elevations.

An alternative method is to mark the locations of the data points on a strip of paper for working by hand (Fig. 7.11a) or on a line drawn on top of the section line in a drafting program. After marking, the line of data locations is rotated to be parallel to the cross-section horizontal (Fig. 7.11b) In a drafting program, group the points before rotating. Then project the data points from the line onto the cross section (Fig. 7.11c). The points can be projected with a right triangle as in the previous method, or the overlay can be moved to the correct elevation on the section and each point marked at the appropriate distance from the end of the section.

After compiling the data onto the profile, it should be checked. Then the profile is constructed by connecting the dots (Fig. 7.12). If the correct shape of the profile is not clear, points can be added by interpolation between contours on the map. A topographic profile should be inked at this stage because it should not be changed. Next the locations of the bedding attitude measurements and contact locations should be projected onto the section.

All dips shown on the cross section must be the apparent dips in the plane of the section. A profile across a structure contour map automatically shows the apparent

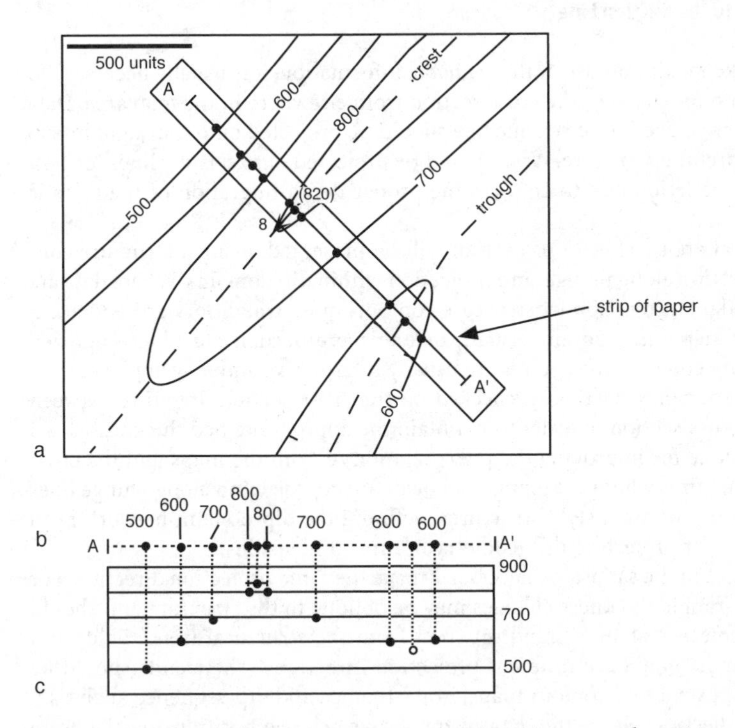

Fig. 7.11. Transferring data from map to cross section using an overlay. **a** Data points are marked on the overlay. **b** The overlay is aligned with the section. **c** Points are projected onto the section (*dotted lines*). *Filled circles* are projected from known elevations; the *open circle* is an interpolated elevation

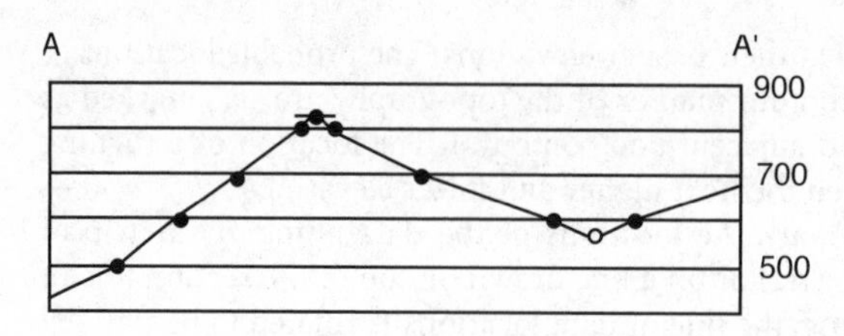

Fig. 7.12. Cross section *A-A'* from the structure contour map of Fig. 7.11. Vertical section, no vertical exaggeration. The *short horizontal* line is the apparent dip from the bedding attitude on the map

dips but measured attitudes must be converted using Eq. (2.20) or (2.21). In Fig. 7.10 and 7.11, the bedding dip is at a right angle to the cross section and so the apparent dip in the line of section is zero (Fig. 7.12). On an exaggerated profile, the dip must be the exaggerated dip from Eq. (7.3) or (7.4). Apparent dips are exaggerated if the section is exaggerated.

When the data have been transferred to the cross section, the section should again be checked against the data seen along the line of section on the map. One of the most common mistakes is to produce a cross section that fails to match the map along the line of section. All elevations, geological contacts, attitudes (apparent dips), and attitude locations must match exactly along the line of section.

7.4.2
Projecting Data to the Section Line

In order to make maximum use of the available information, it is usually necessary to project data onto the plane of the cross section from elsewhere in the map area. Data from a zig-zag cross section or seismic line should be projected onto a straight line to correctly interpret the structure. Wells should be projected onto seismic lines for best stratigraphic correlation and to confirm the proper depth migration of the seismic data.

Three general approaches to projection will be presented: using a structure contour map, projection along plunge, and projection within dip domains. Where data are relatively abundant, projection by structure contouring is straightforward and accurate. Computer mapping programs usually take this approach. It must be recognized that the structure contours themselves are usually interpretive and may not be correct in detail until after they have been checked on the cross section. Iterating between maps and the cross section in order to maintain the appropriate bed thicknesses is a powerful technique for improving the interpretation of both the maps and the cross section. For structures where the plunge can be defined, projection along plunge lines is effective. For dip-domain style structures, defining the dip-domain network is an efficient method for projecting the geometry in three dimensions.

Not all features in the same area necessarily have the same projection direction. For example, stratigraphic thickness changes may be oblique to the structure and should therefore be projected along a trend different from the structural trend. Folds and cross-cutting faults may have different projection directions. The trends should be determined from structure contour maps, isopach maps and dip-sequence analysis.

Incorrect projection places the data in the wrong relative positions on the cross section and renders the interpretation incorrect or impossible. Projection along strike, down the dip, normal to the cross section, and parallel to or perpendicular to a fault, are all poor projection techniques unless they also happen to coincide with one

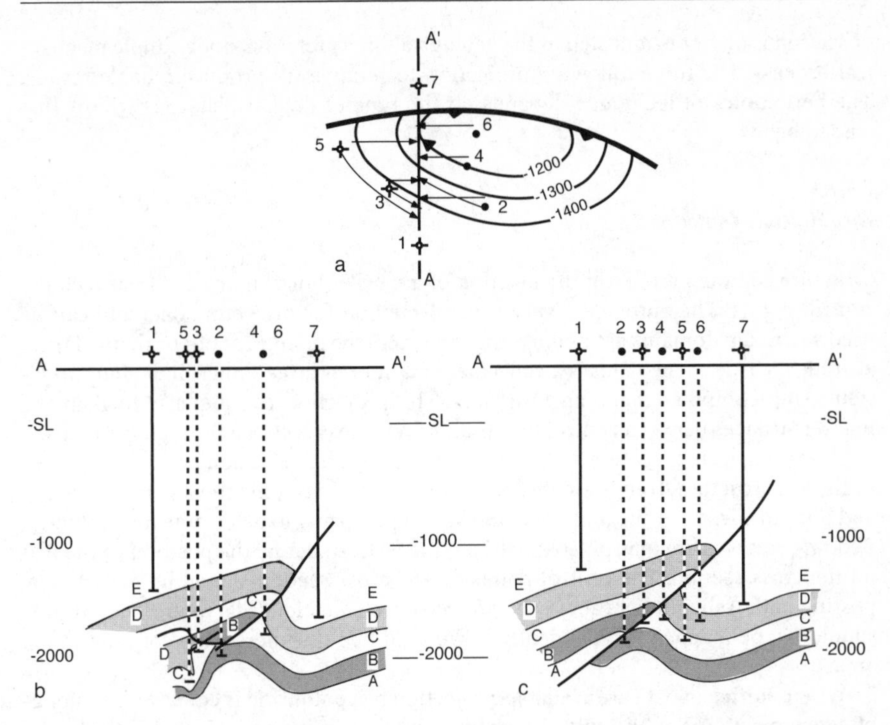

Fig. 7.13. Different cross sections obtained by different methods of data projection. **a** Structure contour map of horizon *E*, showing the alternative projection directions. **b** Cross section produced by projecting wells along structure contours. **c** Cross section produced by projecting wells along the plunge of the fold axis. *Solid lines* are parallel to the fold axis; *dashed lines* are parallel to structure contour; *SL* sea level. (After Brown 1984)

of the three methods just mentioned. Correct projection within a domain of uniform dip, for example, could be done along strike or down the dip. The effect of the projection technique is illustrated with an example (Fig. 7.13) presented by Brown (1984). The cross section of the fault in Fig. 7.13a might be constructed by projecting the wells onto the line of section along the strike of the structure contours (Fig. 7.13b). Well 6 cannot be projected by this method because it does not lie on a contour that crosses the line of section. The resulting profile (Fig. 7.13b) is poor in terms of structural style. The cross section shows multiple small faults instead of a single smooth fault. Note that no well shows more than one fault, yet the cross section shows locations where a vertical well should cut two faults. Changing the type of the projection significantly improves the cross section. The north half of the structure has a uniform cylindrical plunge to the north and so projection along plunge produces a reasonable cross section (Fig. 7.13c). Only one fault is present and it is relatively planar, as expected.

The additional data that are obtained by projection from the map to the line of section help constrain the interpretation of the cross section and help ensure that the interpretation is compatible with the structure off the line of section. The lesson from figure 7.13 is that an incorrect projection technique can make the interpretation worse and incompatible with the surrounding structure. Projection of data to the line of section is an important step in the geological interpretation, not a simple mechanical process. The three following projection techniques all produce reliable results. The best choice of technique depends on the type of data available, as will be discussed below.

7.4.2.1
With Structure Contours

Structure contours represent the position of a marker horizon or a fault between the control points. They provide a very general method for projecting data and can be used where dip domains are not present and where the plunge cannot be defined from attitude measurements. This is a convenient method in three-dimensional interpretation using computer mapping programs. The projection technique is to map the marker surfaces between control points and draw cross sections through the maps.

In Fig. 7.14a, cross section A-A′ is to be drawn through a region occupied by three wells. The structural trends are different from the stratigraphic trends and so the top and bottom surface of the unit are mapped independently to determine the geometry. Bedding surfaces are interpolated between the wells to define the position of the bed on the cross section. The control points on the cross section should be recorded by posting both well numbers above the projected well location (Fig. 7.14b). The surfaces should *not* be mapped independently, however, if bed thicknesses are constant in the area.

Where sufficient data are available, projection by contouring is equivalent to along-plunge projection. For example, the well data in Fig. 7.13 can be projected to the line of section by contouring without knowing the plunge amount or direction (Fig. 7.15). To demonstrate this, the fault in Fig. 7.12a is contoured. The wells (Fig. 7.13b) are inspected individually to be sure that each one shows only one fault cut, indicating that each well may cut the same fault. The fault contours generated from the elevations of the fault cuts (Fig. 7.15a) are smooth, as expected for a single fault. The fault contours superimposed on the original map (Fig. 7.15b) give the projected elevations of the fault along the line of section. The contours trend almost north–south, parallel to the plunge direction of the fold. A cross section of the fault along A-A′ in Fig. 7.15b would be nearly identical to that in Fig. 7.13c. Each map horizon could be similarly projected into the line of section to give a complete cross section like that in Fig. 7.13c.

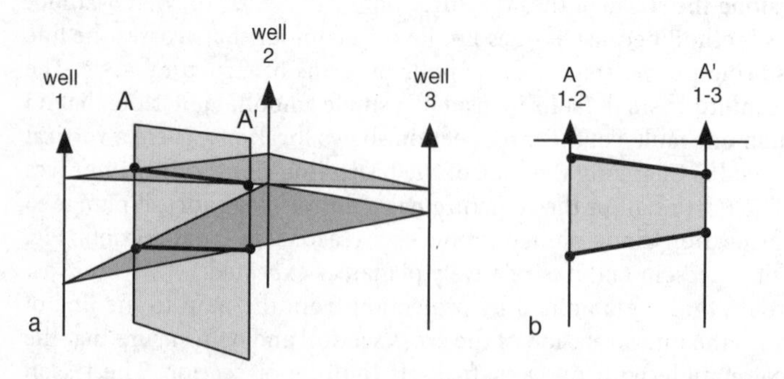

Fig. 7.14. Projection between wells by structure contouring. *Heavy lines* mark intersections between the maps and the cross section. **a** Perspective diagram. **b** Cross section *A–A′*

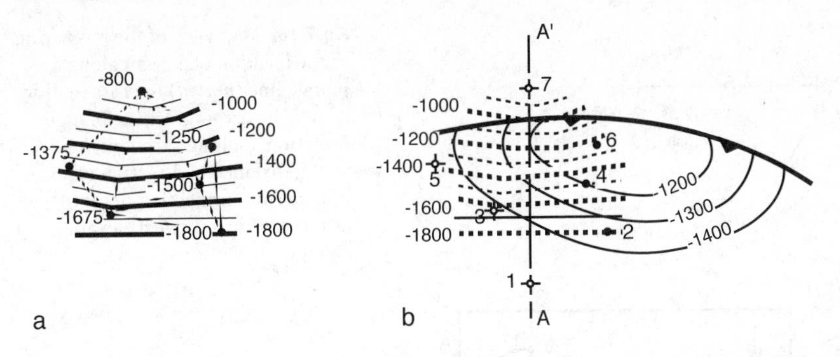

Fig. 7.15. Fault-surface map based on the wells in Fig. 7.13 that cut the fault. **a** *Points* give well locations and depths of fault cuts; *dotted lines* connect the nearest neighbors for contouring. Structure contours (*solid lines*) are derived by triangulation. **b** Structure contours on the fault (*dotted lines*) from a, superimposed on the structure contours of horizon E (*solid lines*) to show their parallelism to the fault

7.4.2.2
Along Plunge

Projection of information along plunge is most appropriate where the data are too sparse to generate a structure contour map, but where the plunge can be determined from bedding attitudes. For example, the plunge can be determined from a dipmeter and the well can be projected along plunge into the cross section. It is also possible to project all the outcrop data along plunge to generate a cross section. Projection along plunge is conveniently done using plunge lines (Sect. 4.3), which are lines in the plunge direction, inclined at the plunge amount (Wilson 1967). The orientation of a plunge line can be found from the composite stereogram or tangent diagram of bedding attitudes across the fold (Sect. 4.2) or from dip-sequence analysis (Sect. 4.6 and 5.3).

The projection of a point, such as a formation top in a well, along plunge to a vertical cross section (Fig. 7.16) is done using Eq. (4.5), which is

$$v = h \tan \phi, \tag{7.8}$$

where v = vertical elevation change, h = horizontal distance in the direction of plunge from projection point to the cross section, and ϕ = plunge. For example, if h = 1 km and the plunge is 15° (Fig. 7.16), the elevation of the projected point is 268 m lower on the cross section than in the well.

The concept of projection along plunge is the basis of the graphical cross-section construction technique of Stockwell (1950), presented below. This method makes it possible to project data onto cross sections that have steep dips, such as vertical sections or sections normal to gently dipping fold axes. An analytical version of the technique will be given at the end of the section. The graphical method is given by the following steps. Refer to Fig. 7.17 for the geometry.

1. Create the graph on which the cross section will be constructed at the same scale as the map and align it perpendicular to the plunge direction. Draw the line AC parallel to the plunge direction at the edge of the map. AC represents a horizontal line on the plunge projection. The plunge projection will be constructed from this line.

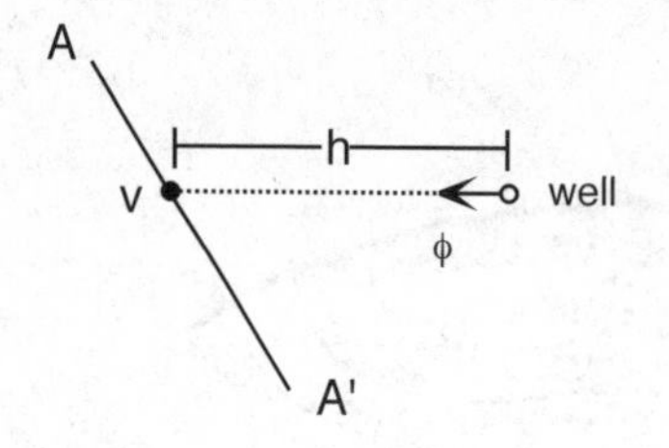

Fig. 7.16. Map view of the projection of the location of a point along a plunge line (*dotted*) to cross section *A–A′*. The *arrow* gives the plunge direction; ϕ plunge amount; *h* horizontal distance in direction of plunge from projection point to cross section; *v* vertical elevation change

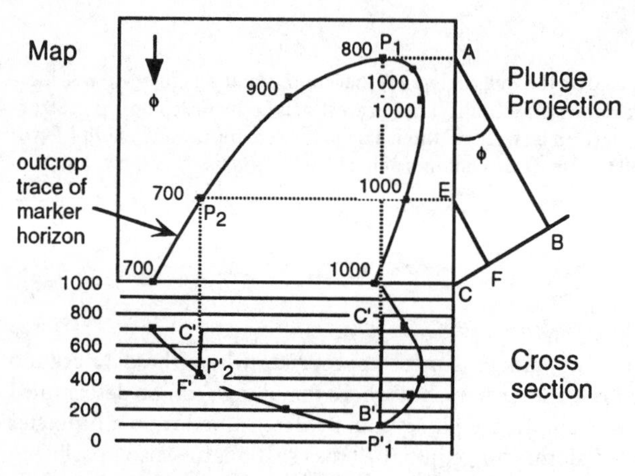

Fig. 7.17. Projection of map data onto a cross section normal to plunge. The plunge is ϕ to the south. *Numbers* on the map (*square box*) are the topographic elevations of the points to be projected

2. Begin the projection with the point that is to be projected the farthest (P_1). Project this point along the dotted line P_1A, perpendicular to the line AC to point A. From point A, draw the line AB at the plunge angle, ϕ, from AC.
3. Draw the orientation of the plane of the cross section (line CB) at the desired orientation to the vertical (angle ACB). In Fig. 7.17 the plane of the cross section has been chosen to be perpendicular to the plunge (angle ABC = 90°). The plane of ABC represents a vertical cross section in the plunge direction through point P_1.
4. The length CB is the distance in the plane of the east-west section from the map elevation of P_1 to its location P_1' on the cross section. Draw a vertical construction line (dotted line P_1C') through P_1 onto the line of section and measure the length C′B′ (solid line) = CB down from the map elevation of the point to find P_1'.
5. Repeat step 4 to project all other points, for example P_2 is projected to P_2'.

The ratio of the vertical projection length AB to the cross section projection length CB is constant for all points, CB:CA = CF:CE = sin ϕ. Projection by hand is very rapid if a proportional divider drafting tool is used. Set the divider to the ratio CB/AB; then as the projection length AC is set, the divider gives the required length CB. If attitude information is available at the points to be projected, the apparent dip can be shown at the projected point on the cross section. See Section 2.5.2 for equations for finding the apparent dip.

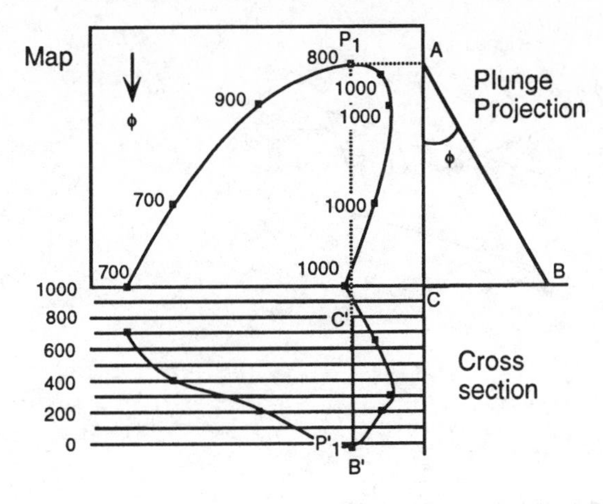

Fig. 7.18. Projection along plunge onto a vertical cross section. *Numbers* on the map (*square box*) are the topographic elevations of the points to be projected

The method can be modified for other cross section orientations by changing the orientation of the line of section on the plunge projection (Fig. 7.18). The orientation of the plane of section on the plunge projection is CB. In Fig. 7.18, CB is at 90° to AC, making the section plane vertical. The ratio CB:CA is constant for all points. Follow steps 1 to 5 above, changing the orientation of the line CB.

The plunge of a fold typically changes along the axis. Cylindrical fold axes may be curved along the plunge and cylindrical folds will change into conical folds at their terminations. Projection along straight plunge lines should be done only within domains for which the geometry of the structure is constant. Variable plunge can be recognized from undulations of the crest line on a structure-contour map or as excessive dispersion of the bedding dips around the best-fit curves on a stereonet or tangent diagram. If the plunge is variable, then the geographic size of the region being utilized should be reduced until the plunge is constant and all the bedding points fit the appropriate line on the stereonet or tangent diagram. If the sequence of plunge angle changes along the fold axis direction can be determined, then the straight line AB (figs. 7.17, 7.18) could be replaced by a curved plunge line.

If many points are to be projected and especially if the fold is conical, it is quicker to calculate the projected point locations using the method of De Paor (1988) than to use the hand-drawing techniques given above. An individual point P (Fig. 7.19), given by its x, y, z map coordinate position, can be projected along plunge to its new position P′ (x′, 0, z′) on the cross section plane (defined by y′ = 0). Select the map coordinate system such that x is parallel to the line of cross section and y is perpendicular to the line of section. Choose y = z = 0 to lie in the plane of the cross section (Fig. 7.19). The sign convention requires that the positive (down) plunge direction be in the negative y direction. The elevation of a point is z. The dip of the plunge line = ϕ, the angle between the plunge line and the normal to the cross section = α, and the dip of the cross section = δ. The plunge line is constant in direction in a cylindrical fold but may be different for every location in a conical fold.

The general equations for the projected position of a point P′, derived at the end of the chapter (Eqs. D7.18, D7.23), are:

$$x' = x + y \tan\alpha + \tan\alpha\,(z\cos\alpha - y\tan\phi)\,/\,(\tan\phi + \tan\delta\cos\alpha); \tag{7.9}$$

$$z' = (z\cos\alpha - y\tan\phi)\,/\,(\tan\phi\cos\delta + \sin\delta\cos\alpha). \tag{7.10}$$

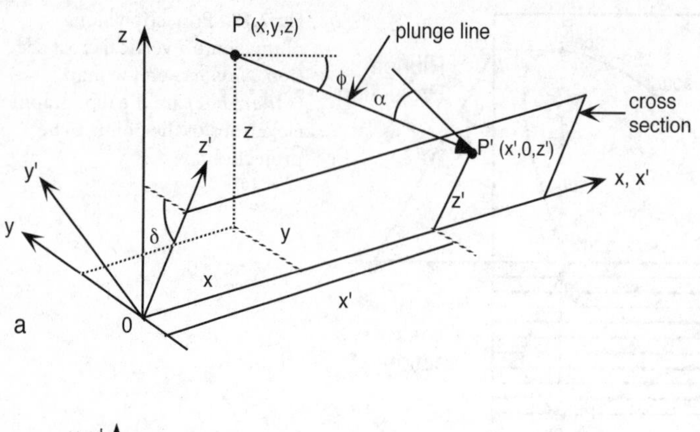

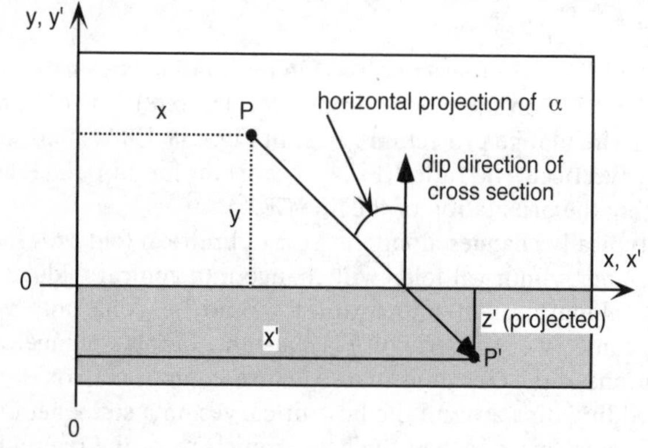

Fig. 7.19. Projection along plunge into the plane of the cross section. **a** Perspective diagram. **b** Horizontal map projection. The projection of the section plane onto the map is *shaded*

For a vertical cross section, from Eqs. (D7.24) – (D7.25),

$$x' = x + y \tan \alpha; \tag{7.11}$$

$$z' = z - (y \tan \phi) \,/\, \cos \alpha. \tag{7.12}$$

A cross section perpendicular to the fold axis is possible only for a cylindrical fold. The equations for projection to the normal section are (D7.26 - D7.27):

$$x' = x; \tag{7.13}$$

$$z' = z \cos \phi - y \sin \phi. \tag{7.14}$$

In a conical fold the plunge amount and direction changes with location. A cross section perpendicular to the crestal line will be closely equivalent to a normal section in slightly conical structures. Substitute the plunge of the crestal line in Eqs. (7.13) and (7.14) to approximate a normal section. The shorter the projection distance, the better the approximation.

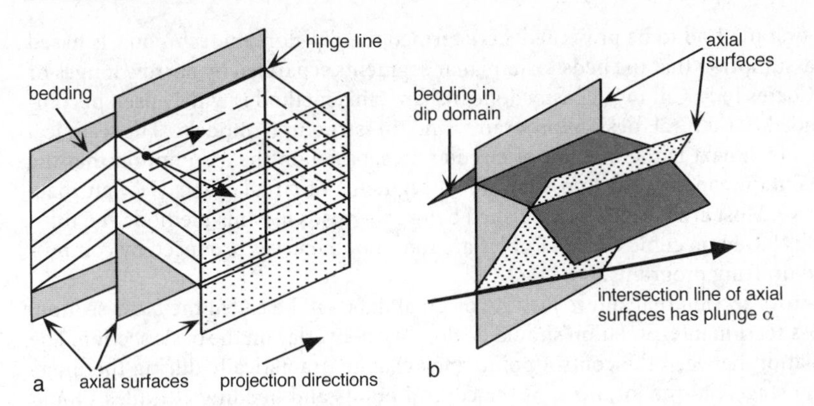

Fig. 7.20. Projection within a dip domain. Domain boundaries are axial surfaces. **a** Bedding dip domains. The *solid point* is the location of an observation. **b** The intersection line between axial surfaces has plunge α

7.4.2.3 *Within Dip Domains*

Because a dip domain is a region of uniform dip, bedding attitudes may be projected anywhere within a single domain (Fig. 7.20a). Bedding surfaces can be projected throughout the entire domain from a single observation point. Some types of information, for example fracture density, may be related to the proximity of the observation point to the fold hinge and so should be projected parallel to the hinge line.

Projection within a dip domain requires determining the locations of the boundaries of each domain. Domain boundaries are axial surfaces (Fig. 7.20). A hinge line (Fig. 7.20a) represents the line of intersection of an axial surface with bedding and forms the limit of a dip domain within a single bed. Axial-surface intersection lines (α lines) are parallel to fold hinge lines in non-plunging cylindrical folds (Fig. 7.20a) but differ from hinge lines in plunging and conical folds (Fig. 7.20b). The methods for finding the orientations of the axial surfaces, axial-surface intersections and the hinge lines have been covered in section 4.5. An example in which the three-dimensional network of axial surfaces is determined will be given after the dip-domain cross-section interpolation technique is described.

7.5 Section Drawing Techniques

Drawing a cross section requires interpolation between data points and perhaps extrapolation beyond the data points. The best method depends on the nature of the bed curvature and on the nature of the bed thickness variations, or the lack of bed thickness variations. Where closely spaced control points are available, interpolation based on any method will produce essentially the same result. Where the data are sparse, the model that best fits the structural style will provide the basis for drawing the most internally consistent cross section. See Section 1.5.1 for a discussion of fold styles.

The first method to be presented, here termed the dip-domain technique, is based on the assumption that the beds form planar segments separated by narrow hinges or faults (Coates 1945; Gill 1953). Easily done by hand, this method is widely used in computer-aided structural design programs and in structural models. The classical method given next is the method of circular arcs, based on the assumption that the beds maintain constant thickness and form segments of circular arcs (Hewett 1920; Busk 1929). Most cross sections published before the 1950s use this method. The third method is based on cubic curves. Cubics are the smooth curves produced by computer-aided drafting programs.

The cross section that shows just the original data will be called the *basis* section. The cross section interpretation should be done as an overlay on the basis section. The interpolation between the control points may change dramatically during the interpretation stage, but the locations of the control points and bedding attitudes should remain the same. Including the basis section along with the final interpretation separates the data from the interpretation, a fundamental distinction that should always be made. The data may be subject to revision, of course. For example, inconsistencies in the cross section may indicate that a geologic contact has been mislocated or that a fault is required. This is one of the important reasons for constructing a cross section. In the best scientific procedure, the original data and the interpreted result are both presented in the final report.

Regardless of the formal interpolation technique to be used, a good first step after the data have been compiled on the basis section is to make a freehand sketch of the tentative interpretation. If working from a map, the sketch should be compared to the down-plunge view of the structure (see Sect. 7.6) to insure that the interpretation agrees with the structure along strike. It is possible to get lost in the details of section design and misconnect horizons, misidentify fold hinges, and forget faults. A careful preliminary sketch will bring the big picture into focus.

It is useful to summarize the stratigraphic thicknesses in a "stratigraphic ruler" that will greatly speed up the drawing of an unexaggerated cross section. Draw the stratigraphic section at the scale of the cross section on a narrow piece of paper. This can be used as a ruler to mark off the stratigraphic units on the cross section and provides a quick check to see if the thickness of a unit is consistent with its dip. This only works on unexaggerated cross sections in the dip direction.

Many structural interpretations are based on seismic reflection profiles which are already displayed in the form of a cross section. A seismic line that is to be interpreted structurally should satisfy the same criteria with respect to the choice of the plane of section and vertical exaggeration as a geological cross section. If geological data are available, projecting the data from maps to the seismic line will provide constraints that will help in the construction or validation of the depth interpretation. A successfully depth-converted seismic line must follow the same geometric rules as a geologic cross section.

7.5.1 Planar Dip Domains

The dip-domain method is based on the assumption that the beds occur as planar segments separated by narrow hinges. This was originally called the method of tangents by Gill (1953). Some interpreters refer to this as the kink-band method. Although the

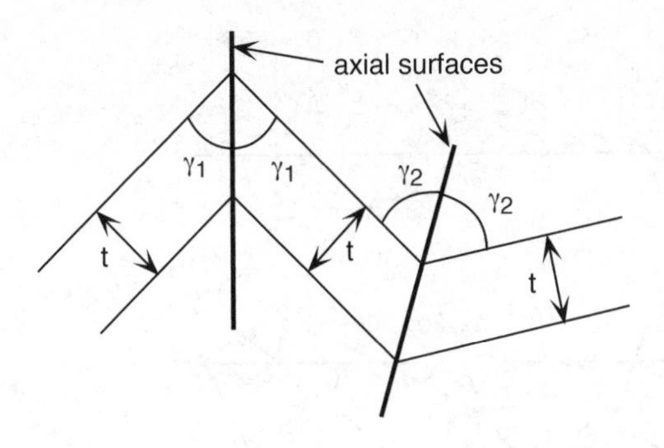

Fig. 7.21. Dip-domain fold hinges in a constant thickness layer. t_1 Bed thickness; γ_i half-angles of the interlimb angle

geometry can resemble that of kink bands, the term kink band has mechanical implications that may not be appropriate; hence the purely geometric term dip domain is used here. The basis for the technique lies in the relationship between the bedding thickness and the symmetry of the hinge. For constant thickness beds, the axial surface bisects the interlimb angle between adjacent dip domains (Fig. 7.21). This maintains constant bed thickness. If the beds change thickness across the axial surface, then the axial surface cannot bisect the hinge. The technique is described here in the context of constant thickness beds. The technique is the same for beds that change thickness except that the axial surfaces do not bisect the hinges. See Eq. (4.10) for a method to calculate the axial surface orientation in folds that do not maintain constant bed thickness (see also Gill 1953).

7.5.1.1
Method

The following steps outline the dip-domain technique.

1. On the map or cross section, define the dip domains and locate the boundaries between domains as accurately as possible (Fig. 7.22a). A certain amount of variability from constant dip is expected in each domain (perhaps a 5–10° range).
2. Define the axial surfaces between domains (Fig. 7.22b). If bed thickness is constant, the axial surfaces bisect the hinges, but if bed thickness changes are known, use Eq. (4.10) to find the dips of the axial surfaces. Where axial surfaces intersect, a dip domain disappears and a new axial surface is drawn between the newly juxtaposed domains (for example, X in Fig. 7.22b). Note that a single fold is likely to have multiple hinges, as in Fig. 7.22.
3. Draw a key bed through the structure, honoring the domain dips and the stratigraphic tops (Fig. 7.22c). Sometimes the data do not allow a single key bed to be completed across the whole structure. Shifting up or down a few beds to a new key bed will usually allow the section to be continued. Note that axial-surface intersections do not necessarily coincide with named stratigraphic boundaries. It is usually helpful to draw an horizon through the axial-surface intersection points (Fig. 7.22c).
4. Complete the section by drawing all the remaining beds with their appropriate thicknesses.
5. If desirable on the basis of the structural style, round the hinges an appropriate amount using a circular arc or spline curve.

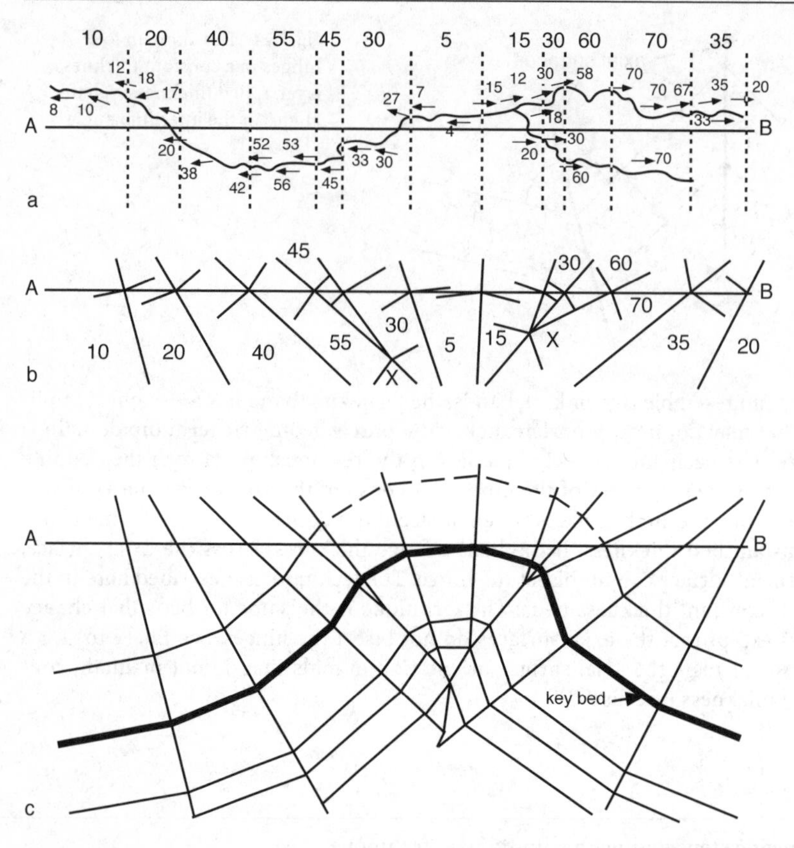

Fig. 7.22. Dip-domain cross-section construction technique. **a** Map of dips measured along a stream traverse and the boundaries (*dotted lines*) between interpreted dip domains. **b** Initial stage of cross section construction showing domain dips and hinge locations with axial surfaces that bisect the hinges. *X* axial-surface intersection points. **c** Completed cross section. (After Gill 1953)

7.5.1.2
Cylindrical Fold Example

The steps in building a cross section and interpolating the geometry using the constant bed thickness dip-domain method is illustrated with the Sequatchie anticline (Fig. 7.23). The map is characterized by domains of approximately constant dip, making it a good candidate for a dip-domain style cross section. The fold is cylindrical within the map area and so the geometry of the structure should be constant along the axis. The crestal line is horizontal (Fig. 4.17), making a vertical section the most appropriate. Prior to drawing the section, the stratigraphic thicknesses are determined and summarized in a stratigraphic ruler at the same scale as the map (Fig. 7.24).

The first step is to transfer the data from the map to the cross section. The line of section is drawn on the map (Fig. 7.23), at right angles to the fold axis. The topography is drawn using the method of Fig. 7.11, with the vertical scale equal to the map scale

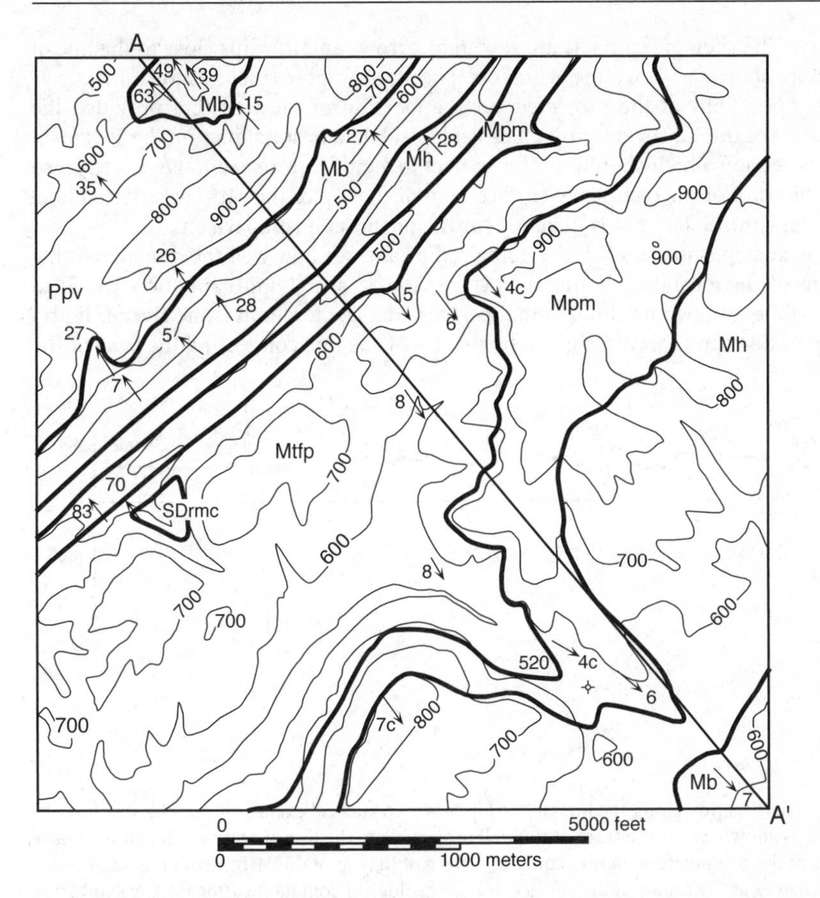

Fig. 7.23. Geologic map of a portion of the Sequatchie anticline at Blount Springs, Alabama, showing the line of cross section. Geologic contacts *wide lines*, topographic contours (ft) *thin lines*, measured bedding attitudes are shown by *arrows*. *c* Attitude computed from three points

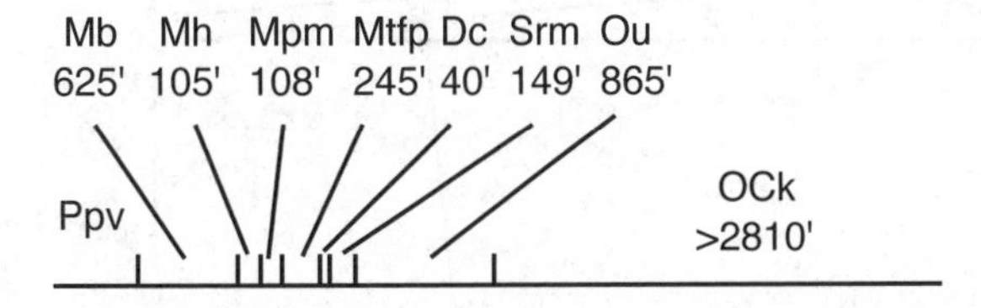

Fig. 7.24. Stratigraphic column for the Sequatchie anticline map area at the same scale as the map, to be used as a stratigraphic ruler. Thicknesses are in feet. Thicknesses of Ppv through Mpm are from outcrop measurements. The top of the Ppv is not present in the map area Thicknesses of Mtfp through OCk are from the Shell Drennen 1 well (Alabama permit No. 688) interpreted by McGlamery (1956), and corrected for a 4° dip. The well bottomed in the OCk and so the drilled thickness is less than the total for this unit

(Fig. 7.25). The geologic contacts are shown by arrows and the dips close to the line of section are shown as short line segments. The stratigraphic ruler is shown intersecting the topography at the projected surface location of the well that provided the thicknesses of the subsurface units. The geological data in solid lines on Fig. 7.25 form the basis section which should not be subject to significant revision. Figure 7.25 also shows the trace of the composite surface map of the top of the Mtfp as a dotted line. This is not control data but is included for the purpose of comparison.

The next step is to establish the domain dips and see how well the domains fit the locations of the formation boundaries (Fig. 7.26). As a first approximation, the fit to one backlimb and two forelimb domains is tested. (The forelimb is the steeper limb.) The dip of domain 1 (3NW) is given by the dip of the line connecting the base of the

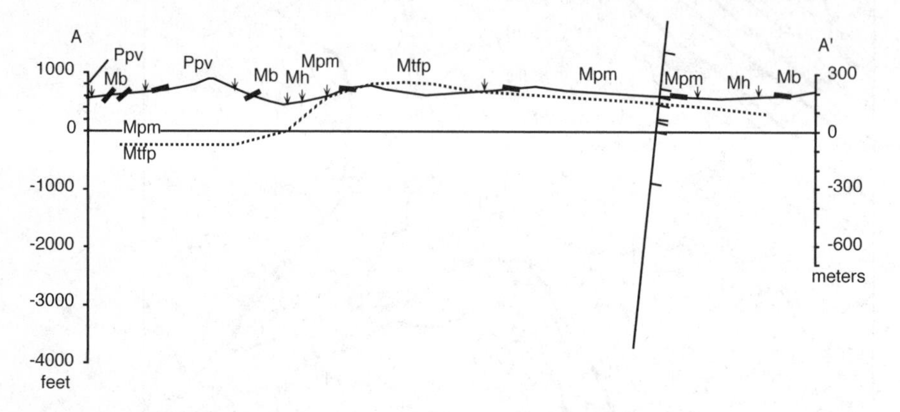

Fig. 7.25. Basis section along the line *A–A′* (Fig. 7.23). No vertical exaggeration. The stratigraphic column is shown where the trace of the well projects onto the line of section. The *dotted line* is the trace of the composite structure contour surface of the top of the Mtfp (from Fig. 3.29). *Short vertical arrows* at the topographic surface are the geological contact locations. *Wide short lines* are bedding dips

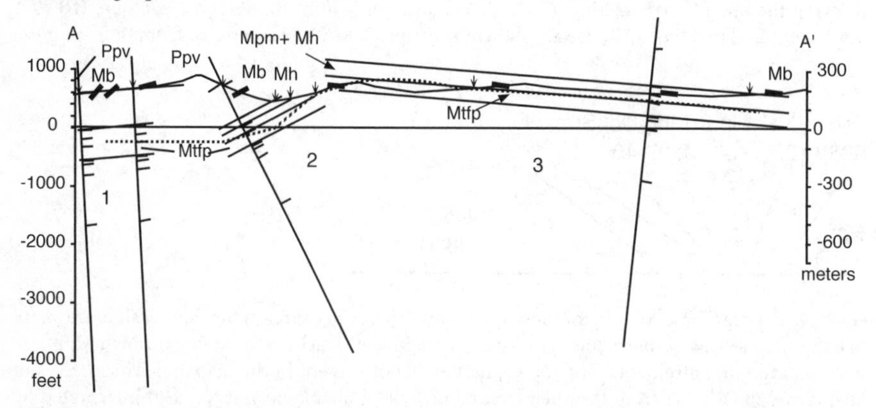

Fig. 7.26. Comparison between domain dips, stratigraphic thicknesses, and contact locations. The *dotted line* is the trace of the composite structure contour surface on the top of the Mtfp. *Short vertical arrows* at the topographic surface are the geological contact locations. *Wide short lines* are bedding dips. Dip domains are *numbered*

Ppv on opposite sides of the Mb inlier. The domain 2 dip is the 27NW dip seen at the surface. The domain 3 backlimb dip of 6SE is seen in outcrop but is selected primarily because with this dip the unit thicknesses match the contact locations. Portions of the beds are drawn in with constant bed thickness to compare with the contact locations. The domain 2 dip fits both contacts of the Mh, even though this information was not used to define the dip. The dotted composite-surface line of the top of the Mtfp is a good approximation to the fold shape, although it does not fit the data in the forelimb of the fold quite as well as do the dip domains. The shape is close enough to that controlled by the additional dips and thicknesses to indicate the value of the composite surface technique as a good first approximation.

The axial surface orientations are determined next (Fig. 7.27). Following the relationship in Fig. 7.20 for constant bed thickness, the axial surfaces bisect the hinges. The interlimb angles are measured, bisected and the axial surfaces drawn between each domain. Two dip domains (2 and 4) are added to those shown in Fig. 7.26 so that the dips can be honored at the ground surface. It is tempting to insert a fault at the location of domain 4, but the map (Fig. 7.23) shows a vertical to near-vertical domain to the southwest in the same position as the vertical dip on the cross section. Not far to the southwest of the map area, the units are directly connected across the two limbs (Cherry 1990) with no fault present. The positions of the axial surfaces in Fig. 7.27 are only approximate; the next step is to determine their exact locations.

The locations of the axial surfaces are now adjusted until the dip domains match the stratigraphic contacts (Fig. 7.28). The dip change of the Ppv at location 1 must be ignored and the corresponding axial surface between domains 1 and 2 removed in order to match the locations of the stratigraphic contacts. A new axial surface dip is determined as the boundary between the two domains in contact (1 + 2 and 3) after the incorrect axial surface is removed. The vertical dip selected for the forelimb provides a good match to all the contacts except for the top of the Dc at location 2. A slight rounding of the contact at this location will provide a match to the map geometry. The internal consistency of the section based on constant thicknesses, planar domain dips and the mapped contact locations and depths in the well is strong support for the interpretation.

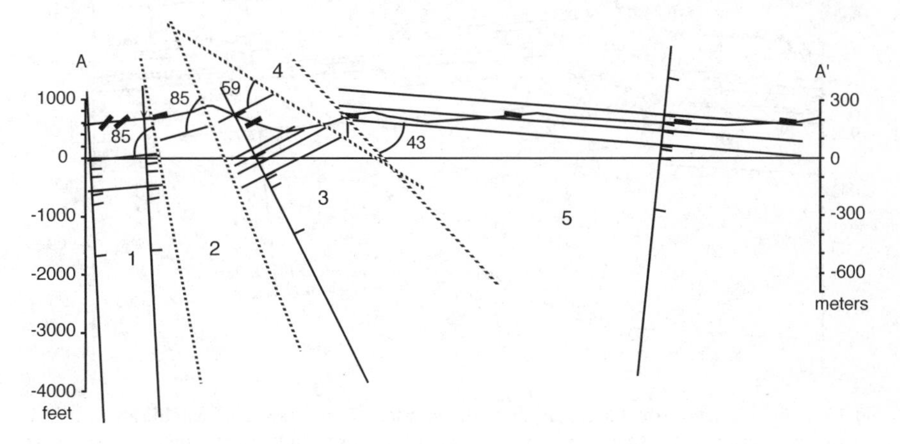

Fig. 7.27. Axial surface traces (*dotted lines*) that bisect the interlimb angles. Exact locations of the axial surfaces are not yet fixed in this step. Dip domains are *numbered*

Axial surfaces are shown as crossing in Fig. 7.28, an impossibility. Where two axial surfaces intersect, the dip domain between them disappears and a new axial surface is defined between the two remaining dip domains (Fig. 7.29). The final cross section (Fig. 7.29) is an excellent overall fit to the dips and contact locations. Locations 1 and 2 are the only misfits. The misfits are quite small. At location 1, the base of the Mh does not match the mapped outcrop location which could be caused by a second-order fold at that point or by the mislocation of a poorly exposed contact. A very small domain of thickened bedding is required at location 2 in order to keep the top of the Dc below the surface of the ground and so that the contacts of the Dc and the Sm meet across

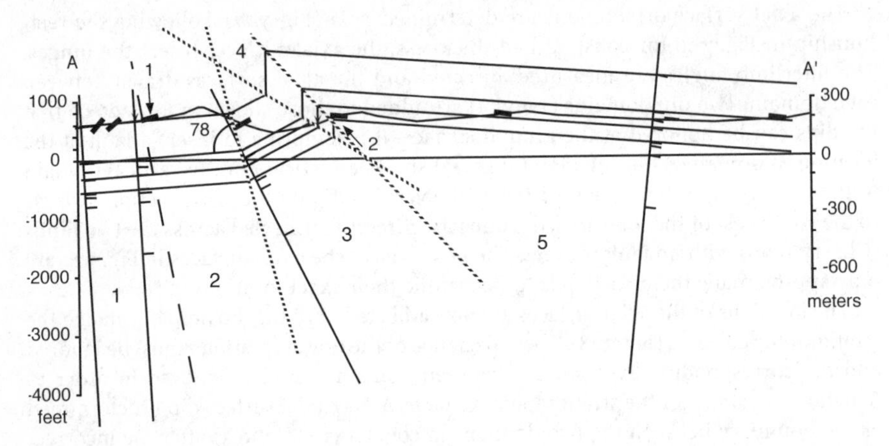

Fig. 7.28. Dip-domain cross section with axial surfaces (*dotted lines*) moved so that the dip domains match the stratigraphic contact locations. The *dashed* axial surface will be deleted and domains *1* and *2* combined

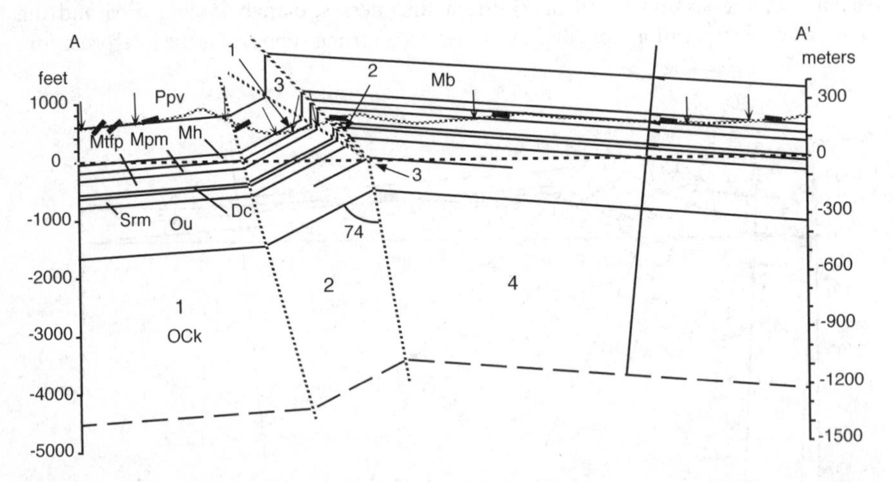

Fig. 7.29. Final constant-thickness, dip-domain cross section across the Sequatchie anticline. No vertical exaggeration. The *large numbered arrows* are explained in the text. *Unnumbered arrows* mark the contact locations. The *dashed line* is the level of the deepest horizon drilled. The *dotted line* is the topographic profile

the axial surface. It is no surprise that bed thickness is not perfectly constant in such a tight hinge. The surprise is that such a small region of thickening is required in the hinge. The effect of the thickening of the Mtfp is to round the hinge, a feature that might continue upward along the axial surface as well, but is shown as ending within the Mtfp. Both the vertical domain and the thickened domain disappear at point 3 where a new axial surface bisects the angle between the remaining two domains (2 and 4). The match of the top of the OCk across this axial surface is an additional confirmation of the cross section geometry because the location of the axial surface is defined by intersection point 3, not by projection of the OCk contact.

This example illustrates the importance of the cross section to structural interpretation. The rule of constant bed thickness allows a few dips and the formation contact locations to tightly constrain the geometry of the cross section. The rule works well even though there is a small amount of thickening in the tightest hinge. The cross section can, in turn, be used to revise the geologic map and the composite structure contour map. The cross section provides the needed control for mapping the deeper geometry. Extrapolation to depth using the composite-surface technique breaks down if vertical lines through the control points pass through axial surfaces, as happens in the forelimb of the Sequatchie anticline (Fig. 7.29). Composite surface maps (Sect. 3.4.2) provide a good first approximation, but the final interpretation should be controlled directly by cross sections based on multiple horizons.

7.5.1.3
Non-cylindrical Fold Example

The non-cylindrical fold example illustrates the use of the three-dimensional axial-surface geometry developed by map interpretation to draw cross sections across a non-cylindrical, dip-domain anticline. The map (Fig. 7.30) is a horizontal slice through

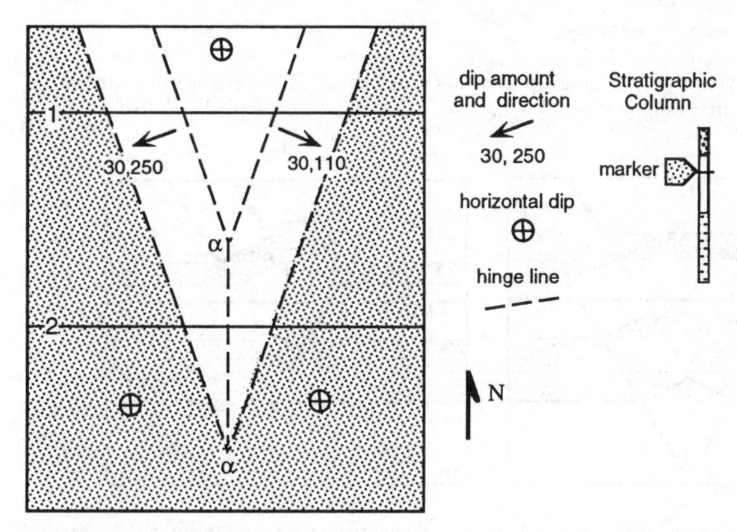

Fig. 7.30. Dip domains within a non-cylindrical, plunging anticline. The *shaded* portion of the map is the outcrop of the top of the marker horizon. Points of intersection between axial surfaces occur at α. Lines of cross sections *1* and *2* are indicated

the anticline that coincides with the sand marker horizon in the shaded region of horizontal dip. The hinge lines are the traces of the axial surfaces and form the boundaries between dip domains. The methods for finding the orientations of the axial surfaces, axial-surface intersections and the hinge lines from the domain dips have been covered in Section 4.5. The two axial surfaces on the east are oriented 60, 340 and the two on the west are oriented 60, 020. The third axial surface strikes north–south and is vertical. The plunge of the axial-surface intersection lines (at α, Fig. 7.30) is 30° north. The three-dimensional shape of the structure is illustrated in Fig. 7.31.

The apparent dip of both axial surfaces in the plane of the cross section is 58° and the apparent dip of bedding is 27° (dashed lines, Fig. 7.32a). The stratigraphic contacts are projected from where they are known, in this case from the outcrop of the sand marker, into the adjacent dip domains. The apparent dip of bedding is used in the limbs

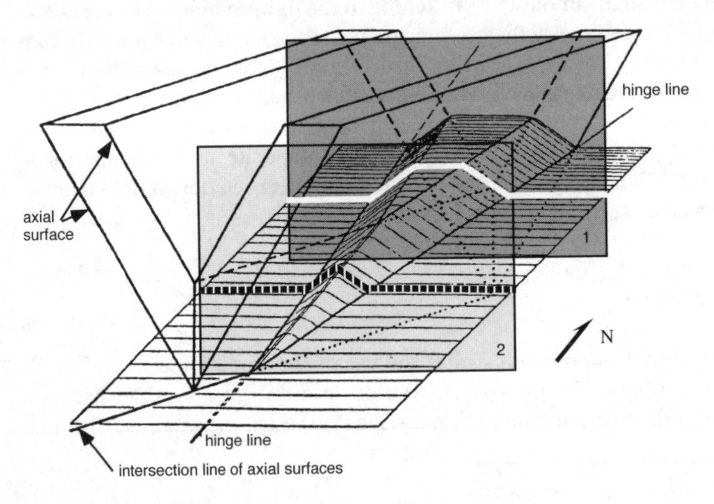

Fig. 7.31. Non-cylindrical, plunging dip-domain fold. Planes of cross sections *1* and *2* are shown. (From Fig. 4.27; after Faill 1973)

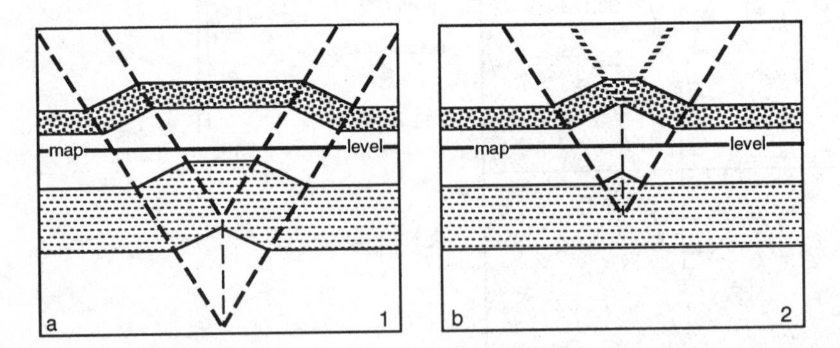

Fig. 7.32. Vertical cross sections through the plunging anticline in Figs. 7.30 and 7.31. No vertical exaggeration. *Dashed lines* are the axial surfaces that intersect the map at the line of section. **a** Cross section *1*. **b** Cross section *2*. *Dotted lines* are the axial surfaces that must be projected in the air onto the line of section

and connected to the horizontal dip at the crest of the structure. In cross section 2, only the outer two axial surfaces intersect the map surface (Fig. 7.32b), and the inner two must be projected onto the section. The line of intersection of the inner two axial surfaces is located at point α on the map. The α line is projected onto cross section 2 just like a plunge line (Eq. 7.8). The intersection point is used to place the inner axial surfaces onto the cross section. Then section 2 is completed by drawing in the beds using the appropriate apparent dips and stratigraphic thicknesses. Note that the axial surfaces do not bisect the hinges and so the beds are thicker on the fold limbs (Sect. 4.5.3).

7.5.2 Circular Arcs

The method of circular arcs is based on the assumptions that bed segments are portions of circular arcs and that the arcs are tangent at their end points (Hewett 1920; Busk 1929). This type of curve can be drawn by hand using a ruler and compass. The resulting cross section will have smoothly curved beds. The method of circular arcs produces a highly constrained geometry in which both the shape of the structure and the exact position of each bed within the structure are predicted. When these predictions fit all the available data, the cross section is very likely to be correct. If the stratigraphic and dip data cannot be matched by the basic construction technique, as often happens, dips can be interpolated that will produce a match. The basic method is given first, then two techniques for dip interpolation.

7.5.2.1 *Method*

If the dips are known at the top and bottom of the bed (Fig. 7.33), the geometry of a circular bed segment is constructed by drawing perpendiculars through the bed dips (at A and B), extending the perpendiculars until they intersect (at O) which defines the center of curvature. Circular arcs are drawn through A and B to define the top and bottom of the bed.

This process is repeated for multiple data points to draw a complete cross section (Fig. 7.34). The first center of curvature (O) is defined as the intersection of the normals to the first two dips (A and B). The marker horizon located at point A is extended to the bedding normal through B along a circular arc around point O. The next center of curvature is located at O^1. The marker horizon is extended to the normal through C as a circular arc with center O^1. The same procedure is followed across the section to complete the key horizon A-A (Fig. 7.34). The remaining stratigraphic horizons are drawn as segments of circular arcs around the appropriate centers. Constructed in this fashion, the beds have constant thickness. To maintain constant bed thickness, the beds form cusps in the core of the fold.

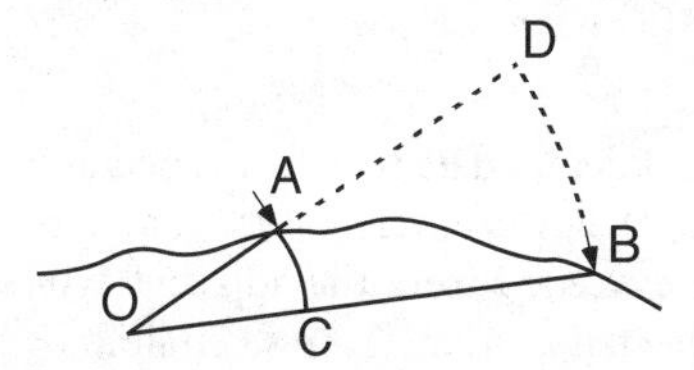

Fig. 7.33. Cross section of a bed that is a portion of a circular arc. (After Busk 1929).

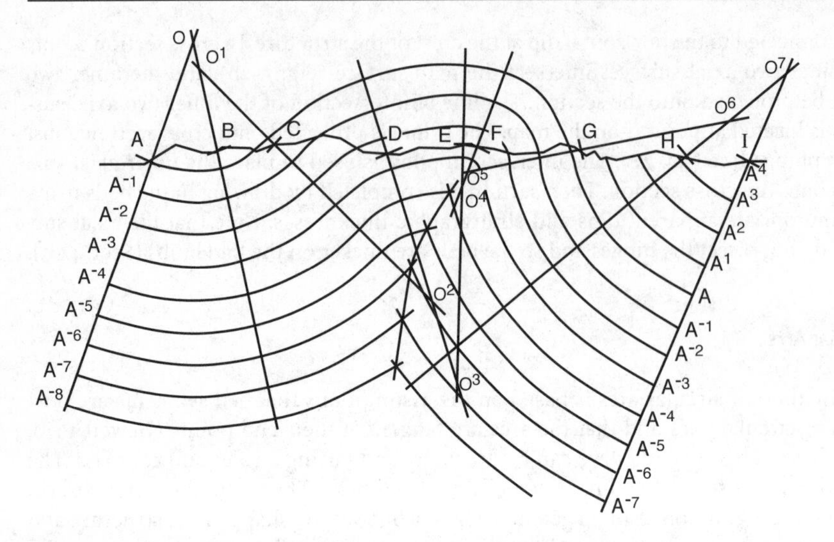

Fig. 7.34. Cross section produced by the method of circular arcs. *A–I* outcrop dip locations; A^i marker horizons; O^i centers of curvature. (After Busk 1929)

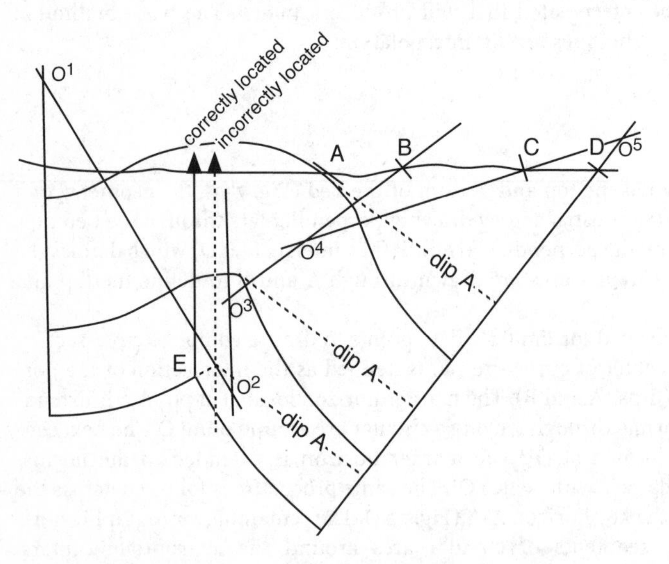

Fig. 7.35. Sensitivity of the crest location on a circular-arc cross section to the dips in the adjacent syncline. *A* dip on surface anticline used for linear projection of the fold limb; *B–D* dips in adjacent syncline used for circular-arc construction of the limb; O^i centers of curvature. Wells attempting to drill the lowest unit at the crest are shown. (After Busk, 1929)

To properly control the geometry of a cross section at depth, data may be needed at a long distance laterally from the area of interest (Fig. 7.35). For example, in order to correctly locate the crest of an anticline at depth, dips are needed from the adjacent synclines. If the last dip in the anticline (Fig. 7.35) was collected at A, then the steep limb of the structure would be drawn with the long dashed lines and the crest on the lowest horizon would be at the location of the incorrect well. Using the dips at B, C, and D, the structure is drawn with the solid lines, and the crest is found to be at E (Fig. 7.35). This is a general property of cross-section geometry and also applies to dip-domain constructions.

7.5.2.2
Dip Interpolation

Frequently the predicted geometry and the contact locations do not agree. The predicted location of horizon A (Fig. 7.36) on the opposite limb of the anticline is at B, but the horizon may actually crop out at B′ or B″. This result means that insufficient data are available to force a correct solution. It is necessary to modify the data or to interpolate intermediate dip values between A and B in order to make the horizon intersect the section at B′ or B″. Two methods of dip interpolation will be given; the first is to interpolate a planar dip segment and the second is to interpolate an intermediate dip.

The simplest method is to insert a straight line segment (AY, Fig. 7.37) between the two arc segments that produce the disagreement. This method is usually successful and provides an end-member solution. The procedure is from Higgins (1962):

1. Extend the dips at A and B so that they intersect at X.
2. On AX locate point Y such that YX = XB.

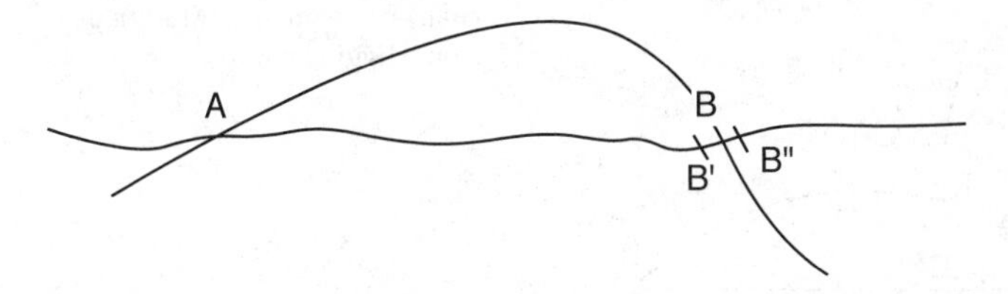

Fig. 7.36. Cross section showing the mismatch between the predicted location of the key bed at *A* and its mapped location (*B*′ or *B*″) at *B*. (After Busk 1929)

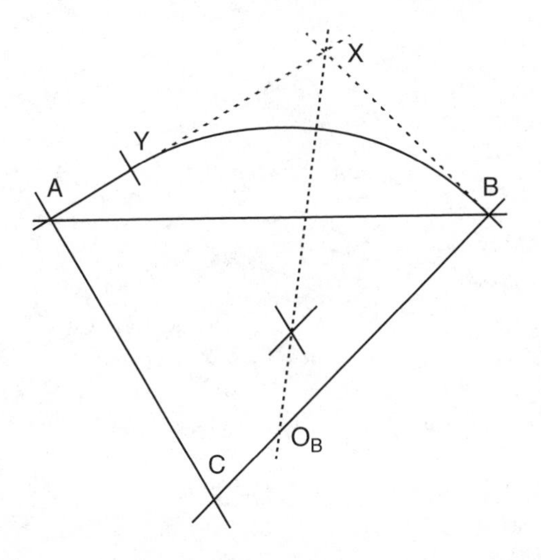

Fig. 7.37. Interpolation using linear domains and circular arcs. (After Higgins 1962)

3. Bisect angle YXB. The bisector will intersect BC, the normal to B, at O_B.
4. With center O_B and radius BO_B, draw the arc from B to Y. This arc is tangent to AY, the straight-line extension of the dip from A.

The second method is to insert a dip such that the two data points are joined by two circular arcs that are tangent at the data points and at the interpolated dip. The result is a cross section with continuously curving beds. This method is given by Busk (1929) and Higgins (1962). Beginning with the two dips A and B (Fig. 7.38):

1. Draw AA′ perpendicular to the lesser dip at A; draw BB′ perpendicular to the greater dip at B.
2. Draw the chord AB. Angle CAB must be greater than angle CBA; if not, switch the labels on points A and B.
3. Erect the perpendicular bisector of AB. This line intersects AA′ at Z.
4. Choose point O_A anywhere on line AA′ on the opposite side of Z from A. (If the length O_A–Z is very large, it is equivalent to drawing a straight line through A.)
5. On BB′ locate point D such that $BD = AO_A$.
6. Draw DO_A connecting D and O_A.
7. Erect the perpendicular bisector of DO_A. This line intersects BB′ at O_B.
8. Draw O_AE' through O_A and O_B, intersecting AB at E.

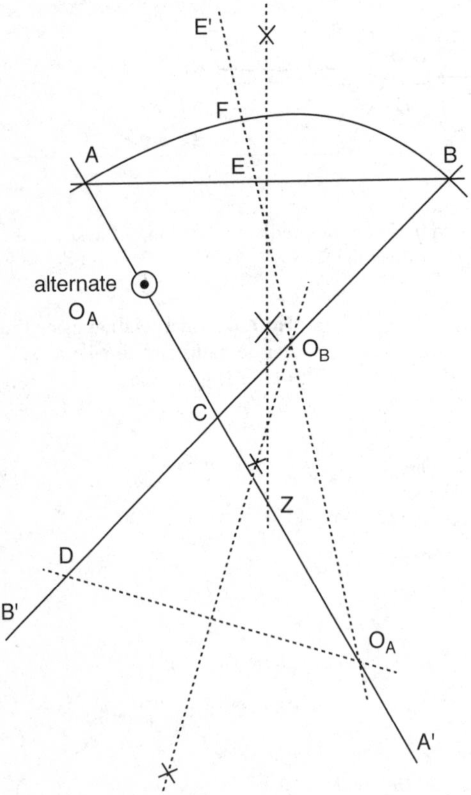

Fig. 7.38. Interpolation using circular-arc segments. (Modified from Higgins 1962)

9. With center O_A and radius AO_A, draw an arc from A, intersecting O_AE' at F.
10. With center O_B and radius BO_B, draw an arc from B intersecting O_AE' at F. This completes the interpolation.

If a correct solution is not obtained, it may be because the sense of curvature changes across an inflection point, causing the centers of curvature to be on opposite sides of the key bed. Modify step 4 above by using the alternate position of O_A (Fig. 7.38: alternate O_A), located between C and A.

7.5.3 Cubic Curves

Interactive computer drafting programs provide several different tools for drawing smooth curves through or close to a specified set of points. Typically they are parametric cubic curves for which the first derivatives, that is the tangents, are continuous where they join (Foley and Van Dam 1983). In this respect the curves are like the method of circular arcs, for which the tangents are equal where the curve segments join, but cubics are able to fit more complex curves than just segments of circular arcs. Two different smooth curve types are widely available in interactive computer drafting packages, Bézier and spline curves. The two curve types differ in how they fit their control points and in how they are edited. Both types are useful in producing smoothly curved lines and surfaces (Foley and Van Dam 1983; De Paor 1996).

A Bézier curve consists of segments that are defined by four control points, two anchor points on the curve (P_1 and P_4, Fig. 7.39a) and two direction points (P_2 and P_3, Fig. 7.39a) that determine the shape of the curve. The curve always goes through the anchor points. The shape is controlled in interactive computer graphics applications by moving the direction points. In a computer program the direction points may be connected to the anchors by lines to form handles (Fig. 7.39b) that are visible in the edit mode. At the join between two Bézier segments, the handles of the shared anchor point are colinear (Fig. 7.39b), ensuring that the slopes of the curve segments match at the intersection.

A spline curve only approximates the positions of its control points (Fig. 7.40) but is continuous in both the slope and the curvature at the segment boundaries, and so the curve is even smoother than the Bézier curve (Foley and Van Dam 1983). The shape

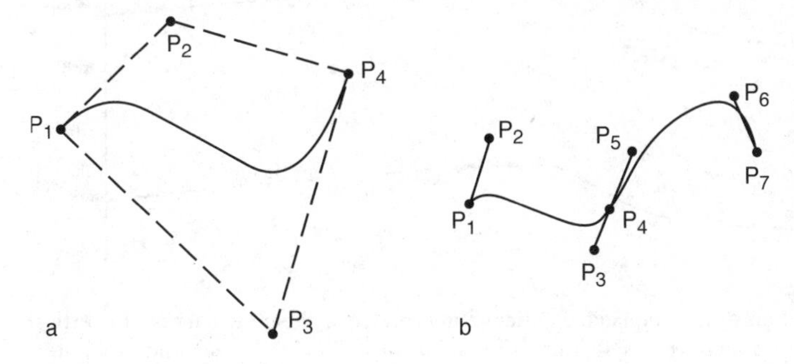

Fig. 7.39. Bézier curves. **a** The four control points that define the curve. **b** Two Bézier cubics joined at point P_4. Points P_3, P_4, and P_5 are colinear (After Foley and Van Dam 1983)

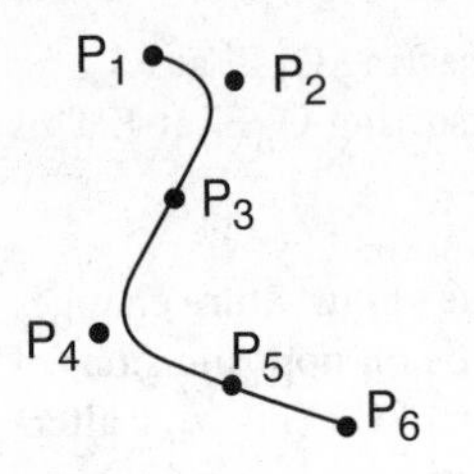

Fig. 7.40. Spline curve and its control points

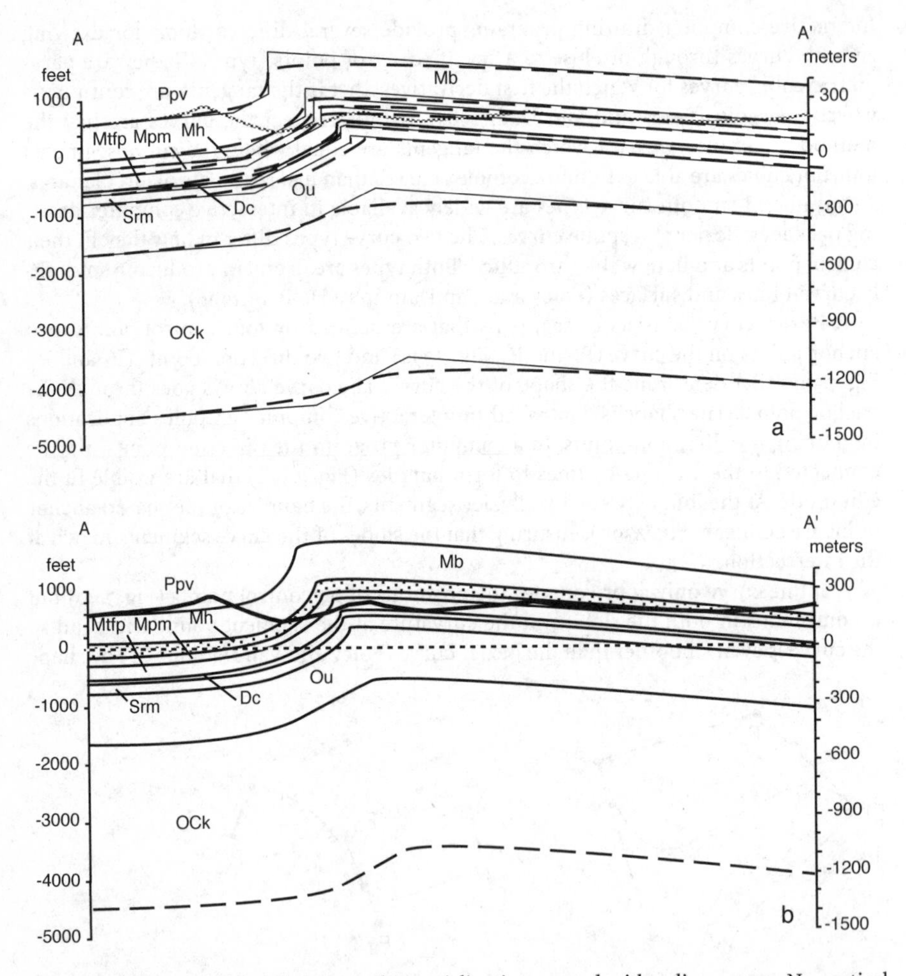

Fig. 7.41. Cross section of the Sequatchie anticline interpreted with spline curves. No vertical exaggeration. **a** Dip-domain cross section (*thin solid lines* from Fig. 7.29) and computer-smoothed spline interpretation (*thick dashed curves*). **b** Spline curve section edited to more closely resemble the dip-domain section

is controlled in interactive computer graphics applications by moving the control points that are visible in the edit mode. This curve type should be drawn separately from the actual data points because editing the curves changes the locations of the points that define the curve. The control points can be manipulated until the match between the curve and the data points is acceptable.

Drawing a cross section (or a map) using curve smoothing techniques requires care to maintain the correct geometry. Constant bed thickness, for example, is not likely to be maintained if the section is drawn from sparse data. The appropriate bed thickness relationships can be obtained by editing the curves after a preliminary section has been drawn. The cross section of the Sequatchie anticline illustrates the problems. The original section (Fig. 7.29) was redrawn with spline curves in a computer drafting program. The resulting cross section (Fig. 7.41) may be more pleasing to the eye than the dip-domain cross section, but it is less accurate. The unedited spline-curve version (Fig. 7.41a) is much too smooth. Each bedding surface is defined by 4 to 6 points, a data density that might be expected with control based entirely on wells. Bedding thicknesses are not constant as in the dip-domain version, and the amplitude of the structure is reduced. These are the typical results of analytical smoothing procedures, including the smoothing inherent in gridding as used for map construction. Editing the spline curves produces a better fit to the true dips (Fig. 7.41b). A more accurate spline section can be produced by introducing many more control points, which is the appropriate procedure for producing a final drawing of a known geometry. The addition of control points to improve an interpretation based on a sparse data set requires additional information, such as the bedding dips, or the requirement of constant bed thickness.

7.6 Changing the Orientation of the Section Plane

Only a cross section perpendicular to the plunge (the normal section) shows the true bed thicknesses. If a cross section is not a normal section, it will be helpful for structural interpretation to rotate the section plane until it is a normal section. Within a domain of cylindrical folding, changing the orientation of the section plane is equivalent to changing the vertical or horizontal exaggeration (Fig. 7.42). This relationship

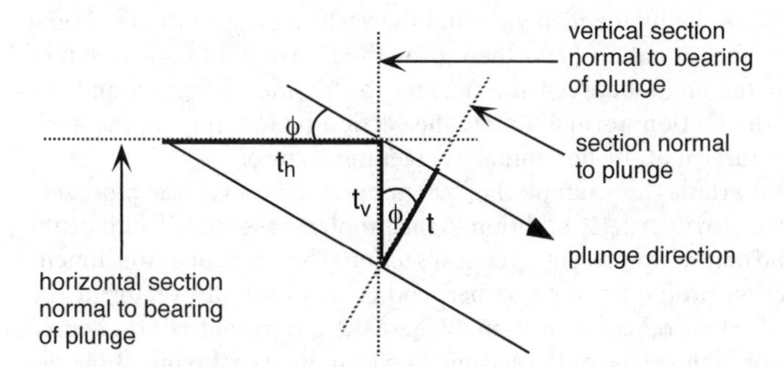

Fig. 7.42. Vertical exaggeration in cross section parallel to the plunge direction, caused by a plunge angle of ϕ. The true thickness is t; the exaggerated thickness is t_h in the horizontal plane and t_v in the vertical plane

is the basis of the map interpretation technique of down-plunge viewing. The map pattern in an area of moderate topographic relief represents an oblique, hence exaggerated, section through a plunging structure. Viewed in the direction of plunge, the map pattern becomes a normal section (Mackin 1950).

The plunge of a cylindrical fold is the orientation of its axis, which can be found from the bedding attitudes using the stereonet or tangent diagram techniques given in Section 4.2. A conical fold does not have an axis and so, in the strict sense, there is no normal section. The orientation of either the crestal line or the cone axis is an approximate plunge direction for a conical fold. On a structure contour map the trend and plunge of the crestal line is readily identified. The plunge angle is given by the contour spacing in the plunge direction (Eq. 4.8).

The down-plunge view of a fault should give the correct cross-section geometry and the sense of the stratigraphic separation. The plunge direction of a fault is parallel to the axis or crest or trough line of ramp-related or drag folds. If the fault is listric or antilistric, the plunge direction should be the axis of the curved surface, just as if it were a folded surface. If the fault is planar and there are no associated folds, the appropriate plunge direction is parallel to the cutoff line of a displaced marker against the fault (Threet 1973). The sense of separation given by the down-plunge view is not necessarily the slip direction, as discussed in Section 5.5.

In a plunging structure, it can be convenient to construct a vertical cross section normal to the trend of the plunge and then to remove the vertical exaggeration due to the plunge angle. The same approach can be used to convert a map view into a normal section.

From the geometry of Fig. 7.42, the exaggeration on a vertical section, t_v / t, (Eq. 7.1) is:

$$V_e = t_v / t = 1 / \cos \phi . \tag{7.15}$$

The exaggeration on a vertical section due to the plunge is removed by multiplying the vertical scale of the section by the reciprocal of the vertical exaggeration, $\cos \phi$.

The exaggeration on a horizontal section (map view), t_h / t, (Fig. 7.42) is:

$$V_e = t_h / t = 1 / \sin \phi . \tag{7.16}$$

The exaggeration on a horizontal section due to the plunge is removed by multiplying the vertical scale of the section by the reciprocal of the vertical exaggeration, $\sin \phi$.

The same procedure can be used to rotate the plane of a cross section around a vertical axis. Treat Fig. 7.42 as being the map view and the vertical exaggeration as being a horizontal exaggeration. Equation (7.15) then gives the horizontal exaggeration of the profile, with ϕ = the angle between the normal to the line of section and the desired direction of the section normal. Rotate the section by multiplying the horizontal scale by the reciprocal of the horizontal exaggeration, $\cos \phi$.

A seismic reflection profile (for example, Fig. 7.8) appears to be a vertical cross section, but this is not necessarily true. In addition to the problems associated with depth conversion, the profile may not represent a vertical section. The plane of a two-dimensional seismic reflection profile tends to be perpendicular to the dip vector of the reflectors (Fig. 7.43). The true reflector location is where the wavefront from the source first meets the reflector. If the plane of the seismic line is in the true dip direction, the reflections can, in principle, be migrated to their true locations and dips as a function of their apparent dips. Migration always steepens and deepens the reflections. If the seismic line is parallel to the strike direction, however, the dips on the seismic line will

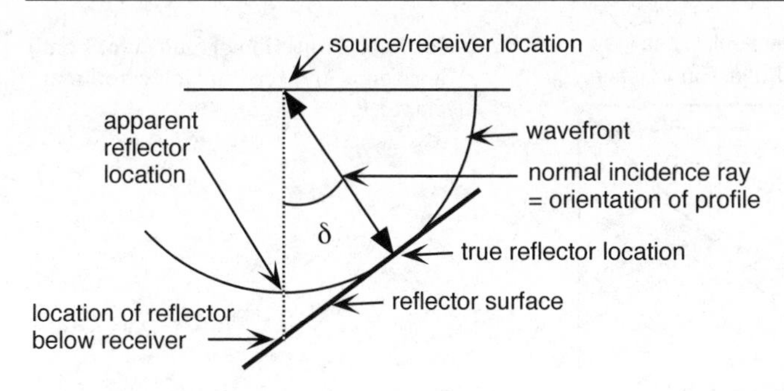

Fig. 7.43. Location of a dipping bed imaged on a seismic reflection profile. δ Dip of the reflector

be zero and no migration is possible. A dip component oblique to the seismic line (Fig. 7.43) causes the true reflector locations to be up-dip of the vertical plane, normal to the dip. This causes a problem in mapping points from the seismic line based on the assumption that the points are directly below the source/receiver locations, but is the preferred direction for a cross section. The vertical exaggeration required to remove the effect of the profile tilt can be estimated by dividing the vertical dimension on the profile by cos δ. A vertical slice through a time-migrated three-dimensional seismic reflection data set should more nearly represent a true vertical section.

7.7 Fault Cutoff Maps

A fault cutoff map is a map of the traces of beds where they are truncated against a fault. Maps can be made of either or both hangingwall and footwall cutoffs. Superimposed hangingwall and footwall cutoff maps are known as Allan diagrams from their use by Allan (1989) in the prediction of the potential pathways and traps for hydrocarbons migrating in the vicinity of a fault zone. Fault cutoff maps also provide one of the few unambiguous methods for determining the slip on a fault.

7.7.1 Construction

The first step in producing a fault cutoff map is to project the truncated horizons into the fault surface and find the lines of intersection between the horizons and the fault. One way to do the projection is with plunge lines (Sect. 7.4.2.2). This is done separately for both sides of the fault and the two maps are superimposed to see the relationships at the fault surface.

Fault cutoff maps are usually projections onto a vertical or horizontal plane. Projections of the fault cutoffs onto either vertical or horizontal planes are convenient when the section is made directly from a completed structure contour map. A vertical map plane is suitable where the fault is steeply dipping (Fig. 7.44) and will show the throw on the fault. If the fault is curved in plan view, a vertical projection foreshortens the length of the fault. A horizontal map plane is suitable where the fault is gently dipping and will show the heave on the fault (Fig. 7.44). If the fault is curved in cross

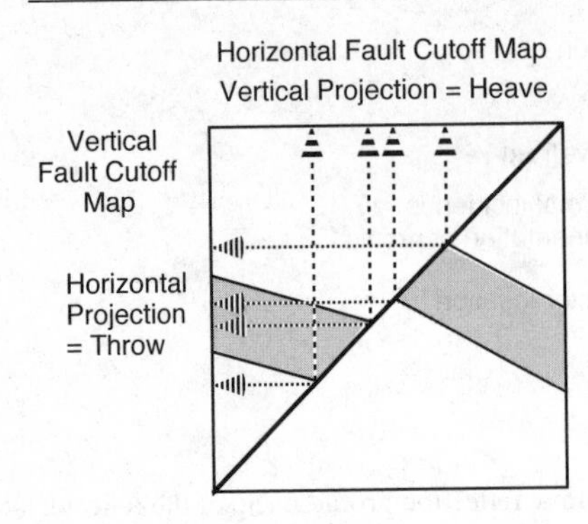

Fig. 7.44. Projection of fault cutoffs onto horizontal and vertical fault cutoff map planes

section, a horizontal projection foreshortens the width of the fault. Three-dimensional computer graphics techniques allow maps to be made directly on the fault surface, but this alternative is generally too complex to be clear on a flat sheet of paper, and so only horizontal and vertical projections are considered here.

The construction of a fault cutoff map is illustrated for a fault seen in an underground mine along two coal seams in the Black Warrior basin of Alabama. The fault dips about 60° to the northeast and forms one side of a full graben (Fig. 7.45). To the southeast it dies out by transferring its displacement to a parallel fault. To the northwest the fault zone continues along multiple parallel branches. The first step in constructing the cutoff map is to choose the plane of the map. In this example the fault dips steeply and so projection onto a vertical plane is suitable. The surface trace of the map plane is taken parallel to the trend of the fault and with a view direction to the southwest, from the hangingwall to the footwall (Fig. 7.46). The vertical scale is exaggerated by a factor of three because the maximum displacement on the fault is so small relative to its length. The elevations of the cutoffs of the top of both coal seams against the fault are transferred to the cutoff map along lines perpendicular to the trace of the map, which is the method of Fig. 7.11. The splays at the northwest end of the fault necessitate a choice of which elevations to map. The entire zone is included here so that the cutoff map (Fig. 7.46) shows the total displacement across the fault zone.

The cutoff map (Fig. 7.46) shows that the fault separation dies out to the southeast and has a maximum at a map location of about 6000 ft on the profile. The northwest component of dip shown by the footwall in both seams is opposite to the regional trend and is caused by a relay zone of displacement transfer to a fault that is off the map to the west. Both seams have similar throws, with the maximum decreasing downward from 90 ft in the America seam to 80 ft in the Mary Lee seam. The dip separation on the 60° dipping fault is 104 ft in the America seam and 92 ft in the Mary Lee seam (obtained from $L = T / \sin \phi$, where L = dip separation, T = throw, and ϕ = fault dip). Assuming that the point of maximum dip separation is close to the center of the fault, the length/displacement ratio is about 8000/98 or 82 to 1.

7.7.2 Determination of Fault Slip

The net slip on a fault is given by the offset of geological lines that pierce the plane of the fault. Geological lines may be formed by original stratigraphic features, such as

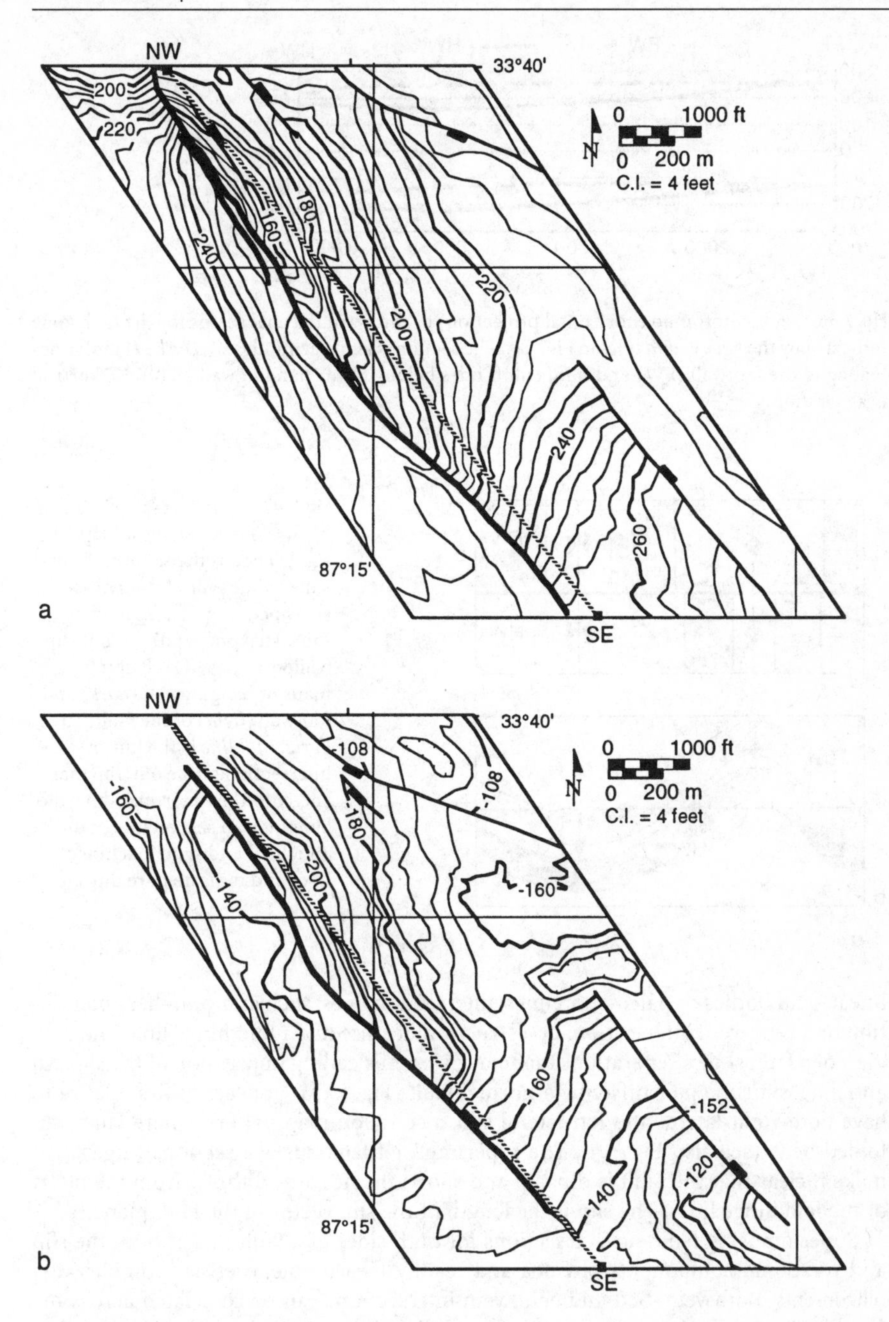

Fig. 7.45. Structure contour maps on the tops of two coal seams in the Drummond Goodsprings No. 1 Mine. The line of the fault cutoff map is the NW–SE *dotted line*. The fault zone represented on the cutoff map is indicated by *heavy solid lines*. **a** Top of America coal seam. Elevations are in feet above sea level. **b** Top of Mary Lee coal seam. Elevations are in feet below sea level. *C.I.* Contour interval. (After Hawkins 1996)

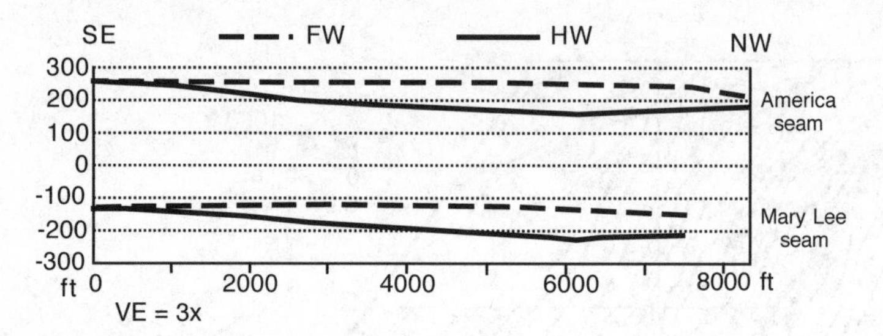

Fig. 7.46. Fault cutoff map (horizontal projection) of throw on the coal seams for the fault zone indicated by the *heavy solid lines* in Fig. 7.45. The section is along azimuth 321°. *Dashed* cutoff lines belong to the footwall (FW) and *solid* cutoff lines belong to the hangingwall (*HW*). *VE* Vertical exaggeration

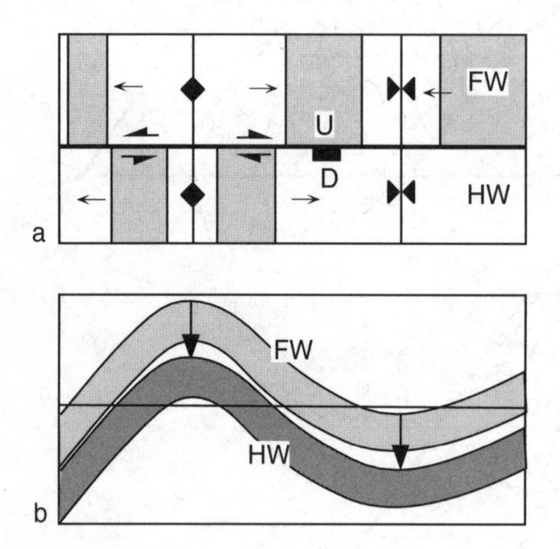

Fig. 7.47. Anticline–syncline pair cut by a normal fault. **a** Map. The *shaded* bed is the same on both sides of the fault. *Full arrows* show bedding dips, *half arrows* show strike separations, not slip. **b** Superimposed fault cutoff maps of hangingwall (*dark*) and footwall (*light*) of the fault. The *horizontal line* is the line of intersection of the outcrop map and cutoff map. *Arrows* show the hangingwall slip vectors of the anticlinal and synclinal hinge lines and indicate pure dip slip

linear sand bodies or paleo-shorelines; intersection lines, such as a vein–bed intersection or a vein–vein intersection; or a structural feature like a fold hinge line. The map view of a fault shows separations that may give a misleading impression of the slip. An anticline–syncline pair offset by a normal fault (Fig. 7.47a) appears at first glance to have both right-lateral and left-lateral slip, a common map pattern where faults cut folded beds (see also Fig. 6.3). The superimposed fault-surface sections (Fig. 7.47b) make it clear that the fault is dip slip and shows the heave and throw from the offset of the fold hinge lines. The slip is the length of the slip vector in the fault plane.

Superimposing fault-surface sections for both sides of a fault makes both the slip and rotational components obvious and readily measurable, even for complex displacements. Both vein–bed and vein–vein intersections can be correlated across the fault in Fig. 7.48. As a result of the rotational displacement on the fault, each correlated point has a different net slip.

Structures may develop independently across some fault zones and so may not directly correlate across the fault. In this situation, the offset of primary stratigraphic features provides the only opportunity to find the true slip on the fault.

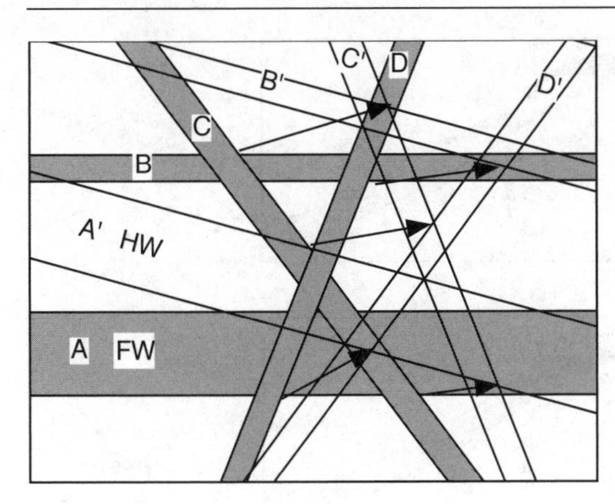

Fig. 7.48. Rotational fault displacement on a fault cutoff map. Superimposed fault-surface sections of hangingwall (*unshaded*) and footwall (*shaded*). Horizontal footwall beds labeled A and B rotate to A' and B'; veins C and D rotate to C' and D'. *Arrows* show hangingwall slip of correlated bed–vein intersection lines. Fault slip is a left-lateral reverse translation produced by a 15° clockwise rotation of the hangingwall relative to the footwall

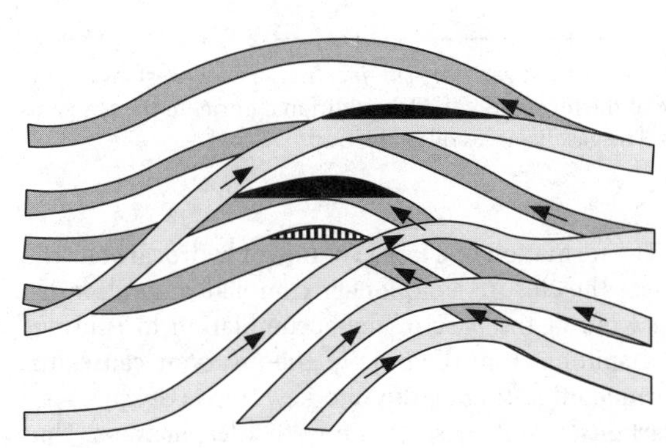

Fig. 7.49. Superimposed fault cutoff maps show fluid migration pathways through permeable beds separated by impermeable beds; footwall *shaded darker* than hangingwall. *Arrows* give migration routes of fluids that are lighter than water. Oil accumulations are *solid black*; gas accumulation is indicated by *vertical lines*. (After Allan 1989)

7.7.3 Determination of Fluid Migration Pathways

Fault cutoff maps are extremely useful in determining the possible routes of fluid migration in fault zones where the fault plane is not a barrier to the migration (Allan 1989). A trap is a closure in a permeable bed that is sealed by impermeable units across the fault. Multiple porous beds in the section can lead to very complex migration paths (Fig. 7.49) in which the migrating fluid crosses back and forth across the fault. Figure 7.49 shows eight fold closures against the fault, only three of which are sealed for the up-dip migration of a light fluid like oil or gas. The spill point at the base of each closure is located where permeable beds are in contact across the fault. Fluids that are heavier than water, such as man-made contaminants, may spiral downward across a fault into synclinal closures at some distance from the original contamination site.

The sequential migration of hydrocarbons through multiple traps as in Fig. 7.49 can lead to a reversal in the expected positions of the oil and gas. The expected sequence in the filling of a trap in place is illustrated in Fig. 7.50. Thermal maturation of the hydrocarbon source leads first to the formation of oil (Fig. 7.50a) which is less dense than water and displaces the water in the trap. Gas forms later in the source bed or in the trap itself and, being less dense than the oil, displaces the oil to form a gas

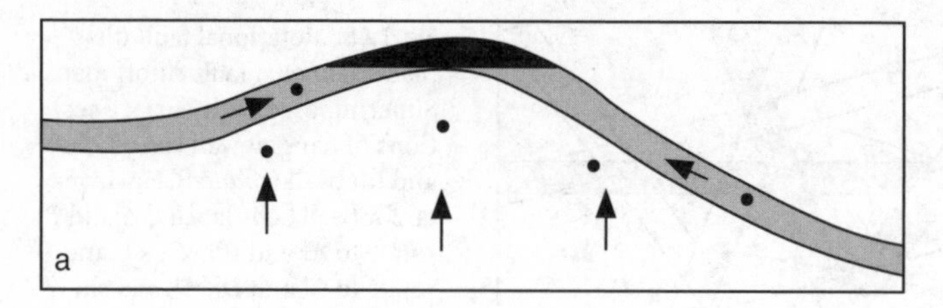

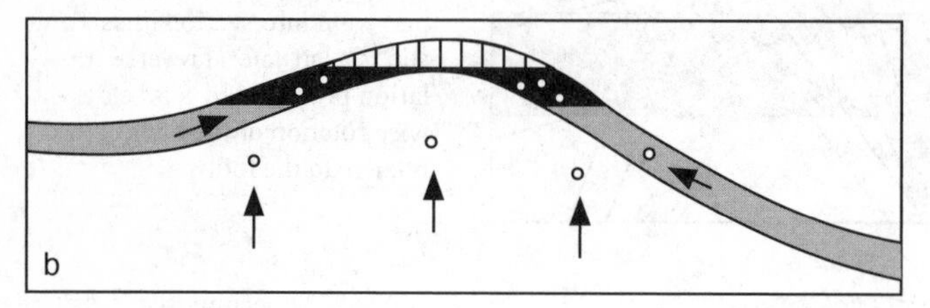

Fig. 7.50. Sequential formation of oil (*black*) and gas (*vertical lines* and *open circles*) and filling of a trap. **a** Burial to the temperature of the formation of oil. **b** Additional burial to the temperature of the formation of thermal gas. The gas displaces oil in the trap

cap on the reservoir (Fig. 7.50b). The formation of a large volume of hydrocarbons can fill the trap to the spill point where the closure is no longer complete and allow displaced hydrocarbons to be forced out at the base of the accumulation to continue migrating up dip. The process of spilling from the base of the reservoir causes the most dense hydrocarbons to continue migrating up dip into new traps (Gussow 1954; Allan 1989). The lightest hydrocarbons, mainly gas, remain in the deeper traps. Thus the presence of deep gas-filled traps and shallow oil-filled traps as shown in Fig. 7.49 can be caused by the upward migration of hydrocarbons across multiple spill points (assuming that oil and gas are both thermally stable at the trap depths).

7.8 Derivations

7.8.1 Vertical and Horizontal Exaggeration

From Fig. D7.1a, the dip of a marker on an unexaggerated profile is

$$\tan \delta = v / h, \tag{D7.1}$$

and the thickness of a unit in terms of its vertical dimension is

$$t = L \sin (90 - \delta) = L \cos \delta, \tag{D7.2}$$

where δ = unexaggerated dip, t = unexaggerated thickness, and L = unexaggerated vertical thickness. Let the vertical exaggeration be $V_e = v_v/v$ and the horizontal exaggeration be $H_e = h_h/h$, where v and h are the original horizontal and vertical scales and the subscripts h and v indicate the exaggerated scale. The equations for the exaggerated dips (Fig. D7.1b) have the same form as Eq, (D7.1):

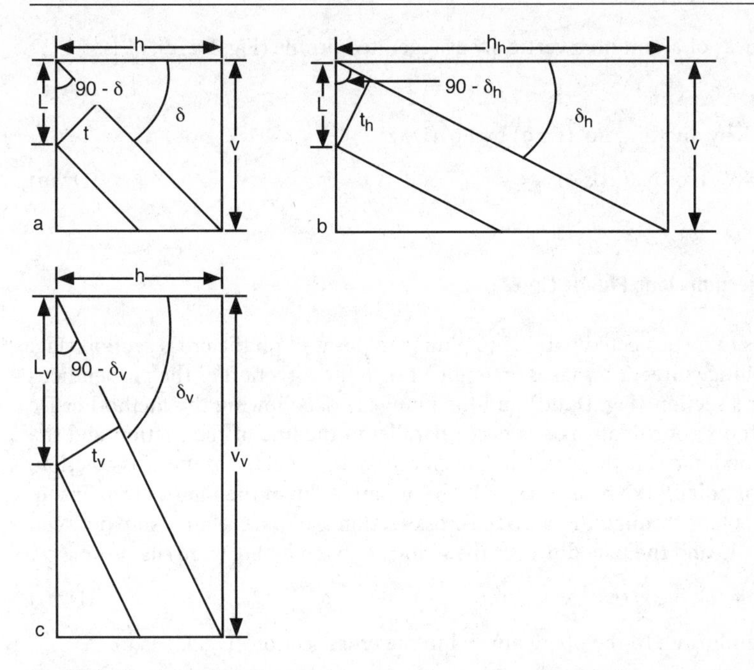

Fig. D7.1. Horizontal and vertical exaggeration. **a** Unexaggerated cross section. **b** Exaggerated horizontal scale; horizontal exaggeration (H_e) = 2:1. **c** Exaggerated vertical scale; vertical exaggeration (V_e) = 2:1

$$\tan \delta_v = v_v / h; \tag{D7.3a}$$

$$\tan \delta_h = v / h_h. \tag{D7.3b}$$

Replace h in Eq. (D7.3a) and v in (D7.3b) with the values from (D7.1) and use the definition of the exaggeration to obtain the relationship between original and exaggerated dips:

$$\tan \delta_v = V_e \tan \delta; \tag{D7.4a}$$

$$\tan \delta_h = \tan \delta / H_e. \tag{D7.4b}$$

To relate the horizontal to the vertical exaggeration, substitute the value of tan δ from Eq. (D7.4a) into (D7.4b) to obtain

$$V_e H_e = \tan \delta_v / \tan \delta_h. \tag{D7.5}$$

To obtain the same exaggerated angle by either horizontal or vertical exaggeration, set $\delta_v = \delta_h$ in Eq. (D7.5):

$$V_e = 1 / H_e. \tag{D7.6}$$

The thickness of a unit on a horizontally exaggerated profile (Fig. D7.1b), t_h, is

$$\sin (90 - \delta_h) = \cos \delta_h = t_h / L. \tag{D7.7}$$

Eliminate L by dividing Eq. (D7.7) by (D7.2):

$$t_h / t = \cos \delta_h / \cos \delta. \tag{D7.8}$$

The thickness of a unit on a vertically exaggerated profile (Fig. D7.1c), t_v, is

$$\cos \delta_v = t_v / (V_e L). \quad \text{(D7.9)}$$

Eliminate L by dividing Eq. (D7.9) by Eq. (D7.2):

$$t_v / t = V_e (\cos \delta_v / \cos \delta). \quad \text{(D7.10)}$$

7.8.2 Analytical Projection along Plunge Lines

The point P is to be projected parallel to plunge to point P′ on the cross section (Fig. D7.2a). The plunge direction makes an angle of α to the direction of the perpendicular to the cross section (Fig. D7.2d) and the plunge is ϕ. Following the method of De Paor (1988), the x coordinate axis is taken parallel to the line of the section and the plane of section intersects the x axis at zero elevation. In the plane of the cross section, the position of point P (x,y,z) is P′ (x′,z′). The apparent dip of the intersection line, q, of the vertical plane through PP′ with the cross section is δ'. The relationship between the apparent dip and the true dip, δ, of the z′ line, is given by Eq. (2.20) as:

$$\tan \delta' = \tan \delta \cos \alpha. \quad \text{(D7.11)}$$

Begin by finding z′. In the plane normal to the cross section (Fig. D7.2a,c),

$$z' = OP' / \sin \delta. \quad \text{(D7.12)}$$

In the plane of the plunge (Fig. D7.2b), using triangles AQP′ and PQR,

$$OP' = q \sin \delta'; \quad \text{(D7.13)}$$

$$\Delta z = L \tan \phi; \quad \text{(D7.14)}$$

$$v = z - L \tan \phi, \quad \text{(D7.15)}$$

and by using the law of sines with angles AQP′ and QP′A in triangle QAP′, along with $\cos \phi = \sin (90 - \phi)$,

$$q = v / (\tan \phi \cos \delta' + \sin \delta'). \quad \text{(D7.16)}$$

In the plane of the map (Fig. D7.2d),

$$L = y / \cos \alpha. \quad \text{(D7.17)}$$

Substitute Eqns. (D7.11), (D7.13), (D7.15), (D7.16) and (D7.17) into (D7.12) to obtain:

$$z' = (z \cos \alpha - y \tan \phi) / (\tan \phi \cos \delta + \sin \delta \cos \alpha). \quad \text{(D7.18)}$$

The x′ coordinate is found from (Fig. D7.2a),

$$x' = x + \Delta x + b. \quad \text{(D7.19)}$$

In the plane of the map (Fig. D7.2d),

$$\Delta x = y \tan \alpha; \quad \text{(D7.20)}$$

$$b = OB \tan \alpha. \quad \text{(D7.21)}$$

In the plane normal to the cross section (Fig. D7.2c),

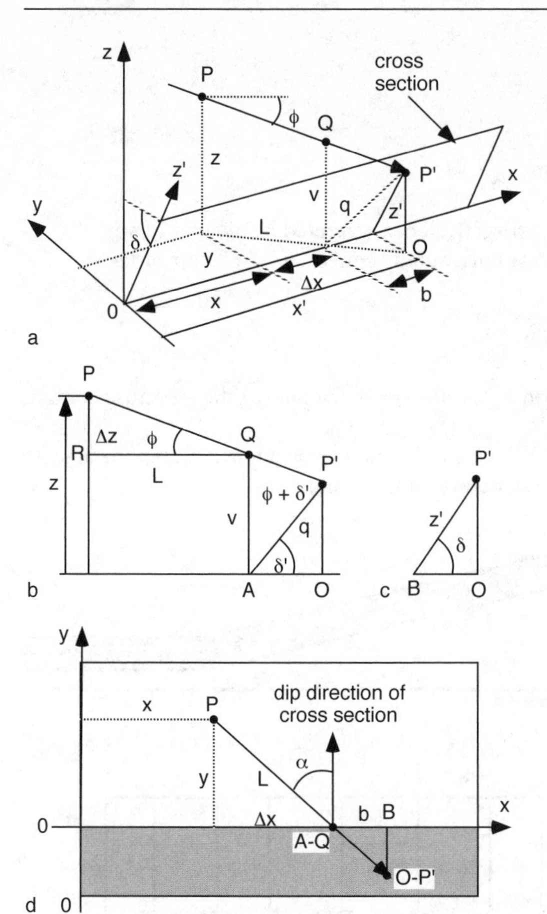

Fig. D7.2. Projection along plunge. **a** Perspective diagram. **b** Vertical plane through plunge line *PP'*. **c** Vertical plane normal to the cross section through line *OP'*. **d** Plan view. Projection of the cross section is *patterned*. Point *Q* is vertically above *A* at *A–Q* and point *P'* is vertically above O at *O–P'*

$$\mathrm{OB} = \mathrm{OP}' / \tan\delta. \tag{D7.22}$$

Substitute Eqs. (D7.11), (D7.13)-(D7.17) and (D7.20)–(D7.22), into (D7.19) to obtain:

$$x' = x + y\tan\alpha + \tan\alpha\,(z\cos\alpha - y\tan\phi) / (\tan\phi + \tan\delta\cos\alpha). \tag{D7.23}$$

For a vertical cross section, $\delta = 90°$ and Eqs. (D7.18) and (D7.23) reduce to

$$x' = x + y\tan\alpha; \tag{D7.24}$$

$$z' = z - y\tan\phi / \cos\alpha. \tag{D7.25}$$

For a cross section normal to the plunge line, possible only for a cylindrical fold, $\alpha = 0$, $\delta = (90 - \phi)$ and Eqs. (D7.18) and (D7.23) reduce to:

$$x' = x; \tag{D7.26}$$

$$z' = z\cos\phi - y\sin\phi. \tag{D7.27}$$

7.9 Problems

7.9.1 Vertical and Horizontal Exaggeration

1. Draw the cross section in Fig. 7.51 vertically exaggerated by a factor of 5:1.
2. Draw the cross section in Fig. 7.51 horizontally squeezed by a factor of 1:2.

7.9.2 Cross Section and Map Trace of a Fault

1. Draw an east–west cross section across the northern part of the structure contour map in Fig. 7.52.
2. Suppose a fault that dips 40° south cuts the structure on a line between the arrows. What would its trace be on the structure contour map?

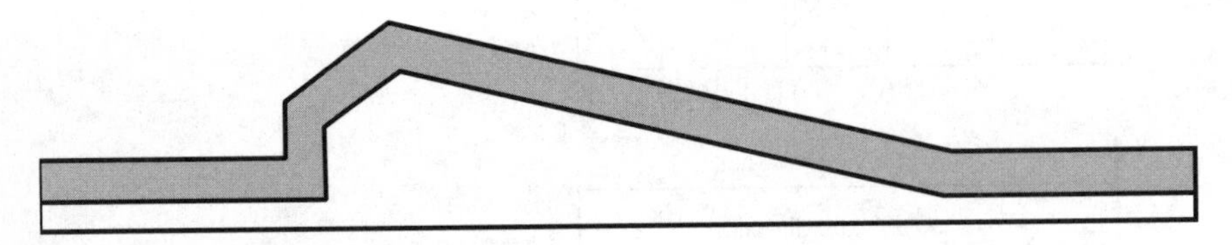

Fig. 7.51. Cross section of a fold

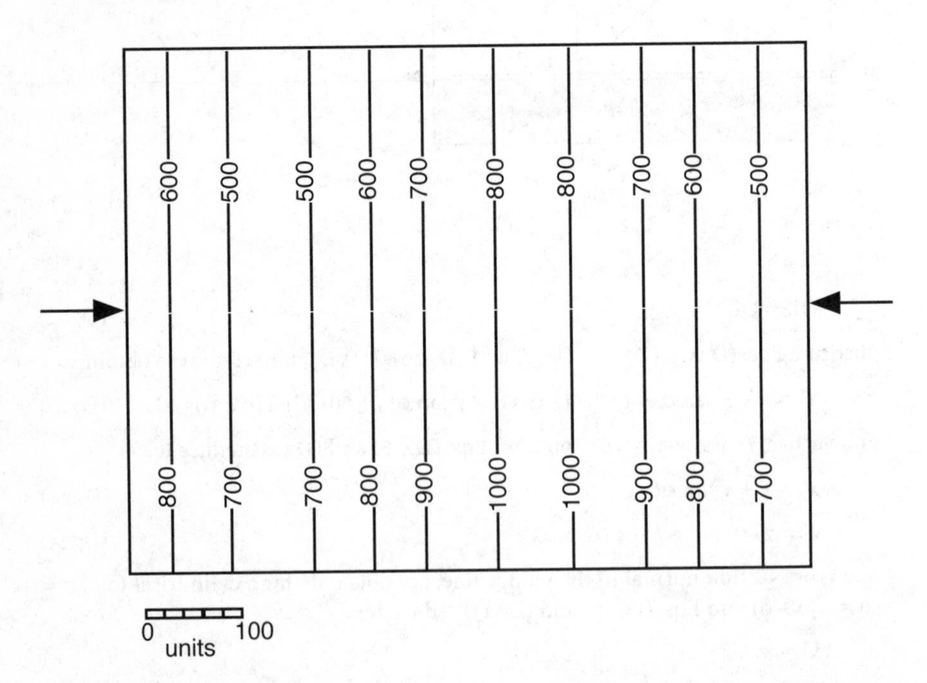

Fig. 7.52. Unfinished structure contour map. *Arrows* indicate the general position of a fault trace

3. Make the fault gap 60 units wide.
4. Adjust the contour elevations on the hangingwall to agree with the throw.
5. Draw a north–south cross section showing the fault.

7.9.3 Cross Section from a Structure Contour Map: Parallel Faults and Folds

1. Draw a cross section perpendicular to the major structural trend in Fig. 7.53. Discuss any assumptions required.
2. What are the dips of the faults?
3. Are the faults normal or reverse?

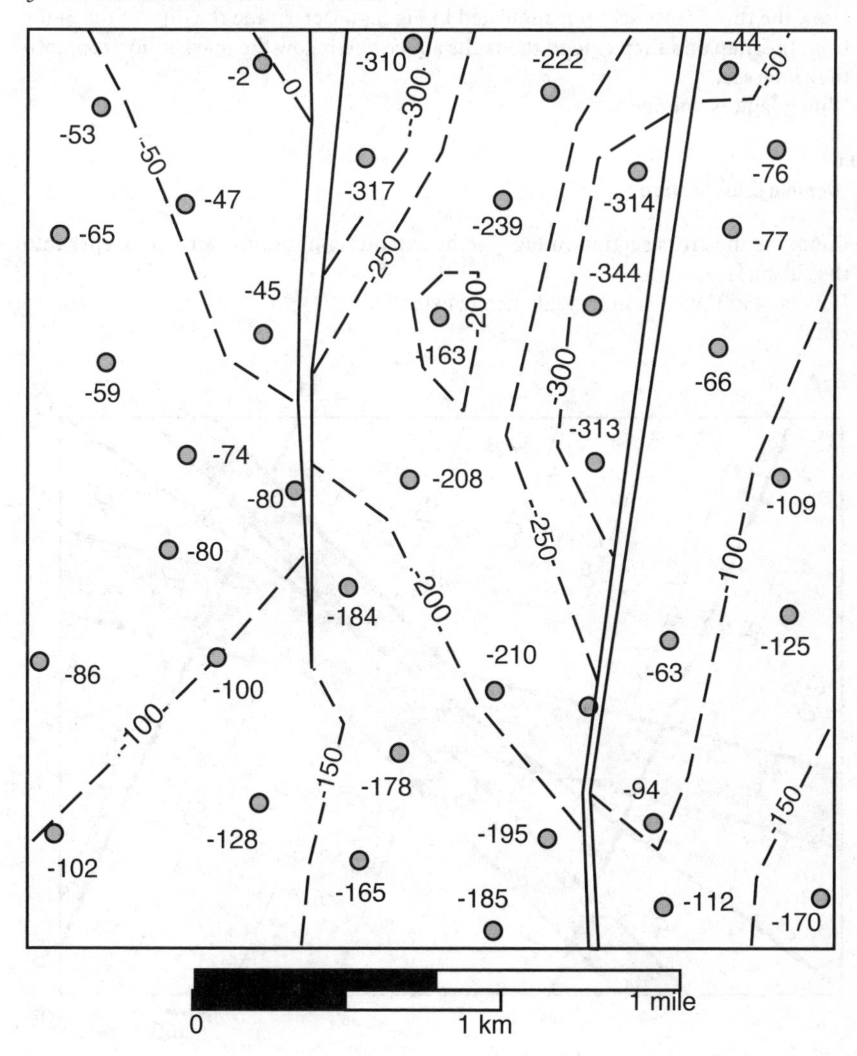

Fig. 7.53. Structure contour map of the top of the Gwin coal cycle (from Fig. 6.16a). Elevations of the top Gwin are posted next to the wells. Units are in feet

7.9.4
Cross Section from a Structure Contour Map: Two Normal Faults

1. Draw the three cross sections indicated on the map of Fig. 7.54. Using the fault dip determined from the map, extend the faults above and below the marker horizon until they intersect.
2. Which fault formed last?

7.9.5
Cross section from a Structure Contour Map: Two Reverse Faults

1. Draw the three cross sections indicated in Fig. 7.55. Determine the dips of the faults from the map and then extend the faults above and below the marker horizon until they intersect.
2. Which fault is younger?

7.9.6
Dip-Domain Cross Section

1. Complete the cross section in Fig. 7.56 by extending it into the air and deeper into the subsurface.
2. How far can the section be realistically extended?

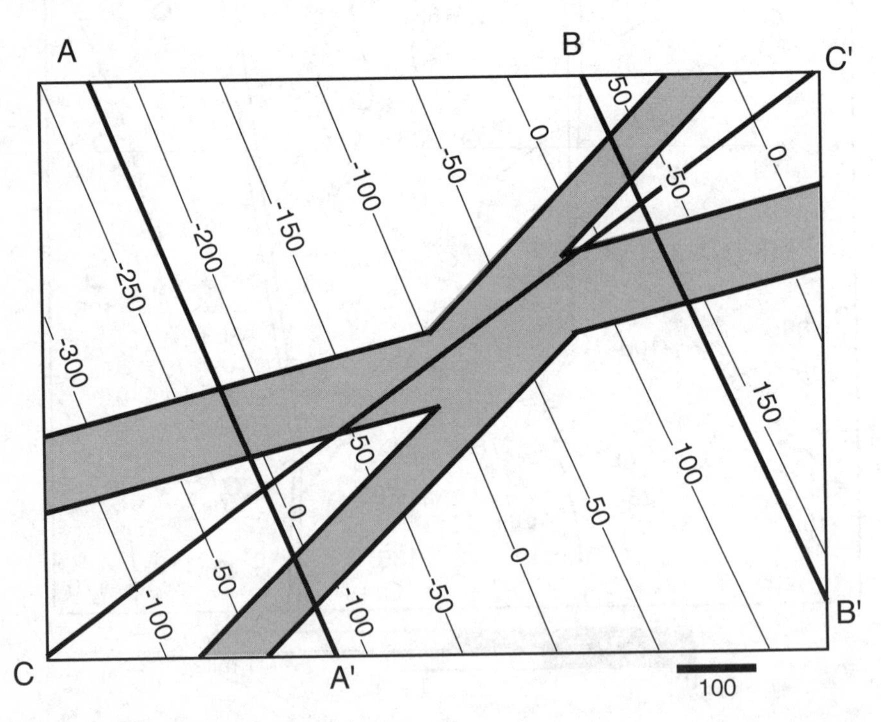

Fig. 7.54. Structure contour map of a normal-faulted surface (from Fig. 6.35). The horizon surface is missing in the *shaded* fault zones. Locations of lines of section *A–A′*, *B–B′*, and *C–C″* are shown

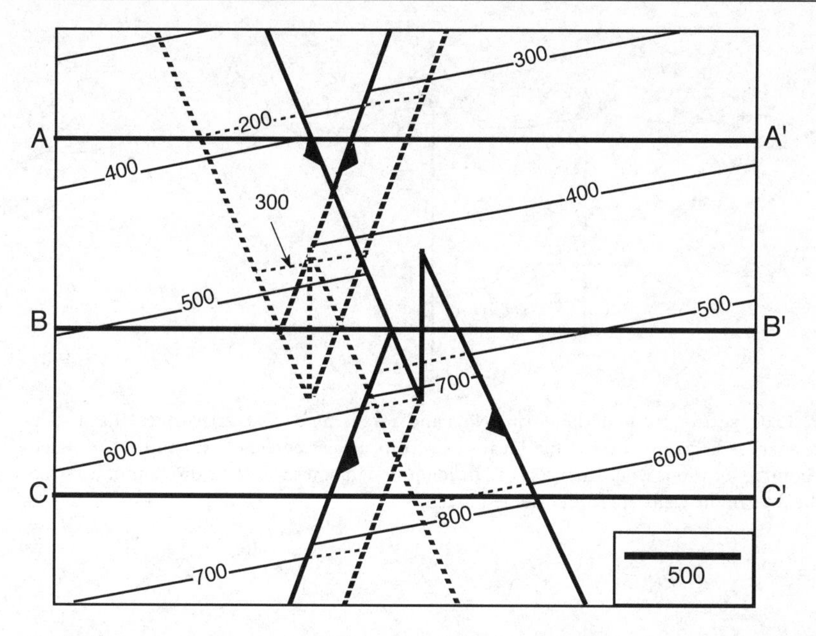

Fig. 7.55. Structure contour map of a reverse-faulted surface (from Fig. 6.41). The horizon surface is repeated by the fault zones. The fault contours are *wider lines*. Hidden contours are *dashed*. The locations of lines of section $A–A'$, $B–B'$, and $C–C'$ are shown

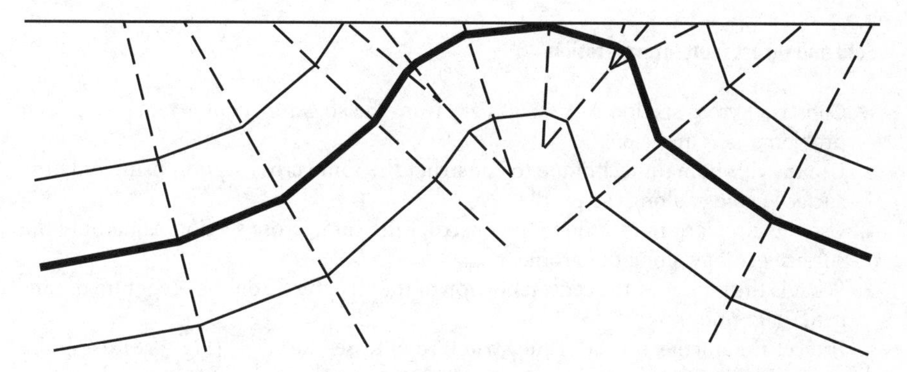

Fig. 7.56. Partially complete dip-domain cross section. *Dashed lines* are axial-surface traces

7.9.7 Cross Sections from Attitudes: Different Styles

1. Complete the cross section in Fig. 7.57, keeping bed thicknesses constant.
2. Use the dip-domain technique.
3. Use the method of circular arcs.
4. Scan the section into a computer and complete using the smooth curves provided by a drafting program.

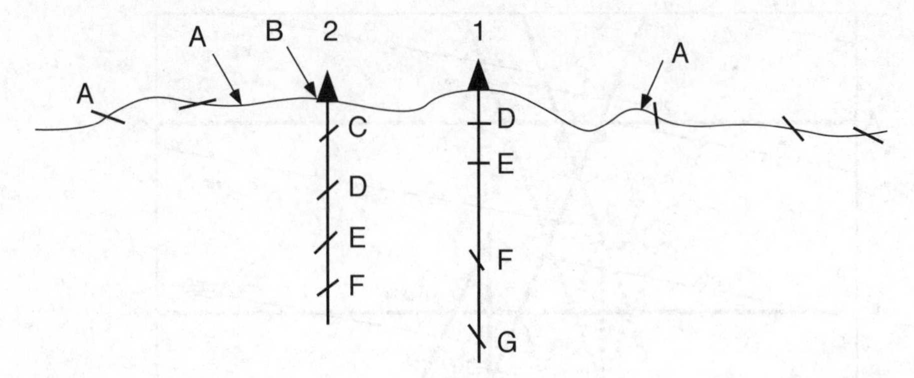

Fig. 7.57. Cross section through the Burma No. 1 and 2 wells. *Short lines* are surface dips. Letters *A–G* are marker horizons seen at the locations of dip measurements that can be correlated. *Arrows* point to locations where markers can be identified in outcrop but the dip cannot be measured. The dips in the wells are from oriented cores

5. Compare the results of the different techniques.
6. Does the cross section resemble one of the examples in this chapter?

7.9.8
Fold and thrust fault interpretation

1. Construct cross section A-A′ (Fig. 3.28) from the structure contour maps made in problems 3.6.5 and 6.8.6.
2. Use the dip-domain technique to construct the same cross section using only the surface geology along the profile.
3. Use the circular arc technique to construct the same cross section using only the surface geology along the profile.
4. What is the plunge of the central portion of the structure from a stereogram or tangent diagram?
5. Project the northern part of the structure onto section B-B′ (Fig. 3.28) using the method of along-plunge projection in Section 7.4.2.2.
6. Compare and contrast the cross sections.
7. The wells to the Fairholme were drilled to find a hydrocarbon trap but were not successful. Use the map and cross sections to determine a structural reason for drilling the wells and a structural reason that they were unsuccessful.

7.9.9
Cutoff Maps: Reverse Faults

1. Draw a fault cutoff map for fault A across the structure contour map in Fig. 7.58.
2. Draw a fault cutoff map for fault B across the structure contour map in Fig. 7.58.

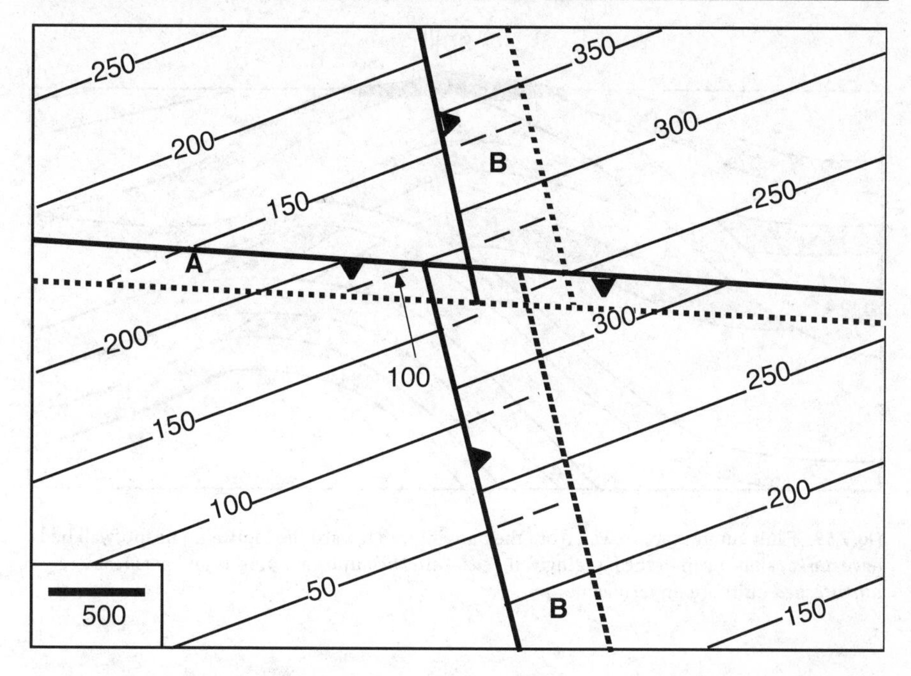

Fig. 7.58. Structure contour map of reverse-faulted marker horizon (*thin lines*). Faults are *thick lines*, hidden contours are *dashed*.

7.9.10 Cutoff Maps: Normal Faults

1. Draw a fault cutoff map on the trend of section C-C′ for both of the faults in Fig. 7.54.

7.9.11 Fluid Migration across a Fault

1. Suppose a toxic liquid that is heavier than water is spilled onto the surface in the center of the structure illustrated by the fault cutoff map in Fig. 7.59. Where will the liquid go?
2. Will the liquid be trapped at a location on the cross section?
3. If the liquid is trapped, will it all be in the same location?

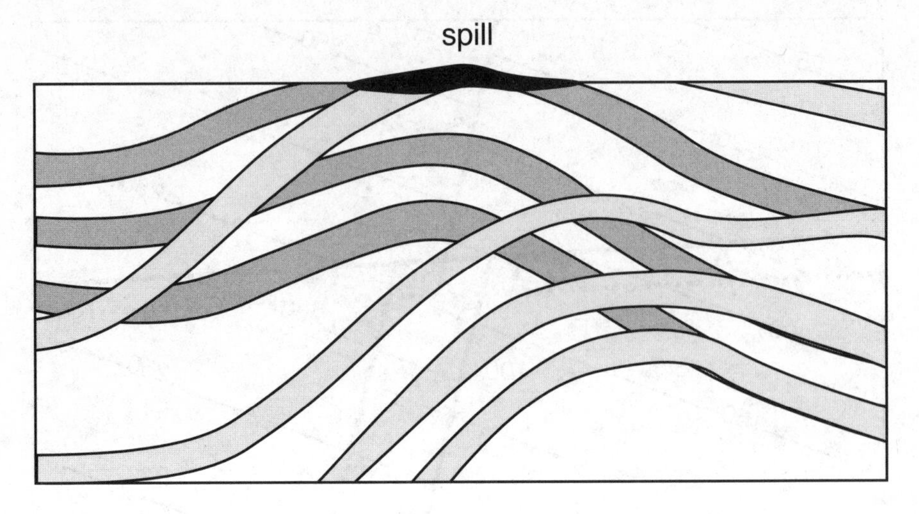

Fig. 7.59. Fault cutoff map, viewed from the hangingwall toward the footwall. The footwall beds have *darker shading* than the hangingwall *beds*. *Patterned* units are porous and permeable; *unpatterned* units are impermeable

References

Adams MG, Mallard LD, Trupe CH, Stewart KG (1996) Computerized geologic map compilation. In: De Paor DG (ed) Structural geology and personal computers. Pergamon, Tarrytown, New York, pp 457–470

Allan US (1989) Model for hydrocarbon migration and entrapment within faulted structures. Am. Assoc. Pet. Geol. Bull. 73: 803–811

Brown W (1984) Working with folds. Am. Assoc. Pet. Geol., Structural Geology School Course Notes, Tulsa, Oklahoma

Busk HG (1929) Earth flexures. Cambridge University Press, London, 106 pp

Cherry BA (1990) Internal deformation and fold kinematics of part of the Sequatchie anticline, southern Appalachian fold and thrust belt, Blount County, Alabama. MS Thesis, Univ. Alabama, Tuscaloosa, 78 pp

Coates J (1945) The construction of geological sections. Q. J. Geol. Min. Metallurg. Soc. India 17: 1–11

De Paor DG (1988) Balanced section in thrust belts. Part I: Construction. Am. Assoc. Pet. Geol. Bull. 72: 73–90

De Paor DG (1996) Bezier curves and geological design. In: De Paor DG (ed) Structural geology and personal computers. Pergamon, Tarrytown, New York, pp 389–417

Faill RT (1973) Kink band folding, Valley and Ridge Province, central Pennsylvania. Geol. Soc. Am. Bull. 84: 1289–1314

Foley JD, Van Dam A (1983) Fundamentals of interactive computer graphics: Addison-Wesley, Reading, Massachusetts, 664 pp

Gill WD (1953) Construction of geological sections of folds with steep-limb attenuation. Am. Assoc. Pet. Geol. Bull. 37: 2389–2406

Groshong RH Jr, Epard J-L (1994) The role of strain in area-constant detachment folding: J. Struct. Geol. 16: 613–618

Gussow WC (1954) Differential entrapment of oil and gas. Am. Assoc. Pet. Geol. Bull. 38: 816–853

Hawkins CD (1996) Thin-skinned extensional detachment in the Black Warrior basin: areal extent. MS Thesis, Univ. Alabama, Tuscaloosa, 45 pp
Hewett DF (1920) Measurement of folded beds. Econ. Geol. 15: 367–385
Higgins CG (1962) Reconstruction of a flexure fold by concentric arc method. Am. Assoc. Pet. Geol. Bull. 46: 1737–1739
Langstaff CS, Morrill D (1981) Geologic cross sections. Internat. Human Res. Development Co, Boston, 108 pp
Mackin JH (1950) The down-structure method of viewing geologic maps. J. Geol. 58: 55–72
McGlamery W (1956) Cuttings log of Shell Oil Company W. E. Drennan #1: records of Permit No 688, Alabama Oil and Gas Board, Tuscaloosa, Alabama, 15 pp
Stockwell CH (1950) The use of plunge in the construction of cross sections of folds. Proc. Geol. Assoc. Can. 3: 97–121
Stone DS (1991) Analysis of scale exaggeration on seismic profiles. Am. Assoc. Pet. Geol. Bull. 75: 1161–1177
Threet RL (1973) Down-structure method of viewing geologic maps to obtain sense of fault separation. Geol. Soc. Am. Bull. 84: 4001–4004
Webel S (1987) Significance of backthrusting in the Rocky Mountain thrust belt. In: 38 th Field Conf.-1987, Jackson Hole, Wyoming. Wyoming Geological Association Guidebook, pp 37–53
Wilson G (1967) The geometry of cylindrical and conical folds. Proc.Geol. Assoc. 78: 178–210

Chapter 8
Restoration and Validation

8.1
Introduction

The previous chapters provide techniques for making the best possible interpretation from the geological data. Because the primary data are always incomplete and may be in part contradictory, the final interpretation should be validated. A powerful and independent test for the validity of a structural interpretation is the restoration of the structure to the shape it had before deformation. A restorable structure can be returned to its original, pre-deformation geometry with a perfect or near-perfect fit of all the segments in their correct pre-deformation order (Fig. 8.1). Restoration is a fundamental test of the internal consistency of the interpretation. A restorable structure is internally consistent and therefore is a topologically possible interpretation. An unrestorable structure is topologically not possible and therefore is geologically not possible. The purpose of this chapter is to give a brief overview of the basic restoration techniques and how to use them. An extended treatment of this subject and related topics in structural modeling will be the subject of a subsequent volume.

In practice, the technique for the restoration of a structure is based on a model for the evolution of the geometry, known as a kinematic model. A valid map or cross section can usually be restored by methods based on more than one kinematic model, and different methods will produce somewhat different restored geometries. It follows

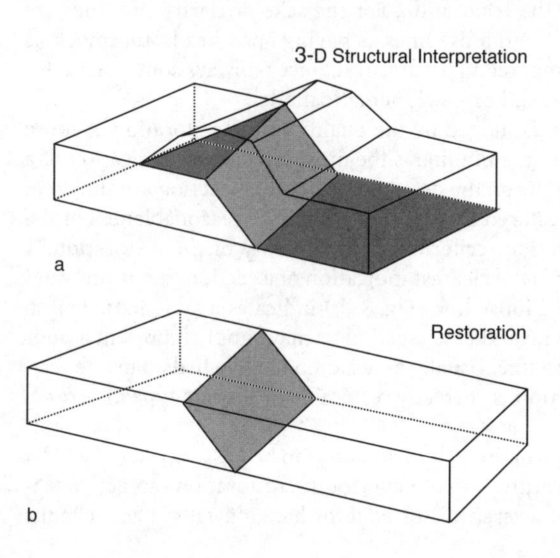

Fig. 8.1. A three-dimensional structural interpretation **a** and its restoration **b**. The *shaded* surface is a fault

that any given restoration does not necessarily represent the exact pre-deformation geometry. The internal consistency of the restoration by any technique constitutes a validation of the interpretation. If a restoration is possible, it shows that the structure is internally consistent even if the restoration technique is not a perfect model for the deformation process. An interpretation based on a large amount of hard data, such as complete exposure, or many wells, or seismic depth sections controlled by wells, is always restorable, whereas interpretations based on sparse data are rarely restorable. This is the primary evidence that confirms restoration as a validation technique. The techniques that will be presented here are rigid-body displacement, flexural slip, planar simple shear, and area restoration. The area–depth graph is presented as a method for judging the internal consistency of a cross section by a technique that is not based on a particular kinematic model. The restoration techniques discussed here are practical for use by hand or with computer drafting programs (Groshong and Epard, 1996). A cross section perpendicular to the main structural trend will provide a good test of the three-dimensional interpretation. The three-dimensional restorability of most structures can be verified by restoring multiple cross sections.

All restorations are based on the initial and final positions of reference lines (Fig. 8.2). On a cross section they are the reference horizon, the pin line, and the loose line (Dahlstrom 1969; Elliott 1980, in Geiser 1988; Marshak and Woodward 1988). The reference horizon is the line that will be restored to the shape it had before deformation. The positions of all other horizons are determined with respect to the reference horizon. The original position of a horizon, including both its shape and elevation, is known as the regional (McClay 1992). The pin line is a straight line on the deformed-state cross section that is required to remain straight on the restored section. A pin line forms a fixed boundary within the cross section or at one end of the cross section. The pin line is always chosen so that its position on the restored cross section can be specified, for example, as being perpendicular to bedding. Pin lines are chosen to be perpendicular to bedding or to lie along an axial surface in the deformed state so that they are likely to be perpendicular to bedding on the restoration. The loose line is a straight line on the deformed-state section that may assume any configuration on the restored section, as required by the restoration. For the sake of clarity, pin lines are shown here as having solid heads and loose lines as having open heads. Inasmuch as pin lines and loose lines are not marked in nature, the choice is always somewhat arbitrary and the best choice may depend on the kinematic model.

The validity of a cross section is judged by the quality of the restoration. A loose line that is straight and parallel to the pin line is the most satisfactory indication of a correct restoration (Fig. 8.2b). Faults on the deformed-state cross section are shown in their restored positions on the restored section (Fig. 8.2b). The reasonableness of the restored shape of the fault is another criterion for the quality of the restoration. A loose line that is highly irregular is the clearest indication of a section that is not valid (Fig. 8.3a). A straight but inclined loose line (Fig. 8.3b) indicates a systematic length difference that may represent an invalid cross section or may simply represent a poor choice of the pin line or the loose line. Usually at least four to five beds must be tested to determine whether the section is correctly restorable or if some type of error is present.

Restoration requires deciding whether a fold or fault is to be passively deformed by the restoration (a pre-existing feature) or is to control the restoration (an active feature). Fault-fold relationships are severely distorted if the hangingwalls of active faults

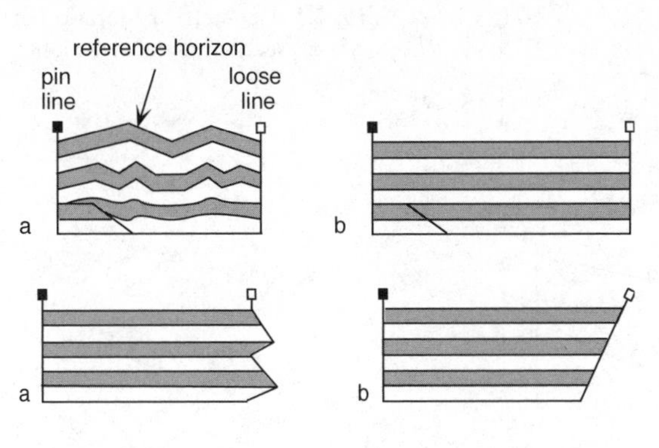

Fig. 8.2. Cross section restoration. *Solid-head pin* pin line, *open-head pin* loose line. **a** Deformed-state cross section. **b** Restored section: equal bed lengths and a straight loose line indicate a satisfactory restoration

Fig. 8.3. Imperfect restorations. **a** Irregular loose line indicates an invalid section. **b** Systematic error: the linear loose line may indicate an invalid section or a poor choice of pin line or loose line

are not restored separately from the footwalls. If the sequencing cannot be determined from the growth history of adjacent beds, as could be the case for a fault in a pre-growth sequence, both possibilities should be considered and perhaps alternative restorations done. Growth structures can be restored sequentially to the original depositional position of each unit.

Restoration is a purely geometric manipulation of the cross section according to a specific set of rules. A geometric restoration (Fig. 8.4) is here defined as a restoration that is not specifically related to the displacements or to the direction of transport that formed the structure. A palinspastic restoration is the restoration of the units to their correct pre-deformation configuration by reversing the displacements that formed the structure. The independence of the restoration from the transport direction is illustrated in Fig. 8.4. Two different cross sections through a fold have been restored. Normally it is assumed that the transport direction is perpendicular to the fold axis, and so the palinspastically restored section would be AB. If the transport is in the CD direction, however, the palinspastic restoration would be in the CD cross section. Either restoration serves to validate the structure but only the palinspastic restoration returns the units to their actual pre-deformation locations.

The only restriction on the choice of the orientation of the line of section to be restored is that the transport direction should be approximately constant in the vicinity of the cross section (Elliott 1983; Woodcock and Fischer 1986). The transport direction need not be parallel to the line of section as long as faults with transport oblique to the section are not crossed. A cross section oblique to the transport direction is the same as a section having vertical or horizontal exaggeration (Cooper 1983, 1984; Washington and Washington 1984). The exaggeration can be removed by projecting the section into the transport direction. Restoration of a cross section that crosses oblique-transport faults will probably result in a structural or stratigraphic discontinuity at the fault. The discontinuity might be removed by a lateral shift of the cross section at the fault.

A balanced cross section or map is one for which the restoration maintains constant area compared to the undeformed geometry and has an acceptably straight loose line. The original criterion, first developed by Chamberlin (1910), was that the deformed and restored cross sections must have the same area and will, if drawn carefully on paper of equal thickness, balance one another on a beam balance, as

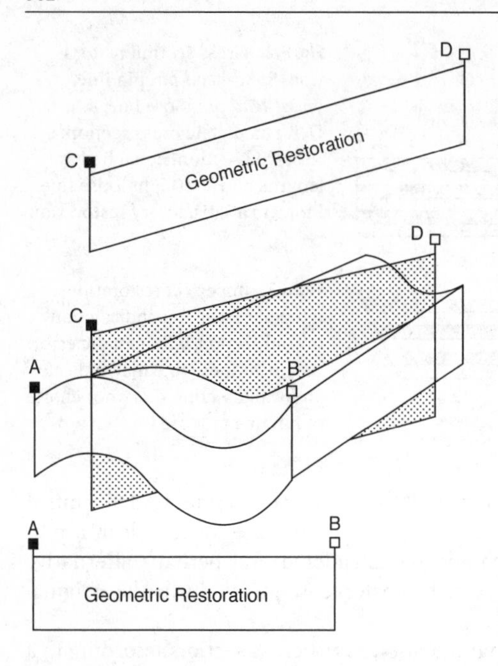

Fig. 8.4. Geometric restoration of a cross section in any direction

Chamberlin did with his cross section. This concept was generalized by Dahlstrom (1969) who observed that the volume should remain constant during deformation. In many structures there is little or no deformation along the axis of the structure, and so in practice the third dimension can often be ignored and constancy of volume can be applied to a cross section as a constant-area rule.

8.2 Rigid-Body Restoration

If all the deformation is by the displacement of rigid blocks (Fig. 8.5), then the section can be restored by rigid-block translations and rotations. Rigid-body displacement preserves all the original lengths and angles within the blocks on the deformed-state cross section. This method is suitable for cross sections consisting of internally undeformed fault blocks, most commonly found in extensional structural styles. It is not an appropriate method for the restoration of folds.

Rigid-body restoration could be done by cutting the cross section apart on the faults, removing the offsets, and reassembling it such that the reference horizon has the required restored shape (usually horizontal). The restoration can be done without cutting by tracing on a transparent overlay (Fig. 8.6). Draw the pin line and the reference line on the transparent overlay. Move the overlay so that the first block is in its restored position and trace it onto the overlay. Then move the overlay so that the second block is in the restored position and trace it. Repeat this step until the section has been restored. If the blocks do not fit together perfectly, it is preferable to leave gaps between them, rather than overlapping the blocks.

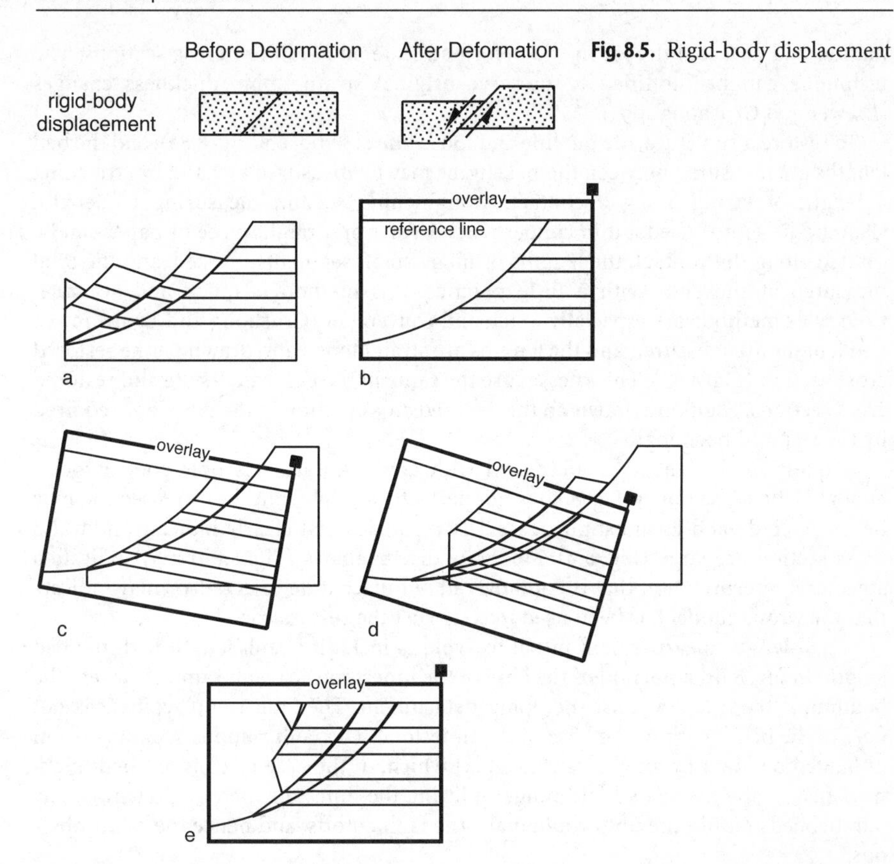

Fig. 8.5. Rigid-body displacement

Fig. 8.6. Restoration of rigid-body displacement by the overlay method. **a** Section to be restored. **b** Preparation of the overlay. **c** Restoration of the first block. **d** Restoration of the second block. **e** Complete restoration

8.3 Flexural-Slip Restoration

Flexural-slip restoration is based on the model that bed lengths do not change during deformation (Chamberlin 1910; Dahlstrom 1969; Woodward et al. 1985, 1989). Internal deformation is assumed to occur by layer-parallel shear (Fig. 8.7). For the area to remain constant, the bed thicknesses must be unchanged by the deformation as well. This is the constant bed length and bed thickness (constant BLT) model. Restoration consists of measuring bed lengths and straightening the lengths while preserving the thicknesses to produce the restored section. If all the bed lengths are the same, the cross section should be valid and restorable. Flexural-slip restoration is appropriate where the beds are folded and structurally induced thickness changes are small, the style of deformation in many compressional structures. The lengths of lines oblique to bedding and the cutoff angles between bedding and faults are changed by the deformation and so are different in the deformed-state and restored cross sections. The technique can be modified to preserve original stratigraphic thickness changes (Brewer and Groshong 1993).

To restore a structure, the pin line and loose line are chosen (Fig. 8.8a), and the bed lengths are measured between them. Lengths may be measured by hand by stretching a length of string along a contact, straightening it, and measuring its length. Chamberlin (1910) used a thin copper wire. A ruler or straight piece of paper can be rotated along the contact, the lengths of many small segments marked, and the total measured at the end. With a little practice, this method is quick and accurate. Computer methods are especially quick and convenient (Groshong and Epard 1996). Bed lengths are measured and the lengths are straightened and drawn on the restored cross section (Fig. 8.8b). Thicknesses are the same in the deformed-state and restored cross sections. Faults are drawn on the restored cross section in the positions required by the restored bed lengths.

A question that must be addressed with any restoration is how good is good enough? The loose line in Fig. 8.8b is not perfectly straight, but the cross section may be considered valid for most purposes. Discrepancies that clearly indicate an invalid cross section are large (Fig. 8.9). Moderate discrepancies (Fig. 8.8b) may indicate a structural interpretation that is good overall but inaccurate in detail, or may indicate that the wrong model has been used to construct the restoration.

A bed-length measurement pitfall to avoid is including fault length with the bed length. In Fig. 8.9a, a portion of the base of the limestone is a fault ramp (f) where the bedding is truncated against the upper detachment. The fault ramp segment is not part of the bed length of the base of the limestone. The fault ramp is separated from unfaulted bedding by an axial surface (ax) which, in the case of constant bed thickness, bisects the associated fold hinge. Splitting the limestone bed by drawing many parallel beds within the unit would make the fault cutoffs, and hence the ramp, obvious.

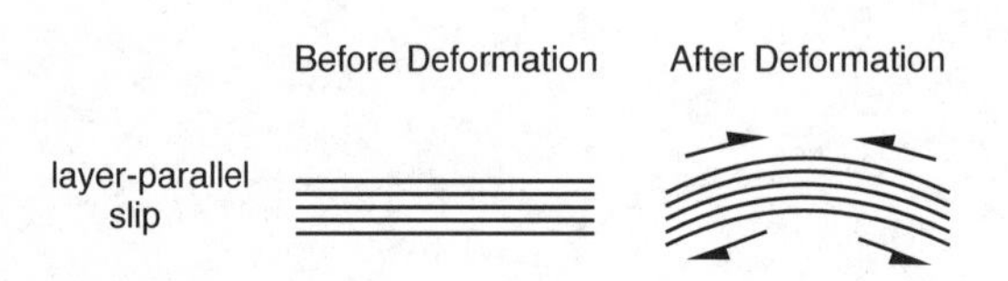

Fig. 8.7. Constant bed length–constant bed thickness flexural-slip model

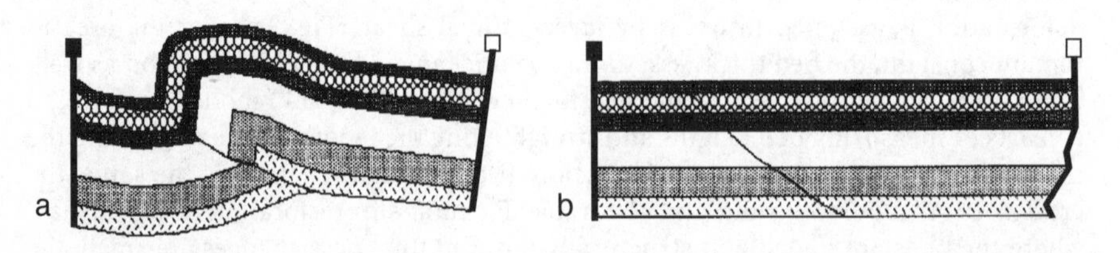

Fig. 8.8. Flexural-slip restoration of a cross section. **a** Deformed-state section. **b** Restored section. (After Kligfield et al. 1986)

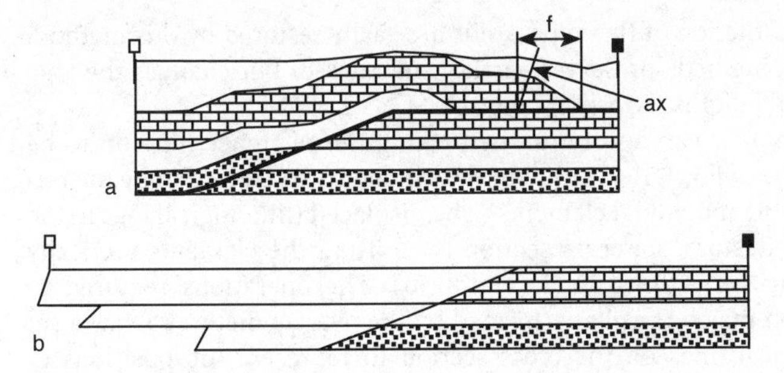

Fig. 8.9. Flexural-slip restoration of an invalid cross section. **a** Deformed-state cross section. Preserved beds maintain constant bed thickness (after a problem in Woodward et al. 1985). *ax* Axial surface; *f* hangingwall fault ramp. **b** Flexural-slip restoration based on length measurements at the top and base of each unit

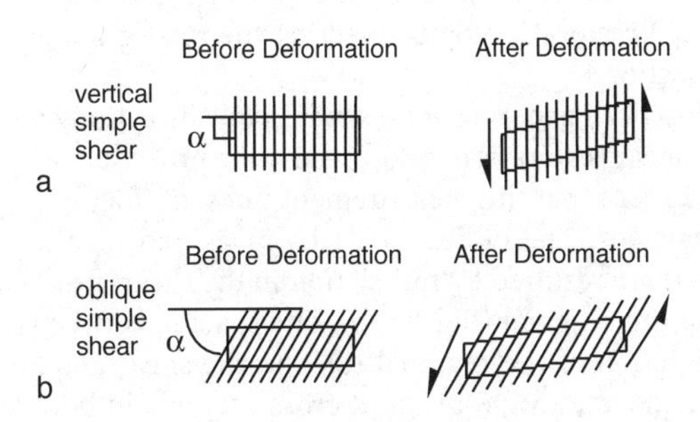

Fig. 8.10. Simple shear oblique to bedding. **a** Vertical simple shear. **b** Oblique simple shear

8.4 Simple-Shear Restoration

According to the planar simple-shear concept, a cross section deforms as if it were made up of an infinite number of planar slices that are free to slip past one another (Fig. 8.10). The shear direction is specified as having a dip α with respect to the regional that does not change during deformation. The special case for which the shear plane makes an angle of 90° to the regional is called vertical simple shear because the regional is usually horizontal (Fig. 8.10a). Oblique simple shear is simple shear along planes at some angle other than perpendicular to the regional (fig 8.10b, $\alpha \neq 90°$). (Some workers measure α from the vertical.) Simple-shear restorations are most widely used for extensional structures, in particular the hangingwall rollovers associated with half grabens.

Restoration by vertical simple shear involves the differential vertical displacement of vertical slices of a cross section in order to restore a reference horizon to a horizontal datum (Verrall 1982). This method has its origins in the well-log correlation technique in which the logs are aligned side-by-side and matched at an easily recognized marker horizon. The alignment makes the correlation of nearby markers much easier. Many computer programs designed for geological or geophysical data interpretation implement a function of this type. Folds are easily restored by this method. The restoration of dipping beds preserves vertical thicknesses but changes the true thicknesses and the bed lengths (Groshong 1990).

Begin a vertical simple-shear restoration by picking the reference horizon to be flattened. In the example of Fig. 8.11a, the top of the sandstone is selected to be the reference horizon. Define the individual elements to be displaced (lithological logs in the example of Fig. 8.11a). Restore the cross section by shifting the elements vertically until the reference horizon is horizontal (Fig. 8.11b). The operations required to restore the cross section may be easily performed by the overlay method. Draw a set of equally spaced vertical lines on the cross section to represent the positions of columns to be restored. The spacing of the vertical lines controls the level of detail in the horizontal direction that will be obtained in the restoration. Draw a horizontal line on a transparent overlay to represent the restored datum. Mark the horizontal position along the datum line of each column to be restored. Shift the overlay to successively bring each column to the restored datum and mark on the overlay the position of each stratigraphic unit. Complete the restoration by connecting the stratigraphic horizons to form a continuous cross section.

The vertical simple shear concept has been generalized to shear in any direction by White et al. (1986). An oblique simple-shear restoration follows the same procedure as the vertical simple-shear restoration, except that the measurement lines are inclined to the regional at an angle other than 90°. The oblique lengths measured on the deformed-state cross section (Fig. 8.12a) are restored by translation in the shear direction to return the reference horizon to the regional (Fig. 8.12b). The spacing between any two measurement lines in the direction of the regional (s), and the shear angle (α), remain constant from the deformed state to the restored cross sections in both vertical and oblique simple-shear restorations. The original distances between the working lines need not be the same between every pair of lines, however.

Fault displacement can be removed by translation parallel to the regional along with the displacement parallel to the shear direction (Fig. 8.13). The following technique is based on the assumption that the fault geometry does not change during deformation. The first step in the restoration is to draw a working line (Fig. 8.13, dotted line) in the shear direction through the hangingwall cutoff point (HCP) of the ref-

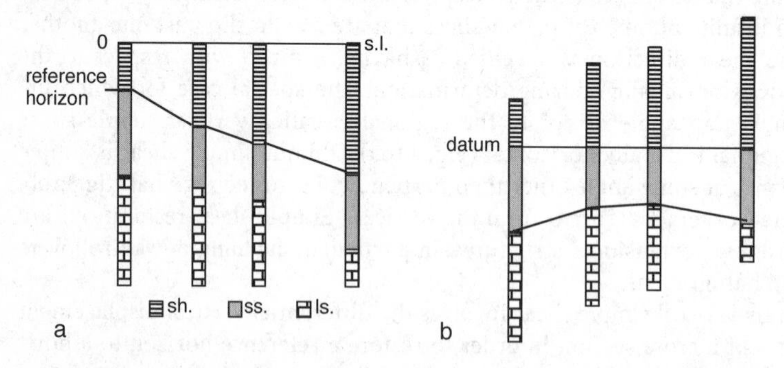

Fig. 8.11. Flattening to a datum by vertical simple shear. **a** Deformed-state cross section based on lithological logs. The zero elevation is sea level (*s.l.*); *sh.* shale; *ss.* sandstone; *ls.* limestone. **b** Cross section flattened to a datum at the top of the sandstone shows a paleo-anticline at the top of the limestone

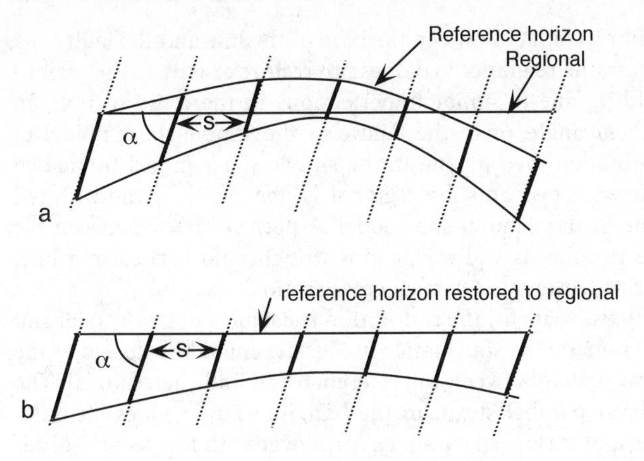

Fig. 8.12. Restoration by oblique simple shear. *Medium-weight solid lines* are marker beds. *Dotted lines* represent the measurement direction and are spaced a distance *s* apart. The *widest lines* are the thicknesses in the shear direction to be restored. The shear angle is α. **a** Deformed-state cross section. **b** Restored cross section

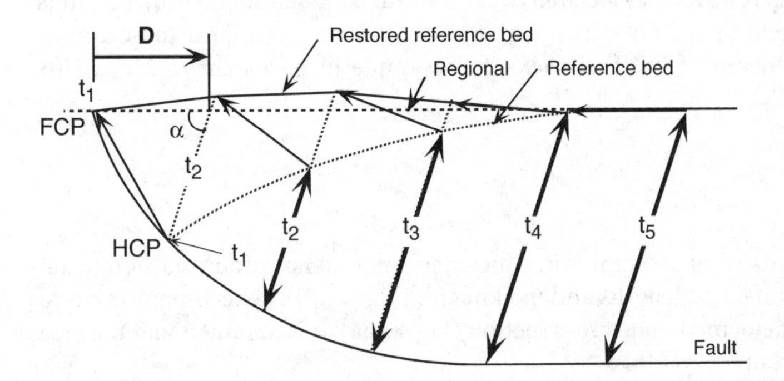

Fig. 8.13. Restoration of the hangingwall rollover on a normal fault by oblique simple shear. *D* Block displacement in the direction of the regional; *FCP* footwall cutoff point of reference bed; *HCP* hangingwall cutoff point of reference bed; α shear angle; t_i hangingwall thickness measured parallel to a shear line; $t_1 = 0$. Working lines are *dotted*. *Arrows* show the restoring displacements at the top of the thickness columns

erence bed. The block displacement (D) to be removed is the distance along the regional from the fault to the intersection of the working line with the regional. Removing the displacement D will restore the hangingwall cutoff point to the footwall cutoff point of the reference bed. Working lines parallel to the shear direction are drawn on the hangingwall. Hangingwall thicknesses measured parallel to the working lines (wide lines in Fig. 8.13) are moved back up the fault the distance D parallel to the regional. The base of the restored thickness column remains in contact with the fault and the top of the thickness column marks the restored position of the reference bed. The thickness, t_1, at the cutoff point is zero. A convenient construction method is to begin by drawing working lines on the cross section evenly spaced a distance D apart, as in Fig. 8.13. Then thickness measured along each working line is restored by shifting it over one working line in the direction that removes the displacement (to the left in Fig. 8.13) while maintaining contact between the base of the line and the fault.

In the example of Fig. 8.13, the reference bed does not restore exactly to the regional. This discrepancy could be due to minor imperfections in the cross section, an incorrect choice of the shear angle, or to the failure of the simple-shear model to exactly describe the deformation mechanism. In the simple-shear model the match between the restored reference bed and the regional is analogous to the restored geometry of the loose line in the flexural-slip model. A perfect match between the restored horizon and the regional is equivalent to a straight and vertical pin line. Small differences are expected, even for a correct cross section.

The shear angle most appropriate for the restoration is usually chosen by trial and error (Rowan and Kligfield 1989). The shear angle in Fig. 8.13 could be selected as the angle that gives the closest match between the reference bed and the regional. The presence or absence of layer-parallel strain in the hangingwall provides an independent indication of the appropriate shear angle. A rollover with few second-order faults probably has less layer-parallel strain than a rollover with many second-order faults. Vertical simple shear produces very little layer-parallel strain, whereas oblique shear produces an increasing amount of strain as the angle between the shear direction and bedding becomes smaller (Groshong, 1990). Vertical simple shear appears to be best for the rollovers associated with natural soft-sediment growth faults (Rowan and Kligfield 1989), whereas shear angles of 60° or less appear to be appropriate for the extension of lithified rocks, for example in basement-involved rifts (Groshong 1990).

8.5 Area Restoration

Area restoration is used for structures in which deformation has produced significant changes in the original bed lengths and thicknesses (Fig. 8.14). The technique is based on the area of the deformed-state cross section (Fig. 8.14a). It is assumed that the area has remained constant:

$$A_o = t_o L_o, \quad (8.1)$$

where A_o = original area, t_o = original bed thickness, and L_o = original bed length. The area between the pin lines is measured and then divided by either the original bed thickness or the original bed length, whichever is better known, and Eq. (8.1) solved

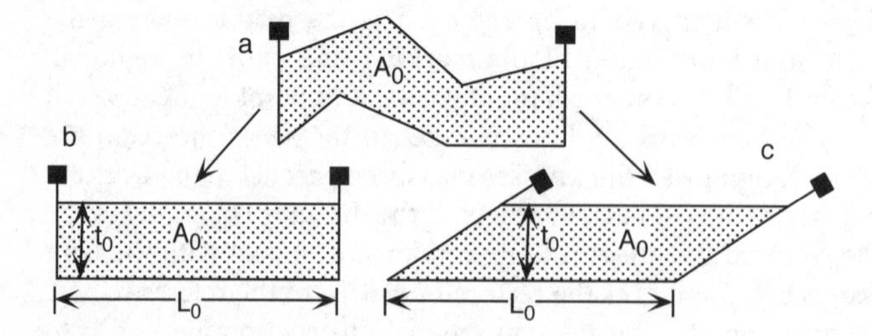

Fig. 8.14. Area restoration. A_o Original area; t_o original bed thickness; L_o original bed length. Shape of the restored area depends on assumed original orientations of the pin lines. **a** Deformed-state cross section. **b** Section restored to vertical pin lines. **c** Section restored to tilted pin lines

for the unknown dimension. The original bed length might be known from that of an adjacent key bed that has not changed thickness (Mitra and Namson 1989) or the original thickness might be known from a location outside the deformed region. Area measurements can be accomplished by dividing the area into simple triangles and trapezoids for which the area can be calculated, measuring with a planimeter, or by measuring on a digital image of the cross section (Groshong and Epard 1996). The shape of the restored area depends on the assumed original shape (Mitra and Namson 1989). Ordinarily the unit is restored to horizontal, leaving only the orientations of the pin lines to be determined. A folded area might be appropriately restored to a rectangular prism (Fig. 8.14b), whereas a block bounded by faults would be restored to fit between the presumed original fault shapes (Fig. 8.14c).

8.6 Area–Depth Relationship

The relationship between the areas of multiple horizons on a cross section and their depths allows a cross section to be tested for area balance and internal consistency without requiring knowledge of the kinematic model that formed it. Deformation causes an amount of the original area to be pushed above the regional (excess area) or dropped below the regional (lost area) as shown in Fig. 8.15. The excess and lost areas (S) are equal to the displacement on the lower detachment (D) times the distance between the regional of a bed surface and the detachment (H), as shown on the right side of the cross sections in Figs. 8.15a,b. It is taken as axiomatic that every structure is detached at some depth, either in the sedimentary cover or in the basement. Measurements are made of the excess and lost areas above or below the regional and the elevation of the regional with respect to the reference level for each horizon. This information is plotted on a graph of area versus elevation (Fig. 8.16a). For many area-balanced structures, the data from multiple horizons give points that define a straight line, the equation of which is:

$$h = (1 / D)\, S + H_e, \qquad (8.2)$$

where h = elevation of a surface above or below the reference level, D = displacement on the lower detachment, S = net displaced area, either excess or lost, and H_e = the ele-

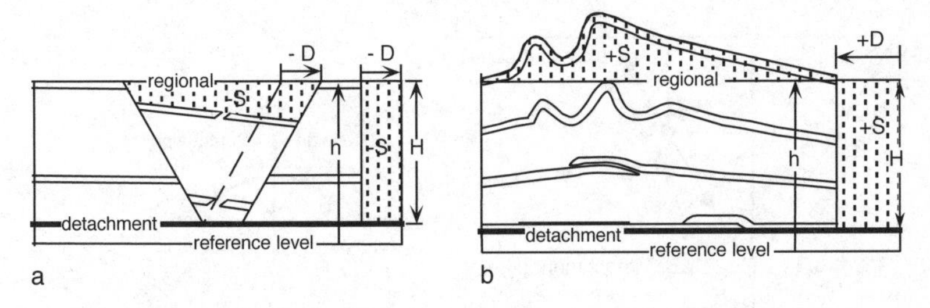

Fig. 8.15. Area-balance terminology. *S* Excess or lost area; *D* displacement on the lower detachment; *H* distance from the lower detachment to the regional; *h* elevation of the regional above or below the reference level. **a** Extension. (After Groshong 1994). **b** Contraction. (After Groshong and Epard 1994)

vation at which the area S goes to zero, which represents the position of the detachment. The elevation of the reference level is arbitrary. If the reference level is at the detachment, $H_e = 0$. If the reference level is chosen to be sea level, then H_e will be the elevation of the detachment relative to sea level. The sign convention for area is that excess area and contractional displacement are positive and lost area and extensional displacement are negative. If deformation causes both uplift and subsidence from the regional, the correct value of S is the algebraic sum or net area (Groshong 1994). The slope of the area–depth line, $1 / D$, is the inverse of the displacement on the lower detachment. To be a valid cross section, the points on the area–depth graph (Fig. 8.16a) must all fall on or close to the area–depth line.

The reference level should be at least approximately parallel to the lower detachment direction and the elevations measured perpendicular to it. Each horizon has its own separate regional. If the regionals are tilted with respect to the reference level, the elevation measurements should be taken at the center of the structure. Structures with displacements on upper detachment horizons within the measured region (for example fault-bend folds, as in Fig. 8.9a) may correctly give area–depth plots that have multiple straight-line segments (Epard and Groshong 1993). Growth units, deposited during deformation, produce curved area–depth lines that trend toward zero area at the depositional surface as well as at the lower detachment.

Equation (8.2) and the graph in Fig. 8.16a are modified from the forms originally given by Epard and Groshong (1993) in order to show elevation as the vertical axis on the graph. This makes the graphical relationship between area and depth more intuitively obvious, with the area on the area-elevation line decreasing downward on the graph to zero at the lower detachment (Fig. 8.16a), as it does in the structure itself (Fig. 8.15). In this form, a vertical area–depth line represents differential vertical displacement (Fig. 8.16b) and the closer the area–depth line is to the area axis, the greater the amount of extension or contraction. A positive slope of the line indicates contraction and a negative slope indicates extension. The slope of the area-depth line (Fig. 8.16a) is the inverse of the displacement, instead of the displacement itself as in the previous formulation (Epard and Groshong 1993). For any given area-elevation data point, the range of h val-

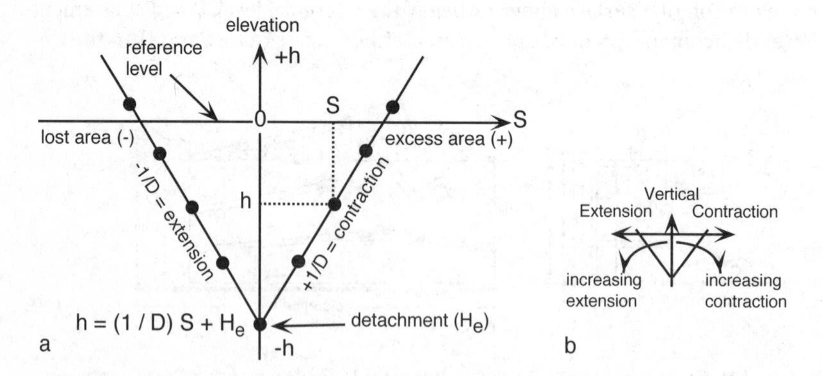

Fig. 8.16. Area–depth relationships for compressional and extensional detachment structures. **a** Area–depth graph. **b** Relationship between the slope of the area–depth line and the amount and type of deformation

ues caused by variations in measurement and interpretation will usually be significantly less than the range of the S values, and so a least-squares fit to the data should be a regression of S onto h, rather than h onto S as the form of Eq. (8.2) might suggest.

The potential pitfall with this technique is in the selection of the correct regional. which is the level of a horizon before deformation. The problem is that this level must be chosen on the deformed-state cross section. If undeformed units are preserved outside the structure of interest, they will record the correct elevation of the regional. The bottoms of synclines are good candidates for the elevations of the regionals in vertical and compressional structures and the elevations of beds in flat horsts are good candidates for the regionals in extensional structures. The effect of an incorrect choice of the regional is to produce an incorrect elevation for the lower detachment. The goodness of fit of the area–depth points to a line on the graph is unaffected by the choice of the regional, and so the method remains valuable in judging the internal consistency of the cross section, even if the regionals are unknown.

Because the area–depth relationship is not dependent on the kinematic model, it applies as an additional constraint on cross sections that can be interpreted by the previous methods as well as on cross sections for which the previous methods are not suitable. For an interpretation that can be validated by one of the previous methods the area–depth relationship will provide additional confirmation and supply additional independent information about the depth to detachment and total displacement.

A typical application of the method is illustrated by a natural example of an extensional structure from the Pennsylvanian coal measures of the Black Warrior basin of Alabama. The cross section (Fig. 8.17a) is based on multiple horizons from closely spaced wells and is relatively accurate. Area–depth diagrams have been constructed for two grabens on the cross section. The lost area is measured for each coal cycle boundary. The positions of the cycle boundaries at their footwall cutoffs coincide with regional dip across the area, from which it is inferred that footwall uplift is negligible. Lines joining the footwall cutoffs are therefore chosen as the regionals. The lost area for a given cycle boundary is the area bordered by the position of the boundary in the graben, the faults, and the regional for that cycle. The reference level was chosen to be at the base of the well control (Fig. 8.17a) and parallel to regional dip. The distance from each regional to the reference level is measured in the center of each graben.

The area–depth curves for both grabens are straight lines with a moderate amount of scatter (Fig. 8.17b). This amount of scatter is normal for an accurate cross section and so the section is validated. The best-fit lines go to zero area at depths below the reference level of -199 and -77 m, the predicted depths of the detachment. The agreement between the depths to detachment, calculated independently for two different grabens of different structural styles, is additional evidence of a valid result. The displacement on the lower detachment that formed the graben is 39 m for the Franklin Hill half graben and 101 m for the Strip Mine full graben.

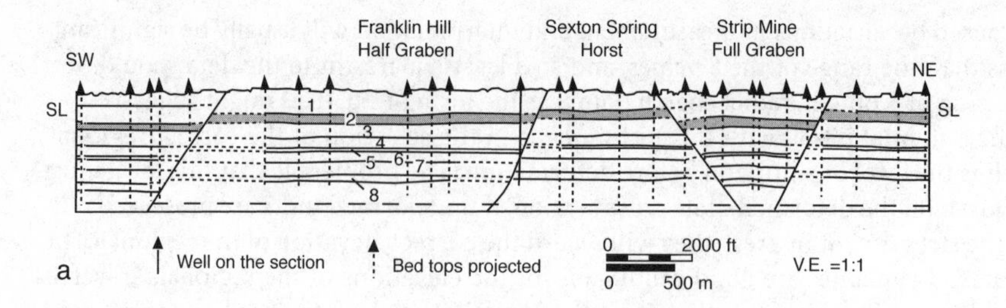

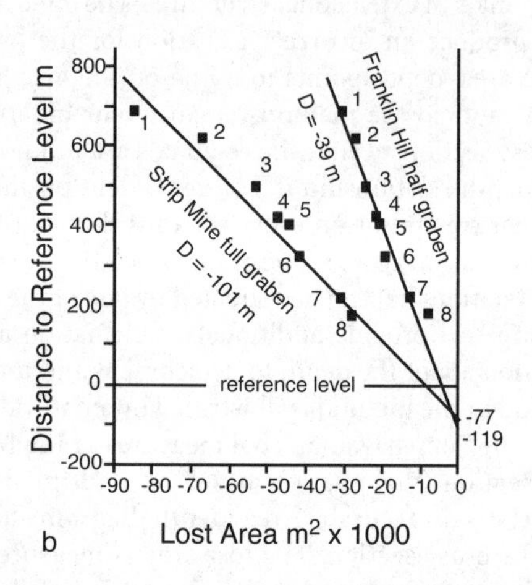

Fig. 8.17. Extensional structure of the Deerlick Creek coalbed methane field. **a** Cross section perpendicular to fault strike showing the calculated lower detachment at the base of the section. The *dashed line* near the base of the section is the reference level. Units *1-8* are coal cycles. **b** Lost-area diagram for the Franklin Hill and Strip Mine grabens. *Points* represent the tops of the coal cycles; *solid lines* are the least-squares best fits; *V.E.* vertical exaggeration. (Modified from Groshong 1994, after Wang 1994)

8.7 Problems

8.7.1 Rigid-Body Restoration

Restore the cross sections in Fig. 8.18.

1. Why is the rigid-body method appropriate?
2. Are the cross sections valid?

8.7.2 Flexural-Slip Restoration 1

Restore the cross section of the Sequatchie anticline in Fig. 8.19.

1. Why is the flexural-slip method appropriate?

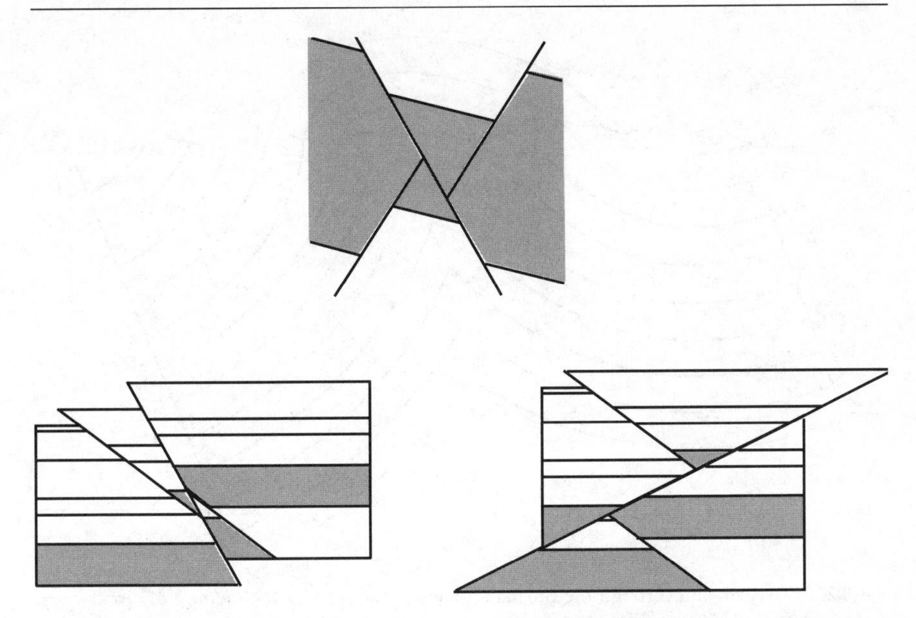

Fig. 8.18. Cross sections of structures formed by rigid-block displacements (from Figs. 6.36, 6.40, 6.43)

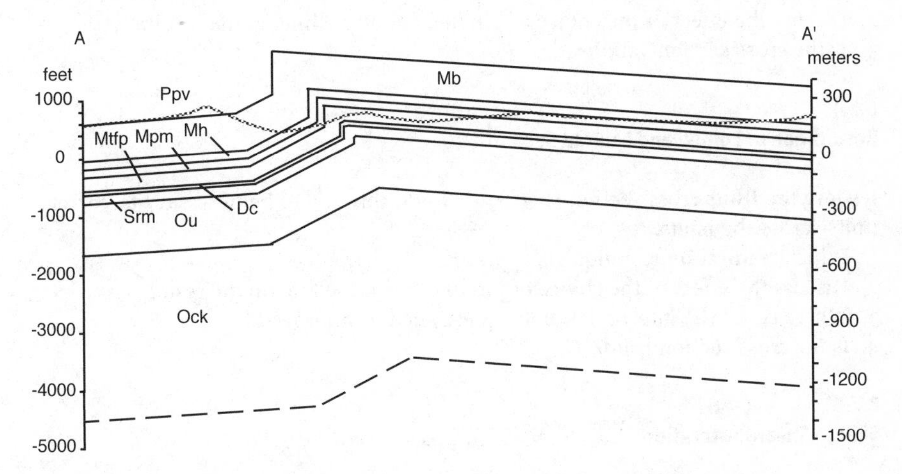

Fig. 8.19. Cross section of the Sequatchie anticline at Blount Springs (from Fig. 7.29)

2. Discuss the effect of the choice of pin line and loose line on the result.
3. Is the cross section valid?

8.7.3 Flexural-Slip Restoration 2

Restore the cross section of the Burma anticline in Fig. 8.20.

1. Why is the flexural-slip method appropriate?

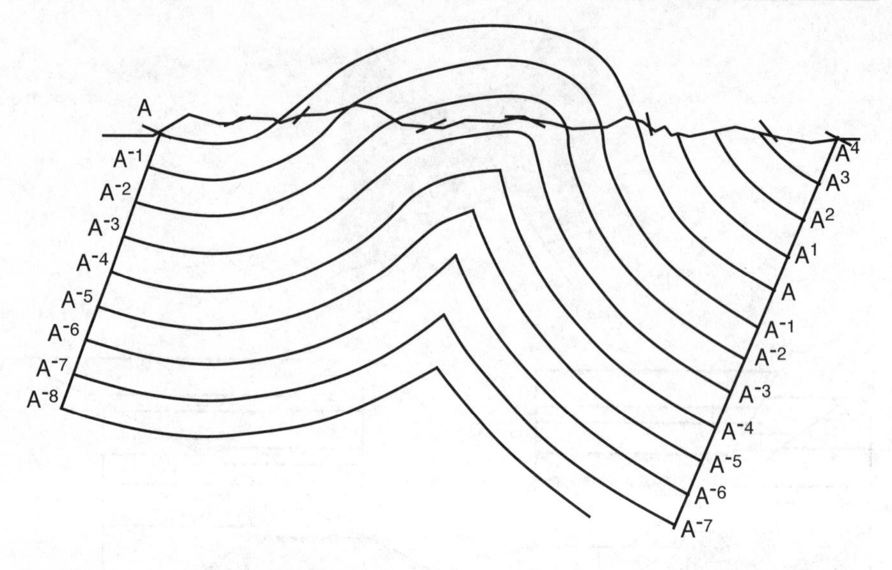

Fig. 8.20. Cross section through the Burma anticline, produced by the method of circular arcs (after Fig. 7.34). A^i Marker horizons

2. Discuss the effect of the choice of pin line and loose line on the result.
3. Is the cross section valid?

8.7.4 Restoration of Folded and Faulted Section

Restore the Rhur cross section (Fig. 8.21). To do this, it will be necessary to correlate units across the faults.

1. Which restoration technique is appropriate? Why?
2. Discuss the effect of the choice of pin line and loose line on the result.
3. Where could the interpretation be questioned or improved?
4. Is the cross section valid?

8.7.5 Simple-Shear Restoration

Restore the growth normal fault in Fig. 8.22. This section contains a growth stratigraphy and can be sequentially restored to the regional for horizons 2 and 3 to show the growth history.

1. Why is the simple-shear method a reasonable choice?
2. What is the appropriate choice of the regional?
3. What is the appropriate shear angle?
4. Is the cross section valid?

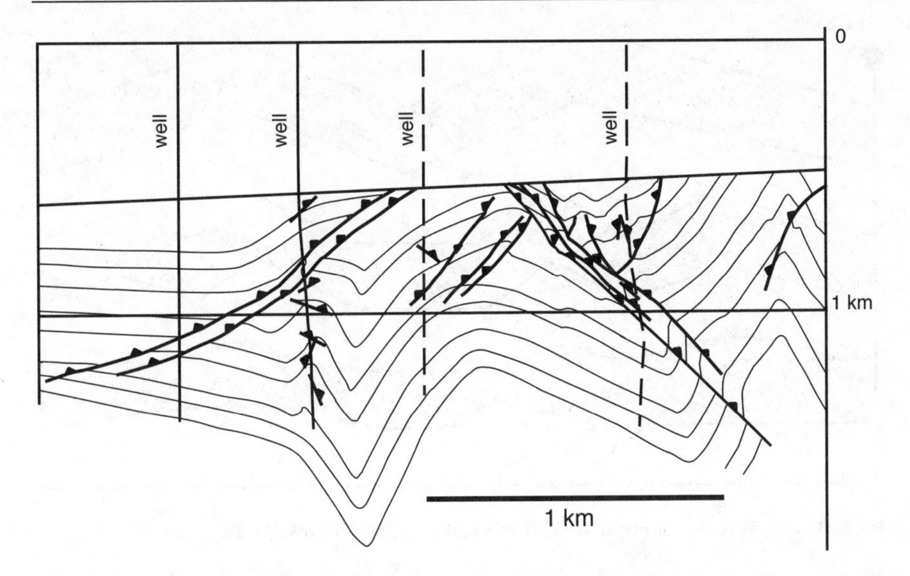

Fig. 8.21. Cross section of a portion of the Ruhr coal district. *Triangles* are located on the hangingwalls of the thrust faults. (After Drozdzewski 1983, and Fig. 5.2b)

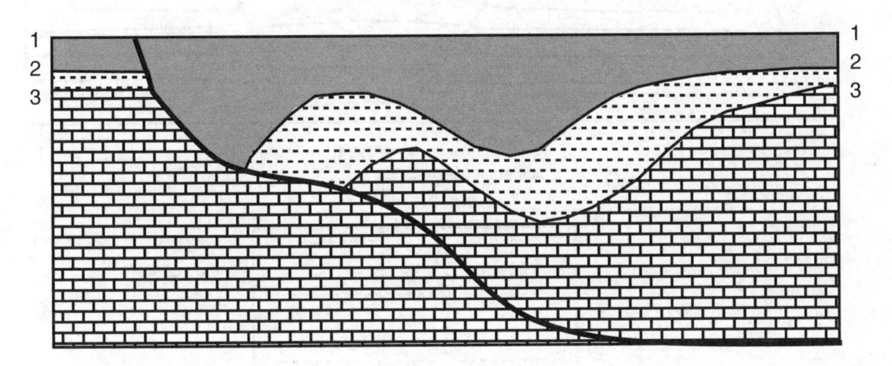

Fig. 8.22. Cross section of a ramp-flat normal fault

8.7.6
Length and Area Restoration

Restore the cross section in Fig. 8.23.

1. Why does this section require a combination of length and area restoration?
2. Which layers probably consist of stiff material and which consist of soft material? Explain your reasoning.
3. Which restoration technique should be used on each bed and why?
4. Explain your choice of the length or the thickness used to determine the shape of the layers restored by area balance.
5. Is the cross section valid?

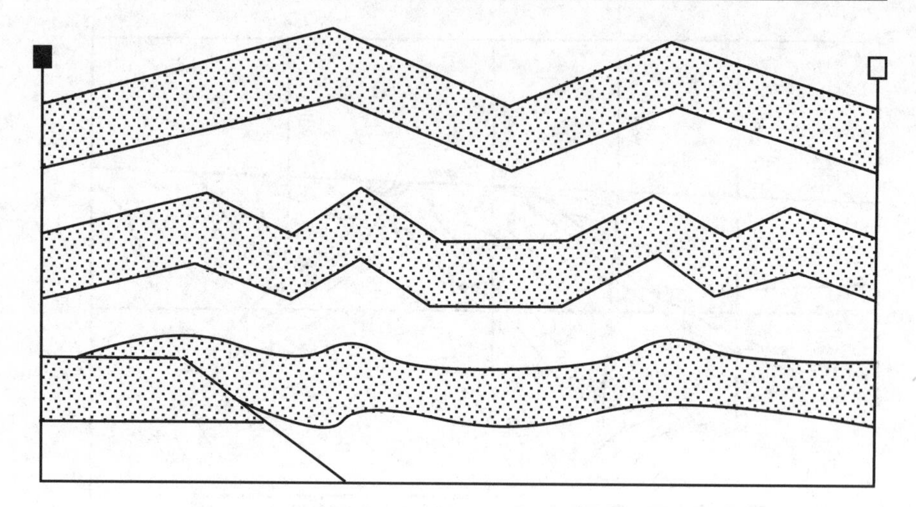

Fig. 8.23. Cross section of buckle-folded and faulted layers (from Fig. 8.2a)

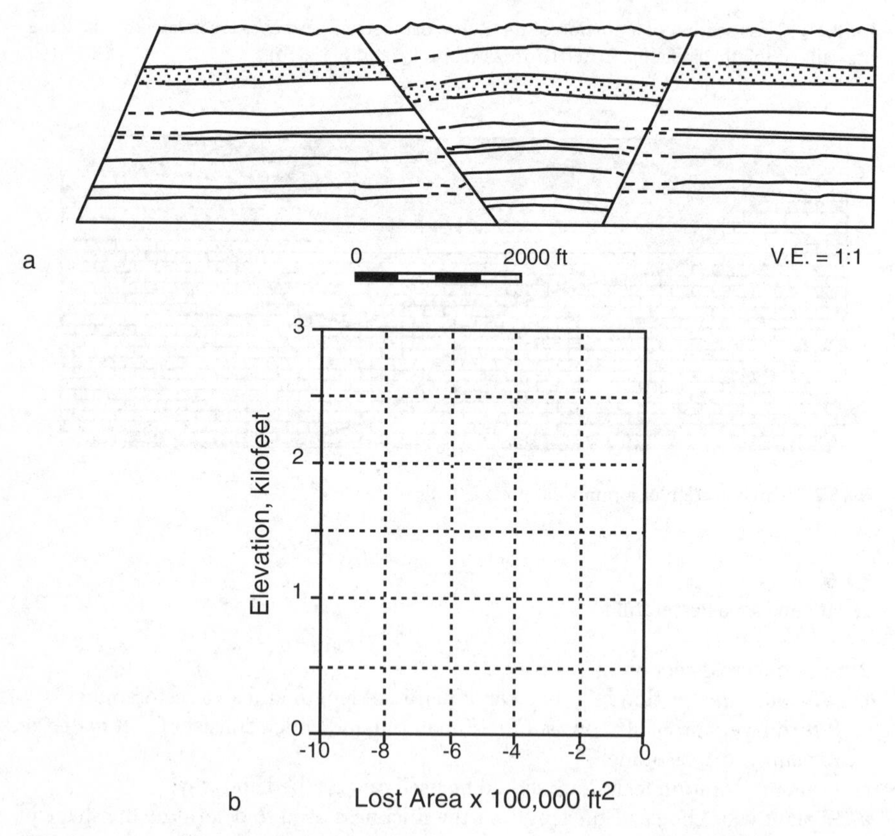

Fig. 8.24. Balance and restoration of the Strip Mine full graben. a Cross section (after Fig. 8.17a); *V.E.* vertical exaggeration. b Area-depth graph

8.7.7 Area–Depth Graph and Area Restoration

Construct an area–depth diagram for the cross section in Fig. 8.24; then restore the section.

1. Is this cross section balanced?
2. What is the location of the lower detachment?
3. What is the displacement on the lower detachment that formed the graben?
4. Is the cross section restorable? What is the best restoration technique and why?

References

Brewer RC, Groshong RH Jr (1993) Restoration of cross sections above intrusive salt domes. Am. Assoc. Pet. Geol. Bull. 77: 1769–1780

Chamberlin RT (1910) The Appalachian folds of central Pennsylvania. J. Geol. 18: 228–251

Cooper MA (1983) The calculation of bulk strain in oblique and inclined balanced sections. J. Struct. Geol. 5: 161–165

Cooper MA (1984) The calculation of bulk strain in oblique and inclined balanced sections, reply. J. Struct. Geol. 6: 613–614

Dahlstrom CDA (1969) Balanced cross sections. Can. J. Earth Sci. 6: 743–757

Drozdzewski G (1983) Tectonics of the Rhur district, illustrated by reflection seismic profiles. In: Bally AW (ed) Seismic expression of structural styles. Am. Assoc. Pet. Geol., Stud. Geol. 15, pp 3.4.1-1–3.4.1-7

Elliott D (1983) The construction of balanced cross sections. J. Struct. Geol. 5: 101

Epard JL, Groshong RH Jr (1993) Excess area and depth to detachment. Am. Assoc. Pet. Geol. Bull. 77: 1291–1302

Geiser PA (1988) The role of kinematics in the construction and analysis of geological cross sections in deformed terranes. In: Mitra G, Wojtal S (eds) Geometries and mechanisms of thrusting with special reference to the Appalachians: Geol. Soc. Am. Spec.Pap. 222: 47–76

Groshong RH Jr (1990) Unique determination of normal fault shape from hanging-wall bed geometry in detached half grabens: Eclogae Geol. Helv. 83: 455–471

Groshong RH Jr (1994) Area balance, depth to detachment and strain in extension. Tectonics 13: 1488–1497

Groshong RH Jr, Epard J-L (1994) The role of strain in area-constant detachment folding J. Struct. Geol. 16: 613–618

Groshong RH Jr, Epard J-L (1996) Computerized cross section restoration and balance. In: De Paor DG (ed) Structural Geology and personal computers. Pergamon, New York, pp 477–498

Kligfield R, Geiser P, Geiser, J (1986) Construction of geologic cross sections using microcomputer systems. Geobyte 1: 60–66, 85

Marshak S, Woodward N (1988) Introduction to cross section balancing. In: Marshak S, Mitra G (eds) Basic methods of structural geology. Prentice Hall, Englewood Cliffs, New Jersey, pp 303–332

McClay KR (1992) Glossary of thrust tectonics terms. In: McClay KR (ed) Thrust tectonics. Chapman and Hall, London, pp 419–433

Mitra S, Namson J (1989) Equal-area balancing. Am. J. Sci. 289: 563–599

Rowan MG, Kligfield R (1989) Cross section restoration and balancing as an aid to seismic interpretation in extensional terranes. Am. Assoc. Pet. Geol. Bull. 73: 955–966

Verrall P (1982) Structural interpretation with applications to North Sea problems. Course Notes No 3, Joint Association of Petroleum Exploration Courses (JAPEC), London.

Wang S (1994) Three-dimensional geometry of normal faults in the southeastern Deerlick Creek coalbed methane field, Black Warrior basin, Alabama. MS Thesis; Univ. Alabama, Tuscaloosa, 77 pp

Washington PA, Washington RA (1984) The calculation of bulk strain in oblique and inclined balanced sections, discussion. J. Struct. Geol. 6: 613–614
White NJ, Jackson JA, McKenzie DP (1986) The relationship between the geometry of normal faults and that of the sedimentary layers in their hangingwalls. J. Struct. Geol. 8: 897–909
Woodcock NH, Fischer M (1986) Strike-slip duplexes. J. Struct. Geol. 8: 725–735
Woodward NB, Boyer SE, Suppe J (1985) An outline of balanced cross sections. Studies in Geology 11, 2nd ed. University of Tennessee Knoxville, 170 pp
Woodward NB, Boyer SE, Suppe J (1989) Balanced geological cross sections: an essential technique in geological research and exploration. Short Course in Geology, vol 6, American Geophysical Union, Washington, DC, 132 pp

Index

Printing (computer to plate): Mercedes-Druck, Berlin
Binding: Buchbinderei Lüderitz & Bauer, Berlin